John James
Audubon

EARLY AMERICAN STUDIES

Series editors:
Daniel K. Richter, Kathleen M. Brown,
Max Cavitch, and David Waldstreicher

Exploring neglected aspects of our colonial,
revolutionary, and early national history and culture,
Early American Studies reinterprets familiar themes and
events in fresh ways. Interdisciplinary in character, and
with a special emphasis on the period from about 1600
to 1850, the series is published in partnership with the
McNeil Center for Early American Studies.

A complete list of books in the series
is available from the publisher.

John James
Audubon

≈≈≈

The Nature of the
American Woodsman

Gregory Nobles

PENN

UNIVERSITY OF PENNSYLVANIA PRESS

PHILADELPHIA

Published by
University of Pennsylvania Press
Philadelphia, Pennsylvania 19104-4112
www.upenn.edu/pennpress

Printed in the United States of America on acid-free paper
1 3 5 7 9 10 8 6 4 2

Library of Congress Cataloging-in-Publication Data

Names: Nobles, Gregory H., author.
Title: John James Audubon : the nature of the American woodsman / Gregory Nobles.
Other titles: Early American studies.
Description: 1st edition. | Philadelphia : University of Pennsylvania Press, [2017] | Series: Early
American studies | Includes bibliographical references and index.
Identifiers: LCCN 2016046873 | ISBN 9780812248944 (hardcover : alk. paper)
Subjects: LCSH: Audubon, John James, 1785–1851. | Naturalists—United States—Biography.
Classification: LCC QL31.A9 N63 2017 | DDC 508.092 [B] —dc23 LC record available at
https://lccn.loc.gov/2016046873

To Phil Terrie,
good writer, good birder, good friend

Contents

Color Plates Follow Page 158

John James
Audubon

Introduction

Creating Art, Science, and Self

Kind Reader,—Should you derive from the perusal of the following pages ... a portion of the pleasure which I have felt in collecting the materials for their composition, my gratitude will be ample, and the compensation for all my labours will be more than, perhaps, I have a right to expect from an individual to whom I am as yet unknown.

    —John James Audubon, "Introductory Address," in *Ornithological Biography*

If anyone had been offering the nineteenth-century version of a MacArthur "genius grant," John James Audubon should have had one. To the extent that genius stems from abundant quantities of imagination, talent, and tenacity, Audubon repeatedly demonstrated all three. He dedicated almost his entire adult life to a remarkably challenging, maybe crazy-seeming commitment to depict, in both paint and print, every bird that flew over and within the United States. He succeeded in that task as no one had done before, and he did so at a great personal sacrifice, with almost no external support. He certainly could have used the MacArthur money.

Today, the main source of Audubon's enduring claim to genius remains the remarkable visual record he left as an artist. His major work—his "Great Work," as he liked to call it—was the famous collection of avian art, *The Birds of America*, which came out in a process of piecemeal publication between 1827 and 1838.[1] Bound together in four huge, heavy volumes, the 435 plates of bird images in the Double Elephant Folio edition define one of the most dramatic achievements in American art. Audubon's massive work has always been, almost by definition, a rare book, and it has recently become the most costly. In March 2000, a Double Elephant Folio sold at auction for $8,802,500, not only setting a record auction price for a book of natural history, but also surpassing earlier sales of the Gutenberg *Bible* ($5,390,000 in 1987) and Chaucer's *Canterbury Tales* ($7,565,396 in 1998) to set a world record amount for a printed book of any sort. Ten years later, in 2010, the price rose even higher, and a complete set of *The Birds of America* fetched $11.5 million.

But whatever its record-setting size and price, *The Birds of America* does not stand alone as the sole measure of Audubon's significance. His bird images are only the most visible—and now, of course, the most valuable—aspects of his larger project. In addition to being a skilled painter, he was also a remarkably prolific writer. Audubon wrote consistently, almost incessantly, throughout his adult years, and his outpouring of words, both published and personal, is especially impressive. He confessed at one point that he thought himself "far better fitted to study and delineate in the forests than to arrange phrases with sensible grammarian skill," but his skill with the pen ranks just a bit behind his mastery with the brush.[2] His major published work, the five-volume, three-thousand-page *Ornithological Biography* (1831–1839), came out as a companion piece to *The Birds of America*, and taken together, the two works make an innovative and interactive reading/viewing experience. In the time of Audubon's life, in fact, when newspapers and other publications reproduced any number of chapters and extracts from his writings, probably as many people read Audubon's words as saw his birds.

We still need to pay attention to his writing. In addition to producing the massive *Ornithological Biography*, Audubon filled thousands more pages as a prodigious journal keeper and letter writer, and those more personal documents provide an immense source of insight into the complicated persona emerging from behind the more famous paintings.[3] Many of those personal writings may not have been so personal after all. He seldom seemed to be writing for himself alone, in the self-reflective, therapeutic sense that many people keep journals today, but always with the notion that someone else would be reading—his wife, his sons, perhaps some broader audience, but certainly someone.

Audubon's sense of audience became more immediately evident in *Ornithological Biography*, in which he repeatedly wrote directly to his "Kind Reader"—a phrase that he used in some variation well over three hundred times—reaching out to embrace the reader as a fellow student of nature, sometimes even as a fictive companion in the search for birds.[4] Audubon sought to create a relationship with the reader—other gentlemen of science, to be sure, but equally important, ordinary people as well—that put himself squarely in the picture as a common man of the American frontier, even as he desperately sought the elevated status of a man of science in the early republic. By looking at Audubon's art *and* science, his painting *and* writing, his elite *and* popular audiences, we can situate his achievements in the larger cultural context of the new nation.

That larger context became a critical element of Audubon's relationship with the reader. His writings went well beyond birds, well beyond scientific description. In the first volume of *Ornithological Biography*, he generously offered to take his reader out of the "mazes of descriptive ornithology . . . by presenting you with occasional descriptions of the scenery and manners of this land."[5] Thus after every five chapters about birds, he would insert an "Episode" about something else altogether, typically an action-packed anecdote about his own experiences of tracking, shooting, and painting birds, or often some tale about other people—and quite often a tall tale at that. Audubon seldom let modesty (or, in many cases, accuracy) stand in the way of a good story, particularly about his own exploits. He once admitted in the pages of *Ornithological Biography* that if he could "with propriety deviate from my proposed method, the present volume would contain less of the habits of birds than of those of the youthful days of an American woodsman."[6] He stuck to his obligations to birds well enough, but, still, *Ornithological Biography* contains so many passages about Audubon himself that it could well be called "Ornithologist's Autobiography."

In whatever form, in fact, whether his published works or personal journals, Audubon almost always wrote about his life, and he was almost always writing a "Life," crafting various parts of the Audubon story that would both shape and reinforce the role he would so brashly embrace. Reading Audubon gives us by far the best way to see the way he created an always expressive, sometimes audacious, and certainly self-conscious sense of himself as an artist, as a scientist, and, above all, as a larger-than-life, self-fashioned symbol, the "American Woodsman."

Audubon's America, America's Audubon

In writing about Audubon's life myself, I do not offer this book as a "Life," as a standard biography, at least not in the sense of providing a chronological, cradle-to-grave account. Over a dozen biographies of Audubon have taken that approach, typically putting *The Birds of America* at the center, even the apex, of the narrative.[7] But chronology goes only so far in exploring a life as richly textured and culturally expressive as Audubon's. Instead, this book follows a different approach. To be sure, the first chapter explores the murky-seeming circumstances of Audubon's birth, and the last opens with transatlantic notices of his death; in between, the fourth chapter focuses on the most famous achievement of his life, the publication of *The Birds of America*. But *The Birds of America* defines only one aspect of Audubon's Great Work. Another was Audubon himself, and his work on that went well beyond his birds.

In its larger trajectory, this book departs from the day-to-day details and offers another way to look at Audubon, exploring more topically, and therefore perhaps more fully, the most meaningful elements of his life. Largely untrained in both art and science, he became one of the most notable men of his era in both, celebrated on both sides of the Atlantic. And in both art and science, it is the process of Audubon's becoming that matters most. Audubon worked for years to make himself a prolific and superb bird artist, and ultimately he succeeded as no one had before or, arguably, since. He also struggled to become a legitimate man of science, doggedly seeking and ultimately taking his place within the transatlantic scientific community. Equally important, though, he also chose to portray himself as an emblematic figure of his era. By devising the guise of the American Woodsman—a highly masculinized amalgam of art and science, but still a friendly denizen of the frontier—he defined a self-promoting notion of national identity that he also perfected in pursuit of transatlantic fame. The American Woodsman wasn't just a catchy nickname; it was a well-chosen role. Audubon didn't just live his life; he performed it.

He seemed perfect for the part. Art and science have many things in common: imagination, creativity, and the attempt to discover and express some fundamental form of information that we commonly call, for want of a more precise term, truth. Today we most commonly find art and science joined together in the modern university, where colleges of arts and sciences give the various academic disciplines an institutional home and a symbolic, if sometimes uneasy, union. In the early American republic, however, universities did not have the depth or diversity they do now. At the beginning of the nineteenth century, the University of Pennsylvania's champion of natural history, Benjamin Smith Barton, complained about the intellectual offerings of American colleges, lamenting that "withered and dead" languages still defined the core curriculum, and, "as yet, little attention is paid to the study of nature in the United-States."[8] The antebellum era defined a time when art and science came together not so much in a complex institution, but more often in a single, albeit exceptional, individual. Audubon was certainly the most prominent of those, a man who stood squarely at the intersection of art and science at a time when natural history was becoming entwined with national history.

Audubon's America was a nation whose ambition seemed boundless. Between the time of the American Revolution and the Civil War, the United States embarked upon a geographically expansive and culturally possessive project, seeking to secure its hold on the continent while spreading over it at the same time. Artists and scientists played almost as much a role in that process as soldiers and settlers. Mapmakers extended the boundaries of the nation on paper, and in the process they helped people imagine the land of the future. Landscape painters depicted the American environment in large-scale, emotion-laden scenes, giving a dramatic, almost daunting power to nature that seemed to require heroic perseverance from its human inhabitants. Naturalists took it as their special mission to give greater emphasis and importance to scientific inquiry that focused on the beauty and abundance of the natural world on the American side of the Atlantic. To do so, they engaged in a massive collective (and quite often competitive) taxonomic attempt to discover, catalog, and classify all the species of America and, in many cases, to define them as distinctly American.

Audubon played his part as well as anyone could. He did not simply paint his birds as stiff specimens for close ornithological examination; he gave them life and location, creating animated images embedded in the American landscape. His unremitting quest to collect and depict hundreds of avian species represented an act of artistic and scientific possession: *The Birds of America* implicitly meant "The Birds of the United States." No one—not even illustrious landscape artists like Asher B. Durand and Thomas Cole or his fellow artist-naturalist Alexander Wilson—could claim a greater degree of engagement with nature in the new nation than Audubon. In an era when emerging mass communications trumpeted a triumphalist insistence on the unique achievements of the United States, Audubon became one of the most adulated artists of antebellum America, and certainly the new nation's first celebrity scientist.[9]

He also became a businessman, an energetic entrepreneur who not only made a very valuable product, but also marketed it aggressively on both sides of the Atlantic. Audubon typically tried to downplay or disguise the overtly commercial implications of his efforts, preferring instead to portray himself as a man completely committed to his artistic and scientific concerns, too lost in his loftier pursuits to bother having a sharp eye on the bottom line. But in the years before he devoted himself fully to producing *The Birds of America*, he sought to make a living by investing both his time and money in a variety of business endeavors—a store, a steam-powered mill, a steamboat—that seemed emblematic of the economy of the early American republic. Even though those ventures ultimately failed—or as he would more readily admit, he himself failed—Audubon apparently did not fail to learn about the need for close commitment to one's work. His own Great Work became both profession and obsession, and he essentially ran the entire enterprise himself, keeping careful track of costs, keeping a close eye on labor, keeping good faith with his subscribers, and often dogging them to keep faith with him. As much as Audubon might resist being seen as a merchant or manager, he could hardly hide the business side of his persona.

But hiding became a central part of his persona. In pursuing his overlapping ambitions in art, science, and business, Audubon developed an elusive and elastic identity. He lived behind a translucent curtain of narrative deception, coy evasion, and outright lies. He could be more than a little loose with the truth about his own life, from his origins ("The precise period of my birth is yet an enigma to me"), to his personal associations ("Daniel Boon . . . happened to spend a night with me under the same roof"), to his exploits in the wild ("Snakes, loathsome and venomous, entwined my limbs"), to his entrepreneurial interests ("I . . . only now and then thought of making any money").[10] The American Woodsman served as a usefully mutable character, shifting from one story to the next to suit the need for a particular narrative effect. Much like Melville's fictional "Confidence Man," the "real" Audubon became a master of personal deception and reinvention.

With that in mind, this book proceeds from the notion that Audubon's frequent address to his "Dear Reader" now applies to us, and it invites us to read Audubon not only with considerable admiration for his literary skill, but also with a measure of skepticism about his meaning. Audubon's writing, exuberant and overblown as it could often be, offers a fascinating entrée into the intersection of American scientific and literary traditions in the first half of the nineteenth century. Many of the stories he told about himself, about other naturalists, or about ordinary Americans cannot be read as factual accounts that might yield a single, demonstrable "truth" about the events in Audubon's life. Quite often, his portrayals of other people create a contrast in character that cast Audubon in struggle against his competitors and detractors, mean-spirited men who would vilify and victimize him but would, in the end, serve as narrative foils for underscoring his eventual success. In other cases, though, some of Audubon's stories—including some of the more unsettling episodes about violence, race, and slavery—put him in a much less flattering light. It does no good to ignore or quickly dismiss these issues, as most earlier Audubon biographies have, or simply give Audubon a polite pass as being "a man of his time." Disturbing though they may certainly be, the unsavory aspects of Audubon's views of society are as important as his views of ornithology, and they have much to tell us about his time, even in ways that Audubon himself might not have intended or even fully understood. In the end, even if we are wise not to take all of Audubon's stories literally, we still need to take them seriously.

Doing so will ultimately be the best way to take Audubon himself seriously. Rather than trying to discern fact from fiction, it makes better sense to accept the various ambiguities and apparent contradictions in Audubon's life as a valuable avenue of approach to understanding his place in antebellum American culture. In the end, as much as we might celebrate Audubon's unmistakable impact in art and science, we must also appreciate the importance of personal ambiguity as a critical element of his identity as the American Woodsman. Taking the full measure of Audubon's "genius" begins with a fundamental point: One of his greatest creations was himself.

Chapter 1

~~~~

Becoming Audubon, Becoming American

The precise period of my birth is yet an enigma to me, and I can only say what I
have often heard my father repeat to me on this subject.

—John James Audubon, "Myself"

In 1827, when Audubon was in his early forties and just beginning to enjoy the
fame that would come his way for the rest of his life, he wrote about his newfound sense of self: "What a curious interesting book a Biographer—well
acquainted with my Life could write, it is still more wonderfull and extraordinary than that of my Father!"[1] Comparisons to his father aside—and they both
lived "curious interesting" lives, albeit in different ways—the problem for any
"Biographer" would be that Audubon made it difficult, nearly impossible, to
become fully acquainted with several critical aspects of his life. He hid some
of his basic biographical information behind a veil of unanswered questions
and even outright deception, shading some of the essential elements of his
life's story from historical view. His written depictions of himself, both published and unpublished, offer enough discernible details for a "curious interesting book," to be sure, but they also suggest a quiet resistance to a deeper level
of personal revelation.

Audubon would not by any means be the first or last person to devise an
incomplete or misleading self-portrait, of course. People often shape their
identities to suit different circumstances, sometimes assuming a dissembling
image for a particular situation, occasionally adopting an enduring disguise
for life. The bookshelves of autobiography have become heavy with self-serving
stories that fabricate various fictions about the writer's life, and the very act of
constructing a narrative requires constructing a selective, sometimes even
secretive, sense of the self.[2] But in Audubon's case, this evasive behavior invites

us to consider what must be the central irony of his life: For a man who spent so much time and trouble depicting birds so carefully and colorfully, presented in scientifically accurate and life-size detail, he left his own self-portrait remarkably incomplete and ambiguous by comparison, often rendered in sketchy contrasts of black and white.

The Ambiguities of Origin

In the introduction to his major written work, *Ornithological Biography*, Audubon offers only one short sentence about his beginning: "I received life and light in the New World." He follows that line with a brief paragraph about his early exposure to nature and "the power of those early impressions," but he provides little more than that. By locating his first "life and light" in "the New World," he vaguely creates the impression that he was born in North America; it remains for the reader to infer. A few pages later, he writes that, as a young man, he "returned to the woods of the New World with fresh ardour, and commenced a collection of drawings," which would eventually become his life's major achievement, a massive collection of avian art, *The Birds of America*.[3] In both instances, "New World" is a geographically elastic and usefully evasive term, allowing him to avoid being more specific about a particular location. We are left to ask, then, when was he born, and where? Almost immediately, we also ask, who were his parents? And what do we know about Audubon's birth and boyhood that would shape his later life as the "American Woodsman"? Audubon could not—or, perhaps more to the point, would not—tell us all we might want to know.

He wouldn't even tell his own family. In a memoir called "Myself," first written initially for his sons, Victor and John, in 1835, and published only posthumously, in 1893, Audubon opens with an evasion: "The precise period of my birth is yet an enigma to me, and I can only say what I have heard my father repeat to me on this subject." He continues with an air of uncertainty: "It seems that my father had large properties in Santo Domingo," he notes, but adds that the elder Audubon also traveled on occasion to Louisiana, where "he married a lady of Spanish extraction, whom I have been led to understand was as beautiful as she was wealthy, and otherwise attractive." She bore three sons and a daughter, Audubon writes, but only he, the youngest of the sons, "survived extreme youth."[4] Audubon enhances the enigma of his nativity by casually but carefully accepting no direct responsibility for the information he gives his own sons about his origins—"I can only say what I have heard" . . . "It seems" . . . "I have been led to understand"—and subtly transferring the only authority for this part of his life story to his long-dead father.

But Audubon goes on, and the story gets better. As an infant, he writes, he accompanied his parents back to "Santo Domingo," or Saint-Domingue, and the family estate, where his mother soon met her doom as "one of the victims during the ever-to-be-lamented period of the negro insurrection on that island." With the help of "some faithful servants," the elder Audubon and his young son escaped the insurrection and made their way back to New Orleans and then back to France, where Audubon's father had a home in the city of Nantes in the Loire Valley. He also had a wife in Nantes, an apparently tolerant and understanding woman who accepted the young boy into the household and raised him as her own; she was, Audubon writes, "the only mother I have ever known." Audubon then spent his early years in Nantes "much cherished by my dear stepmother . . . [and] constantly attended by one or two black servants, who had followed my father from Santo Domingo to New Orleans and afterwards to Nantes."[5] Audubon spins quite an exciting story for his sons: of being born to a Spanish mother on one side of the Atlantic (presumably in Louisiana) and being raised by a French stepmother on the other side, in France; of losing his birth mother in a major slave revolt in a French colony and escaping with his father, thanks to the help of loyal slaves; and coming to safety in a French city, where he found a loving and generous stepmother who immediately cherished him—and he does it all in four short paragraphs. It is, in essence, a broad-reaching tale of the eighteenth-century Atlantic world, beginning with a North American birth, followed by a brief but dramatic West Indian incident, leading to a longer French interlude, and, as Audubon's sons (and we) would know, culminating in a return to America, where he would eventually enjoy a comfortable outcome by gaining artistic and scientific fame. It would have made a promising outline for a popular novel, and perhaps it should have been a work of fiction: Important parts of it were simply untrue.

For that reason, it makes sense to explore more fully the ambiguities of Audubon's autobiography to appreciate the place of the evasions and omissions about his origins in the longer trajectory of his life. It may well be, of course, that the "enigma" of his birth explains it all, that Audubon knew only what he had "been led to understand" by his father. If so, he apparently didn't press his father for details, and if he did, he certainly didn't share them with his own sons—or anyone else, for that matter. Questions about Audubon's origins persisted throughout his lifetime, and he never made much of an effort to answer them in print.

Audubon was always very vague about something as basic as his age, for instance, usually overstating it by a few years, perhaps because he hoped to mislead for some reason or probably because he just didn't know for sure.[6] Some of the contemporary official sources don't offer much help. His 1806 application for American naturalization described him as "a free white person of the Age of Twenty three Years . . . born at Aux Cayes in the Island of St. Domingo sometime in the Year one thousand Seven Hundred and Eighty three." Six years later, in 1812, his record of naturalization referred to him as "John Audubon of the District of Pennsylvania a native of the Island of St. Domingo aged about Twenty Six years." Finally, Audubon's American passport for his second journey to Great Britain in 1830 noted that he was "46 years, 5 feet 8 1/2 inches, common forehead, hazel eyes, prominent nose, common mouth, pointed chin, greyish hair, brown complexion, oval face."[7] The age never seems to come to rest on any definite number—twenty-three in 1806, "about" twenty-six in 1812, and forty-six in 1830—or even on any specific year of birth. The differences may have as much to do with the vagaries of record keeping in various parts of the Atlantic world as with any attempt on Audubon's part to deceive, but the point remains: On such a basic issue as birth date, even birth year, the documentary evidence on Audubon could be, at best, uneven.

Whatever variation there may have been about when Audubon was born, the question of where seems even more salient. By the 1830s, when Audubon had begun to gain decent notoriety on both sides of the Atlantic, the uncertainty surrounding his birth and national identity became a perplexing part of his story in the popular press. Despite the recognition of his birth in Saint-Domingue in his passport and other official documents, the more enduring narrative of his life—the one he created and the one that seemed to stick in the popular press—located his birth in the United States. To be sure, an 1832 essay in the *American Monthly Review* did raise the issue of the ambiguous information in Audubon's "auto-biographical sketch" in *Ornithological Biography*, observing that "Mr. Audubon says that he was born in the New World; but does not inform [the reader] in what part of this wide New World, or at what time the event happened." At about the same time, almost in response, a writer in the *New American Review* sought to clarify the location of his birth, reporting that "Mr. Audubon was born in America, but was descended from a French family, and was sent early in life, to receive his education in France." The following year, a Philadelphia-area newspaper took a different position, trying to rectify an apparently erroneous report that Audubon had been born in Pennsylvania; instead, the author noted that a "gentleman of this vicinity . . . informs us, that France is his native country." A decade later, however, a correspondent for another newspaper, who claimed to have interviewed Audubon in person, stated flatly that "Mr. Audubon was born in New Orleans" and that "nothing would give him more displeasure than to be even suspected of being an European."[8] These differing accounts likewise remained a bit vague on his birth year—"about 1775," "about 1780," "about 1782"—but they came increasingly to reflect a clear consensus around locating Audubon's origins in North America.

By the time Audubon died, his American-based birth narrative had gained considerable currency in the country, and his obituaries increasingly connected him not only to North America, but to the United States, giving him implicit citizenship in the new nation. An 1852 book, *The Homes of American Authors*, located Audubon's origins "on a plantation in Louisiana, then a French possession," and noted that he had been "born the same year the Declaration of Independence was made (1776)," thus rooting his Frenchness in the soil of an emerging Americanness. An 1851 death notice associated Audubon's birth more directly with the new nation, noting that "Mr. Audubon was born about 1775, in the State of Louisiana," thus anticipating by some twenty-eight years Louisiana's acquisition by the United States, much less formal statehood. Another likewise located Audubon's birth in Louisiana, in 1775, and insisted that "Audubon was a native of this country, and not an adopted citizen."[9] Whatever questions might remain about the exact date or location—New Orleans, somewhere else in Louisiana, or in the United States, at any rate—written accounts had made him a fully naturalized native of the nation, and any suggestion of his origins in Saint-Domingue never appeared in print.

Origin Stories

Before we allow Audubon or any of the contemporary sources to tell more of his origins, we might well turn to more recent writers. By the twentieth century, the story of Audubon's background had become clearer, and his modern biographers reached a reasonable measure of agreement on most—but still not all—of the basic details of his birth. In 1917, Francis Hobart Herrick, one of the earliest and most thorough of the twentieth-century Audubon biographers, wrote that Audubon was born on April 26, 1785, in Saint-Domingue; on that there now seems to be no modern scholarly dispute. Unfortunately, there was no birth certificate or baptismal record to document his birth, because Audubon was born out of wedlock—an important issue in itself—and local officials did not give formal sanction to such births. The only documentary evidence for his birth, then, comes from the records of the doctor who attended Audubon's mother, Jeanne Rabin (also spelled Rabine), who was ailing with some sort of tropical sickness during the last days of her pregnancy. On April 26, the physician's entry showed that Jeanne Rabin had delivered a child—but only that, with no name or even description of the child. Still, Herrick safely asserts that the child must have been the future avian artist. Indeed, the baby was her last child; soon after giving birth to the infant son, she died, but as a result of sickness and not, as Audubon would later tell it, at the hands of rebellious slaves. As Herrick explains, "Much other documentary evidence which also has recently come to light is all in harmony with these facts."[10]

Herrick also devotes a good deal of space to the identity of Audubon's father, Jean Audubon: He was a Frenchman, born in 1744 in a village on the Bay of Biscay, who rose from being a cod fisherman to a sailor to a ship captain. But Audubon père's most important career move proved to be his marriage, in 1772, to Anne Moynet, a prosperous widow nine years older (or twelve or fourteen, according to other sources), whose wealth and wifely indulgence allowed him to spend months at a time at sea or, increasingly, on a sugar plantation he purchased in Aux Cayes (now Les Cayes), a port town on the southern coast of Saint-Domingue, which had become the largest sugar-producing island in the world.[11] There he, like so many other Europeans, took full financial advantage of the enormous economic opportunities of the West Indies, turning enough profit to provide quite a comfortable living, albeit one built on the infamous brutality and misery inflicted upon Saint-Domingue's rapidly rising slave population.[12] Jean Audubon came to describe himself as a *négociant*, a merchant, but in addition to trading in sugar and a variety of wares, he also traded in slaves, sometimes dozens at a time. "Great numbers of negroes must have passed through Jean Audubon's hands," Herrick observes, "which strangely reflect the customs of a much later and sadder day on the North American continent." Herrick later adds, with considerable charity, "Jean Audubon, who spent a good part of his life at sea and in a country almost totally devoid of morals, must be considered as a product of his time."[13]

And like other powerful European inhabitants of this "country almost totally devoid of morals," Jean Audubon also took sexual advantage of the dependent women in Saint-Domingue, a highly sexualized society that one scholar has described as a "libertine colony."[14] The elder Audubon maintained a long-standing domestic relationship with a mixed-race woman variously described as a creole or a quadroon and variously called Catharine or Marguerite or, more commonly, "Sanitte" Bouffard, who bore him three children, all girls. Sanitte, the *ménagère* (housewife) of Audubon's household, apparently accepted the entrance of another woman into the domestic scene: Jeanne Rabin, who moved in with Captain Audubon and Sanitte in 1784 and who, two years later, became the mother of Jean Audubon's only son. Or so documentary evidence for Audubon's birth story seemed to suggest. But as Herrick observed, the birth story had long seemed murky: "Much of the mystery which hitherto has shrouded the early life of John James Audubon is involved in the West Indian period of his father's career"—and, one might add, his father's relationships with women in Saint-Domingue.[15]

By 1788, the stability of Jean Audubon's island estate may have begun to seem much less secure. Even though Les Cayes seemed safe for the time being, Jean Audubon sent his three-year-old son, originally called Fougère (French for "fern"), to France for safekeeping. In 1789, the elder Audubon signed up as a soldier in the Les Cayes troop of the National Guard, but before he had to face the prospect of real service, he left Saint-Domingue, first to do some business in the United States, then to go back to France, where his toddler son was awaiting him. He would also have his and Sanitte's daughter, Rose, or Muguet, brought to France in 1791, when the slave insurrection began. He would, however, leave behind Sanitte and the other offspring he had had with her to face their fate on the island.

Back in France, Jean Audubon would find another eruption of unrest, the dramatic revolution that toppled the monarchy, divided the people, and cast the country into carnage. In 1793, his particular part of the country, the Vendée, which included Nantes and its environs, became hotly contested between supporters of the revolution and counterrevolutionary royalists, and the bloodshed continued throughout most of the 1790s. Jean Audubon once again joined the local National Guard, this one organized to defend the revolution against its reactionary enemies in the Vendée. He remained a solid ally of the new French republic, serving on local revolutionary committees and enlisting once again in the navy; as his son later wrote, Jean Audubon "continued in the employ of the naval department of that country" throughout the 1790s and on into the new century.[16] Even with the roiling violence of the revolution, France seemed safer ground for the Audubon family than Saint-Domingue.

It also seemed a safer starting place for family-friendly Audubon biographies. In 1954, for instance, two children's books opened Audubon's boyhood story when he was already in the comparative familiarity of France. Margaret Kieran and John Kieran's *John James Audubon*, a Landmark Book, opens the story on "a warm May afternoon in 1793" in Nantes, which was "beginning to show the beauty of spring." Readers first see young Audubon, still called Fougère at this point, out enjoying the birds and flowers, but as every youngster must know, such freedom has to come to an end with the inevitable call to dinner. "Around the table that night it was a typical family gathering," the story continues, with Fougère and his younger stepsister, Muguet, seated with their kindly and beloved stepmother and the sterner-seeming Captain Audubon. After grilling his young son about his studies and fretting to his wife that the boy "needs careful watching," the captain let his attention drift with "a faraway expression in his eyes . . . thinking, no doubt, about his stay in Santo Domingo where young Audubon was born and where the boy's mother had died not long afterward." Bringing him back to France, he provided Fougère with a new stepmother, Madame Audubon, who "lavished as much affection on him as though he had been her own son," making sure the boy had everything he could want: "his own room, his own nurse, and the finest clothes she could buy." The chapter then concludes on a note of erasure: "Soon he had forgotten all about that far-away tropical island." Whether the "he" in the sentence refers to Captain Audubon or young Fougère remains, however, unclear.[17] Joan Howard's *The Story of John J. Audubon* begins almost a year later, in March 1794, on a "cold bleak day" when the weather seems as ominous as the revolutionary political situation in the Vendée region. As the menace of violence swirls around the Audubon household, Madame Audubon again appears as a source of comfort and kind support. Still, young Fougère "wished he could remember who his real mother was." Staring into a flickering fire, struggling to make sense of stories his father has told him about "the island of Santo Domingo," with its tall mountain, fields of sugarcane, and brightly colored birds, Fougère also flashes on the image of a woman: "There was a lady who was so much different from stout Madame Audubon. She was much younger and prettier. She wore wide satin skirts. Her curls were powdered white and piled on a small, proud head. Was that lady his own mother? Fougère did not know."[18]

Thus 1950s-era children's literature dispensed with the ambiguities of Audubon's origins, perhaps predictably so. At a time when *Father Knows Best* defined the televised standard for the normal postwar family, too much talk of a distant and dead mother, whoever she was, might undermine the integrity of the intact, albeit blended, four-person Audubon household, with its stern but loving father, doting and indulgent stepmother, and two mischievous but charming children, one of them a boy with an incipient penchant for birds and, apparently, vague questions, if not memories, about his birth mother.

But for adult readers, questions about Audubon's mother remained on the table. As late as 1966, another Audubon biographer, Alexander Adams, asked, "Who exactly was Mlle. Rabin?" only to answer, "No one really knows." Jeanne Rabin's background, like that of her famous son, has been shaded with uncertainty, and, as Adams continued, "she may have been . . . a Creole, one of those women of European descent who were famous for their beauty, their charm, and also their competence." But "creole" could be a rather elastic term, especially in an eighteenth-century slave society, sometimes referring to people of African descent born on the western side of the Atlantic or to those with varying mixtures of European and African blood, typically the product of unions between white men and women of color. Herrick, for instance, likewise describes Catharine "Sanitte" Bouffard, Jean Audubon's mixed-race *ménagère*, as a *"creole de Saint-Domingue."*[19] As one of the leading historians of early American slavery has observed, the mixture of so many diverse peoples in slave societies often resulted in fluid forms of self-definition, rendering identity "a garment which might be worn or discarded, rather than a skin which never changed its spots."[20] Race mattered, to be sure, but it also defied the fixed categories so often assigned to racial identity in modern times.

The questions about the category of "creole" have thus contributed considerably to suggestions, even assumptions, that Rabin's identity—and therefore her son's—might well reflect at least some degree of African descent. These notions have become most prominent in African American tradition. In recent times, for instance, *The Negro Almanac* (1989) included Audubon as its first alphabetic entry under "Outstanding Black Artists" and described him as "the son of a French merchant sea captain and planter and his Afro-Caribbean mistress." Similarly, the Ebony Society of Philatelic Events and Reflections has placed Audubon on its list of "African Americans on U.S. Stamps," noting that "Audubon's mother was a Creole (mixed heritage) from Domenica." Among academic institutions, the African American Cultural Center Library at Indiana University of Pennsylvania currently lists Audubon in its Biography File.[21] On the other hand, the *Dictionary of Negro Biography* (1970) contains no entry for Audubon nor does the more recent *Black Biography, 1790–1950* (1991). The *Journal of African American History* (formerly the *Journal of Negro History*) has never carried an article on Audubon since its inception in 1916. The differences among such sources in the African American scholarly community do not indicate a consensus, of course, on the exact biographical accuracy of Audubon's racial identity—or, indeed, his mother's. They do suggest, however, that Audubon has had a persistent place in African American memory.

Questions about Audubon's racial identity apparently never sat well with Alice Ford, the Audubon biographer who wrote the most definitive (and certainly most defensive) account of Audubon's origins—and who has also been the most insistent about his whiteness. In her 1964 biography of Audubon, she sought to erase the uncertainties surrounding Rabin's identity by arguing that she was not a creole woman but a French immigrant to Saint-Domingue, a twenty-five-year-old chambermaid from Les Touches parish in Nantes. According to Ford, Jeanne traveled to the West Indies on the same ship as Captain Audubon, and it was there they first began their relationship—or, at least, within the cramped but ever-exposed social circumstances of an eighteenth-century ocean voyage, their friendship.[22]

Hoping no doubt to close the matter once and for all, Ford offered even more detailed information on the Audubon-Rabin relationship in 1988, in a revised edition of her Audubon biography. As the text on the book's dust jacket put it, "The most startling revelation is the positive identification of Audubon's mother, which leads to clarification of the mystery of Audubon's early life." In this edition, Ford addressed the "mystery" by once again making her case for Rabin's being a French chambermaid, but this time she enhanced the documentation with imagination. She speculated that Jeanne, a young woman of "pious upbringing," might have worried about how Anne Moynet Audubon, Captain Audubon's legal wife back in France, would have taken the news of her husband's various infidelities in Saint-Domingue: "The presence of a black mistress was one thing," Ford fretted for Jeanne, but "the imagined intrusion of a white woman, known by some to be the mother of her husband's expected child, quite another."[23] Getting inside the mind of this young woman of "pious upbringing" was not the only point of that sentence: It was also to assert her racial identity. Indeed, the very crux of Jeanne Rabin's imagined moral quandary hinged on race, the contrast between, on one side, the "black mistress," Sanitte, and, on the other, this pious "white woman," Jeanne, who had nonetheless found herself in a similar situation of extramarital motherhood.

There remains one small, final piece of the parental puzzle. As Ford also notes, Captain Audubon had yet another child with Sanitte, this one a daughter named Rose (or Rosa or Muguet), born on April 29, 1786, almost exactly a year after young Audubon, or Fougère. When Captain Audubon later sent the young girl to France, she was entered on the ship's list as "Demoiselle Rose Bonitte, aged four, natural daughter and orphan of Demoiselle Rabin, white." Ford explained this identity change for the young Rose by suggesting that not mentioning the part-African Sanitte as the girl's mother and instead using Rabin's name served as a "protective shield" for her entry into white society in France; just to make the point clearer, Rose was also given the designation "white" in the ship's list.[24] Ford let the issue go with that, without taking the possible implications one step further: If one child of Captain Audubon could be given a new racial identity by assigning her to a different mother and calling her "white," could not another? But Ford failed to entertain, much less explore, that prospect, and her version of the young boy Audubon's birth still defined him as the white son of a white French father and a white French mother. As one observer has noted, Alice Ford "was the first to bleach Audubon completely."[25]

In fact, Audubon himself was the first. Despite his French–West Indian origins, Audubon would always define himself as decidedly American. Just as he sometimes depicted his birds in odd, even distorted postures in order to fit them within the dimensions of his paintings, so Audubon reshaped and shaded his own identity to conform to the common assumptions of what it meant to be an American—above all, a white American—in the antebellum era. By the time he became famous and, in his own estimation, a superb subject for a "Biographer," he had adopted the identity of the American Woodsman, a free man of the American frontier, an artistic cousin of the Common Man. "America will always be my land," he wrote his wife during his residency in England in the late 1820s. "I never close my eyes without travelling thousands of miles along our noble streams and traversing our noble forests." Wrapping himself both rhetorically and visually in the garb of the American Woodsman, he located himself not just in the natural environment, but also in the social environment of the new nation.[26] Audubon's consistent, sometimes insistent assertion of this selective self-identity became the critical element of his persona, both in his writings and in his more general engagement with Euro-American culture. Being a free, white frontiersman was one of the most valuable identities a man could have.

French Foundation

Before Audubon could become fully American, however, he would freely acknowledge being French. Despite his vague evasions about what he had "been led to understand" by his father about his "New World" birth, he took more ownership of his memory in describing his early life on the far side of the Atlantic. "The first of my recollective powers placed me in the central portion of the city of Nantes, on the Loire River, in France," he wrote, "where I still recollect particularly that I was much cherished by my dear stepmother, who had no children of her own."[27] Madame Audubon, the wealthy widow whose fortune allowed her second husband to have both a sugar plantation and other women in the West Indies, found herself with a three-year-old toddler when young Fougère arrived in France in August 1788. In another three years, she also acquired Fougère's half sister, Rose. Madame Audubon never had children of her own, and she never asked to become a stepmother to these two young children born out of wedlock to two different mothers in the West Indies; still she apparently accepted their presence with a good measure of maternal affection, and she became, Audubon attested, "the only mother I have ever known."[28]

She seems to have been a good one, or certainly an indulgent one, "devotedly attached to me," and, Audubon admitted, "far too much so for my own good." By the time she acquired these two surprise stepchildren from her husband, she was at least in her late fifties, living in the comfortable circumstances first assured by her own inherited wealth and then supplemented by her second husband's sea ventures and slaveholding. She was also happy enough to spend some of the Audubon family wealth on her young stepson, believing, he said, "that fine clothes and filled pockets were the only requisites" to his becoming a gentleman. The only other requisite she insisted upon was that he become confirmed as a Catholic, which he dutifully did at age seventeen; even though young Audubon was "surprised and indifferent" about taking his parents' faith, he learned the catechism and acceded to the confirmation ceremony so that "all was performed to her liking." Audubon himself was much to her liking, too, and she spoiled him, giving him "*carte blanche* at all the confectionary shops in the town," boasting to others of his accomplishments and good looks, and providing him a youth, he said, in which "all my wishes and idle notions were at once gratified."[29]

Among those "idle notions," none proved to be more important than skipping school and heading into the woods with other boys who, like Audubon, "were more fond of going in search of birds' nests, fishing, or shooting, than of better studies." Throughout his youth, Audubon explained, "there existed within me a tendency to follow Nature in her walks." Follow them he did, eventually beginning to make pictures of birds until he had, by his own count and artistic estimation, "upward of two hundred drawings, all bad enough, my dear sons, yet they were representations of birds, and I felt pleased with them."[30] (Audubon would later claim to have studied painting with the great French artist Jacques-Louis David, but there seems to be no definite evidence of his having done so. As in so many other aspects of Audubon's autobiographical "facts," this tutelage under his "honoured Mentor" seems to be a fiction of his self-fashioning.[31])

Audubon's stepmother may have allowed him the latitude to pursue his early passion for nature, but her place in the narrative paled in comparison to an even greater parental influence in Audubon's life: "But now, my dear children, I must tell you somewhat of *my* father." When Audubon wrote his wife, in 1827, about his notion that his own life might be "still more wonderful and extraordinary than that of my father," he acknowledged the standard that he would explore more fully for his sons eight years later, in "Myself." The story Audubon spins there about his father, Jean Audubon, is an almost classic rags-to-riches tale, some of it true, some not, but all of it useful for giving Audubon—and his sons—a way to locate the family lineage in an eighteenth-century success story.

Born into a household of twenty-one children, all of them boys except for one girl, young Jean had to leave home at twelve, Audubon wrote, sent off by his father with "a shirt, a dress of coarse material, a stick, and his blessing." Thus cast into the world, Jean went to sea on fishing boats, soon rising to the level of able seaman at age seventeen, and eventually coming to own several of his own ships by the time he was twenty-five. His fortunes improved dramatically three years later, in 1772, when "at twenty-eight [he] sailed for Santo Domingo with his little flotilla heavily loaded with the produce of the deep."

At that point in the tale, Audubon shifts to a direct quotation from his father: "I did well in this enterprise," Audubon père tells his son, "and after a few more voyages of the same sort gave up the sea, and purchased a small estate on the Isle à Vaches; the prosperity of Santo Domingo was at its zenith, and in the course of ten years I had realized something very considerable." Left unsaid—probably because it hardly needed saying—was that the prosperity of Saint-Domingue as a whole, along with the Audubon estate in particular, reached its zenith on the backs of enslaved black workers on the sugar plantations.[32]

While accumulating wealth in the West Indies, Audubon's father also went to war. While serving with the French forces in the American Revolution, Audubon claimed, the elder Audubon was presented with a portrait of George Washington by the general himself, "a few days only before the memorable battle of Valley Forge." The battle is hardly memorable, of course, because it never took place. More to the point, while Washington and his ragged troops were shivering through the frigid winter at Valley Forge in 1778, Jean Audubon was still basking in the warmth of Les Cayes. He sailed away from there in the spring of 1779, only to be captured and imprisoned by the British. Once French officials had negotiated his release a year later, he did assume command of a French fighting ship, and he happened to be on hand, although presumably on the water, when the combined French and American forces surrounded Lord Cornwallis at Yorktown in 1781 and extracted from him the surrender that effectively ended the British war effort.[33] Perhaps Jean Audubon met George Washington somehow in the process; perhaps he received a gift portrait from the general, but perhaps not. And perhaps his son got his military history confused or, like so many other people of the era, simply invented an encounter with Washington that never took place. For his sons' sake, however, Audubon's story of his father's favor with Washington helped establish an early and certainly valuable-seeming association with the creation of the United States.

Slaveowning, seafaring, and military service made the elder Audubon an absentee parent for much of the time, but when he did come home, he seemed an imposing presence in the eyes of his young son. Indeed, the son saw an impressive image of himself in the father: "In personal appearance my father and I were of the same height and stature," Audubon wrote, "say about five feet ten inches, erect, and with muscles of steel." He also saw something of his own personality: "In temper we much resembled each other also, being warm, irascible, and at times violent; but it was like the blast of a hurricane, dreadful for a time, when calm almost instantly returned."[34]

The contrast of violence and calm in his father's character seemed to reflect the turbulence of the times, and the father's experience in the revolutions of the eighteenth-century Atlantic world seemed to have lodged deep in the son's memory. "The different changes occurring at the time of the American Revolution, and afterward that in France, seem to have sent him from one place to another as if a foot-ball." For all that bouncing around between revolutions, however, the Audubon holdings in Saint-Domingue kept increasing in value— until, Audubon noted, "the liberation of the black slaves there." Audubon even quoted his father directly about the impact of revolutions, which, the elder Audubon told his son, "too often take place in the lives of individuals, and they are apt to lose in one day the fortune they before possessed." In writing to his own sons much later in "Myself," Audubon described the "thunders of the Revolution" in France, when "the Revolutionists covered the earth with the blood of man, woman, and child."[35]

He could just as easily have been writing about the revolution in Saint-Domingue, where, as he had already said in the previous pages, his mother "was one of the victims during the ever-to-be-lamented period of negro insurrection on the island." She wasn't, but his mulatto half sister was: Marie-Madeleine, Sanitte's first daughter with Audubon's father, still lived at Les Cayes with her mother when the slave insurgents ransacked the Audubon estate in 1792, and she died in the violence. Perhaps neither Audubon, father or son, ever learned of her death.[36] At any rate, Audubon the son abruptly dropped the topic of revolutions altogether in writing for his own sons and turned away from saying anything more: "To think of those dreadful days is too terrible, and would be too horrible and painful for me to relate to you, my dear sons."[37]

Still, whatever the "horrible and painful" elements of his father's experience, Audubon seemed to be able to recall the deeds, achievements, and, on occasion, the exact words of his father with considerable clarity. This apparent memory for detail stands in sharp, perhaps surprising, contrast to the vague, second-hand, somewhat offhand information he offered his sons about his own mother and the circumstances of his birth. Again, he claimed to rely on his father for information about the "enigma" of his origins, but in conveying that information he gave it almost no authority, leaving it shrouded in the mystery of hearsay: "I can only say what I have often heard my father repeat to me," or "I have been led to understand." On the whole, in writing "Myself," Audubon used his father as a valuable narrative device. When he needed a way to illustrate the virtues of strong character, the vicissitudes of life, or the violence of revolutionary times, he had his father for that. When he needed a way to evade the personal details and social disadvantages of his West Indian origins, he had his father for that, too.

The Specter of Saint-Domingue

Perhaps most important, he had his father to orchestrate his exit from France and his entrée into the United States, allegedly as a lad from Louisiana. Audubon's father "greatly approved of the change in France during the time of Napoleon," his son wrote, but the elder Audubon's admiration of Napoleon apparently stopped short of committing his teenaged son to the First Consul's military service. Because Napoleon had become determined to reverse the direction of the insurrection in Saint-Domingue—and to restore slavery there in the process—he began pouring upward of thirty thousand troops into Saint-Domingue in late 1801. In short order, about half of them quickly succumbed to the island's most deadly disease, yellow fever, and thousands more fell to the revolutionary forces.[38] Then, needing more and more money to fund this desperate military effort, he decided to sell a huge swath of North American territory he had recently acquired from Spain, and at the end of April 1803, France concluded the Louisiana Purchase with the United States. What those high-level decisions might mean to ordinary people in France seemed a bit unclear at the time, but Jean Audubon knew that he didn't want his son to become cannon fodder (or, equally likely, fever fodder) for Napoleon's West Indian venture. In the summer of 1803, the elder Audubon booked passage for his son to cross the Atlantic on the American brig *Hope*. He also provided his son with another sort of hope, arranging for a passport that described the teenaged boy as a "citizen of Louisiana," thus helping him dodge conscription and the unhappy prospect of being sent back to Saint-Domingue, the real place of his birth, as a soldier.

In the following year, 1804, the "negro insurrection" eventually enabled Saint-Domingue's people of color to gain their freedom from European colonizers. By that time, the movement from insurrection to independence had come to represent a menacing specter of black power for white people on both sides of the Atlantic, and the residual racial anxiety endured well into the century. Audubon had to know that. No matter what he knew about his early

days in Saint-Domingue, he also knew that, as someone with a Haitian association in his background, he clearly needed to establish himself on the white
side of the racial divide. Thus in his memoir, "Myself," Audubon plays loose
with the dates and details. His beautiful "Spanish" mother dies as a victim of
the slaves' violence, he writes, and he and his father have to flee to France to
escape the uprising—along with their black "servants," to be sure. Time is out
of joint in Audubon's story—the slave insurrection did not begin until 1791,
but his mother actually died in 1785, soon after his birth, and his father sent
the three-year-old boy to France in 1788—but chronology is not the issue. The
more important point is to portray his family—and therefore himself—as
white victims of the black unrest.

In the end, the task is not to seek some essential, absolute truth about who
Audubon really was, how "American" he was, or what sort of American he was,
black or white, or whether he might have passed for white. Both in the nineteenth-
century Atlantic world and in twenty-first-century scholarship, questions about
someone's exact identity can often lead back to ambiguity, even with DNA-based
findings. The point, rather, is to accept ambiguity as a possible element of individual identity and to see what the individual does with it. In Audubon's case,
the ambiguity of his background—not to mention his inventive, often evasive,
sometimes duplicitous discussion of his origins—offers a valuable avenue of
biographical approach. As Robert Penn Warren wrote in the preface to his poem
Audubon: A Vision, "By the age of ten Audubon knew the true story, but
prompted, it would seem, by a variety of impulses, including some sound practical ones, he encouraged the other version, along with a number of flattering
embellishments." Whatever the "true story" of Audubon's background might
have been, the real fascination still lies in the "embellishments," including
Audubon's role as, to use Warren's term, a "fantasist of talent."[39]

Audubon kept up the deception throughout his life. In 1837, when he had
become a celebrity on both sides of the Atlantic, he wrote to a close friend, "I
am glad, and proud Too; that I have at last been Acknowledged by the public
prints as a Native Citizen of Louisianna."[40] For his own part, he never took to
the public prints to acknowledge his own West Indian origins, nor did he
discuss the larger geopolitical implications that might be attached to his place
of origin. He offered only a simple explanation in the privacy of his personal
memoir, noting that his father "found it necessary to send me back to my own
beloved country, the United States of America."[41] In that regard, Audubon was
coming "back" to a place he had never been. The United States, his "beloved

country," had indeed acquired Louisiana, allegedly the place of his birth, but he would not set foot in Louisiana until 1819, sixteen years later. Instead, he first came ashore in the United States in late August 1803, when the *Hope* docked in New York City.[42]

It turned out to be a bad time to arrive. Audubon immediately came down with a case of yellow fever, an epidemic disease that swept through the city between mid-July and late October, sickening over 1,600 people and killing upward of 700. At the time, physicians debated the origins of yellow fever, some saying that it stemmed from urban filth, particularly decomposing animal and vegetable matter in hot weather, with others arguing that it came in from outside the country. We now know that the latter explanation makes more sense: Yellow fever is a mosquito-borne illness that first came into North American seaports on ships from the West Indies, including Saint-Domingue.[43] Luckily enough, Audubon survived yellow fever, thus gaining a lifetime immunity. Still, he never noted the obvious irony of his illness. Before this West Indian–born, French-speaking immigrant could begin to fashion a new identity as an American—and eventually, as the American Woodsman—he first had to recover from a malady that stemmed, quite unpleasantly, from the place of his origin, the French West Indies. Try as Audubon might, there was no escaping the specter of Saint-Domingue.

Chapter 2

―――≋―――

Hearing Birds, Heeding Their Call

At a very early period of my life I arrived in the United States of America,
where, prompted by an innate desire to acquire a thorough knowledge of the
birds of this happy country, I formed the resolution, immediately on my landing, to spend, if not all my time in that study, at least all that portion generally
called leisure, and to draw each individual of its natural size and colouring.
—John James Audubon, "Account of the Method of Drawing Birds"

Audubon's father didn't send his son to America to become a bird artist. In
addition to getting him well beyond Napoleon's militaristic reach, he had a
much more prosaic plan in mind. He wanted the teenaged boy to learn to speak
and write better English, which could be a useful skill for another task that
might soon be at hand: helping to manage Mill Grove, a Pennsylvania farm
that the elder Audubon had acquired in 1789. Audubon père could hope that,
with some combination of guidance and experience, his son would someday
become a capable overseer of the Mill Grove operation; he at least assumed
that the boy could benefit from the combination of responsibility and opportunity the place provided. He sent him off to America, then, with a letter of
credit and a connection to a well-trusted local agent, giving him just the sort
of support a young man might need for a promising start in the new nation.

Looking back some years later on that American beginning, Audubon
clearly appreciated his father's sending him to the Pennsylvania farm. He wrote
fondly of his early days at Mill Grove, embracing the place as a "blessed spot"
as he looked upon the work his father had done years before, "the even fences
round the fields, or on the regular manner with which avenues of trees, as well
as the orchards, had been planted by his hand."[1] In fact, that work had been
done by the hand of a tenant, who had also discovered lead deposits under-

ground sometime in the 1790s, giving an additional dimension to Mill Grove. Lead wasn't as good as gold, but in a bullet-hungry hunting country like the United States, such a mineral bonus could certainly be a substantial asset. With the promising combination of land and lead, then, everything seemed quite well laid out for young Audubon, just waiting for him to make it even better.

Unfortunately, however—or, perhaps fortunately, given his eventual artistic success—Audubon had neither head nor heart for farm management, and he never made the most of the opportunity his father first offered. By his own account, he never made much of the various other business ventures he pursued during his first two decades in the United States. All he really wanted, he insisted, was to become a bird artist.

The Precarious Calling of Art

Becoming an artist of any sort has always been a low-percentage career move, particularly in an upstart place like the early nineteenth-century United States. The number of painters who became reasonably prominent at the time could probably be counted on two hands, and those who gained lasting significance on one. We might think immediately of John Singleton Copley (1738–1815) in the second half of the eighteenth century and Washington Allston (1779–1843), Thomas Cole (1801–1848), and Asher B. Durand (1796–1886) in the first half of the nineteenth, all of whom followed different paths to their profession. Copley, for instance, grew up in a poor Boston household headed by a widowed mother, an Irish immigrant who ran a tobacco shop, but he taught himself to paint well enough to make a comfortable living by making exquisite portraits of middle-class New Englanders in the era of the American Revolution. The American colonies, however, had neither the art museums nor the painting masters that could help an aspiring artist rise to the next level, both professionally and financially. In 1774, Copley left America for England, where he studied at the Royal Academy, worked with the British master Benjamin West, and then took off on the near-requisite tour of the Continent to see the treasures of France and Italy. So, too, did Allston, who graduated from Harvard College in 1800 and soon thereafter followed in Copley's footsteps to England, the Royal Academy, Benjamin West, and the Grand Tour. Theirs became the path that other aspiring American artists would hope to follow for the next two centuries.

Cole and Durand came up the harder way. Born into middling backgrounds at best, both of them the sons of merchant-tradesmen, neither had the opportunity to cross the Atlantic for artistic training. Instead, they learned their trade by becoming itinerant artists, slogging along the path of other painters who populated the broader base of the pyramid of American art, most of whom never made it to the loftier reaches of the American art world, such as it was. Itinerants catered to the rising aspirations of clients in the middling range of society, more ordinary folk who nonetheless had the wherewithal to pay for a "correct likeness"—even if the likeness turned out to be not exactly correct. In the first decades of the nineteenth century, the demand for personal pictures grew rapidly as a form of consumption in a commercially expansive society, and itinerant artists carried their brushes from town to town, giving people the images—and emblems of status—they wanted. Largely self-taught, relying more on design books and simple observation than on a formal training, these artist-entrepreneurs did thousands of portraits of individuals, family groups, and sometimes even prized livestock, creating affordable artwork destined to hang on the walls of a family's parlor rather in the staircase of a grand estate or halls of a gallery. They gave their sitters something of vernacular value, and then they moved on to do the same for people in the next town.[2]

Cole and Durand lived the life of the itinerant painter for a while, but both eventually turned their artistic attention to the American landscape and, equally important, attracted the attention of a prominent patron. John Trumbull, the well-born and successful artist, became a valuable advocate for both, helping them attain commissions from wealthy clients, members of the emerging American elite who would willingly pay for a painter's artistry—sometimes if only to satisfy their own vanity. Patronage had a price, and their clients quite often dictated the tone, if not the actual content, of their paintings: In a time of increasing urbanization and industrialization, many patrons preferred images of romantic, even nostalgic, nationalism, paintings of a past American landscape that could be eternally preserved on canvas. Cole and Durand both became famous and prosperous as the twin pillars of the Hudson River School, but they sometimes felt their talents constrained, even squandered, by having to pander to the demands of aristocratic clients who commissioned the work and took it as their right to influence its execution.[3]

Audubon, like Cole and Durand, would work his way upward while still following a career path very much of his own making. Along the way, he would have to undertake the kinds of ad hoc tasks that had become common among

America's itinerant artists—making quick and cheap portraits, painting signs, or teaching art classes—but only as a financial means to a larger artistic end. Audubon never quite used the term "struggling artist" to describe himself, but he did indeed struggle, usually without much help from anyone else. Audubon came from a reasonably prosperous family, and even though he claimed (almost certainly falsely) to have studied with the great French painter Jacques-Louis David, his background in France did little for his art career in America. Friends of his father and, later, father-in-law proved to be useful allies, but none of them became as committed a champion as John Trumbull, much less a financially supportive patron of Audubon's artistic ambitions. Perhaps the best one can say is that Audubon developed such a single-minded dedication to one single task—drawing birds—that he did not have to bow to the artistic expectations or a priori demands of a patron. To be sure, he would spend an enormous amount of time and trouble trying to track down people wealthy enough to pay for his "Great Work," but he did so in pursuit of subscriptions, not commissions. For better or worse, he had no one telling him what to do, and he could follow his own passion.

We now know the remarkable result. In strictly aesthetic terms, he became a better bird artist than anyone else had ever been: The proof is in the pictures. Moreover, no one in nineteenth-century America so successfully combined artistic and scientific ambition or did more to bring nature to the nation. Audubon's birds became emblematic of the fusion of art and science in antebellum American culture.

The Persistent Calling of Birds

Still, for all the fame and lasting success of Audubon's astounding lifetime achievement, *The Birds of America*, there remains a basic question that has to be asked: Why did he do it? Why did he devote the better part of his adult life to producing a huge book about every bird known in North America? How sensible would it be, after all, to make one's calling out of wandering through the woods or along the riverbanks and seashores, looking for birds, shooting them, arranging their decomposing carcasses into lifelike poses, drawing the dead birds, filling in the pictures with paint, working with an engraver and a publisher to make a big and expensive book, and then wandering around some more, essentially from door to door, to try to sell the book to skeptical customers? All sorts of people, from the middling ranks to the upper reaches of society, wanted individual and family portraits, especially in the era before photography made accurate likenesses readily available. Socially prominent patrons might also pay quite well to commission a large landscape, especially if they could exert enough influence over the artist to shape the painting to their vision of a romanticized past. Many devotees of natural history would also pay for one-off paintings of plants and animals—but a whole collection of birds? Of all the many ways a young man might try to make a good living in the energetic but often erratic economy of the early American republic—and as we shall see, Audubon tried more than a few—making and marketing an oversized book of bird pictures might have seemed among the least likely paths to success, an almost crazy career choice even for an artist. What sort of person would devote his life to such a challenging and, at the outset, financially unpromising pursuit?

In Audubon's own account of his life, the answer became clear: someone whose passion began to overshadow any sort of profession, until eventually the two merged into one. It is difficult to say exactly when Audubon determined that painting birds would become his life's work. He began drawing them as a young boy in France, but he apparently destroyed much of that early work. It was only when he came to the United States, he wrote, that he resolved "to draw each individual of its natural size and colouring," and he dated "the *real* beginning of my *present collection*, and observations of the habits of some of these birds, as far back as 1805."[4] He claimed to have been driven by what he called an "innate desire" to know and depict the birds in this new land, an impulse that gripped him as soon as he landed on American shores and that propelled him throughout the rest of his life. Still, writing some three decades later, in the last volume of *Ornithological Biography*, he admitted that had anyone suggested at the outset that he would ever complete "a work comprising five hundred species of birds of the United States and British America, I should have smiled and shaken my head."[5]

He began his artistic career long before it could even be called a career, in fact, first drawing birds as a part-time pastime, in "all that portion generally called leisure," doing something he wanted to do, perhaps *had* to do, while he was doing something else—trying to make a go of his father's farm, running his own store and steam mill, or painting portraits and doing the other sorts of artistic hackwork he hated. Sometimes, he complained, something else got in the way of the one thing he liked most of all: "I have often been forced to put aside for a while even the thoughts of birds, or the pleasures I have felt in watching their movement, and likewise to their sweet melodies, to attend more closely to the peremptory calls of other necessary business."[6] Whatever his other responsibilities, though, Audubon could not help heeding the call of those "sweet melodies," and the art and science of birds would become his sole and certainly most necessary business—with, to be sure, a series of personal and professional missteps that put his eventual success in much sharper contrast.

That contrast, though, stands as an important part of Audubon's life story, an almost necessary element in the larger narrative he created about the artistic and scientific trajectory of his life. The recurring conflict he frequently describes between "peremptory calls of . . . necessary business" and his more pleasurable "thoughts of birds" requires reading with some measure of caution, if not skepticism.

When it came to taking care of business, he tended to depict himself as inconsistent, impatient, and very easily distracted, sometimes making bad decisions, sometimes simply not caring enough to do the work before him. He wrote that he reveled in the delightful diversions of his business trips, which gave him ample opportunities to track down birds while he should have been looking after his merchandise: "Were I here to tell you that once, when travelling, and driving several horses before me laden with goods and dollars, I lost sight of the pack-saddles, and the cash they bore, to watch the motions of a warbler, I should repeat occurrences that happened a hundred times and more in those days."[7] Written some years after the fact in "Myself," Audubon's autobiographical sketch for his sons, this frank confession of failure—what might be called a form of professional attention deficit—did not go on to drive home a fatherly lesson about the need for better commitment to commercial affairs. Instead, Audubon drew a contrast between his fascination with birds and his indifference, even disdain, for day-to-day business.

Just as Audubon used various written accounts of his life—particularly the autobiographical "Myself" but other briefer passages throughout *Ornithological Biography* as well—to complicate, at times even obfuscate, the details of his origins, so, too, did he use his writings to underscore his distaste for, and ultimate failure in, the world of commerce. In the more than two decades between first arriving in the United States in 1803 and then departing for Great Britain in 1826, to devote himself fully to the production of *The Birds of America*, Audubon did indeed have his financial ups and downs. He did well enough in his financial affairs to be able to acquire considerable property—both land and enslaved human beings—to create a comfortable existence, and then he lost essentially everything in the Panic of 1819.[8] The failure was real, and the distaste no doubt just as real.

His later portrayal of his struggles in those years can sometimes appear to be a near-parable about one man's high-minded pursuit of art and science within the less lofty economic context of American society. The story is hardly that simple, though. In an era of ever-expanding economic activity in America, Audubon tried his hand at enough entrepreneurial enterprises to seem just as energetic and resilient as the next small-scale capitalist, and if he turned out to run up against financial failure at one point, so, too, did thousands of others. The point is not to declare his business narrative altogether true or false, but to be aware of the always self-conscious construction of the image he sought to offer the world. Audubon became a self-made man in more ways than one.

The Birds and the Beauty of Mill Grove

Arriving at his father's farm in the autumn of 1803, Audubon alighted in an ornithologically fortunate spot. Situated along Perkiomen Creek just before it flows into a bend in the Schuylkill River, about twenty-five miles west of Philadelphia and less than a hundred miles east of the Kittatinny Ridge, Mill Grove lay within a major American flyway. According to a leading modern-day naturalist, the Kittatinny region is "one of the world's most famous migration corridors," a true "birding paradise" now, and probably even more so back in Audubon's day.[9] Contemporary observers certainly celebrated the avian abundance of eastern Pennsylvania. William Bartram, whose 1791 *Travels* became one of the best works of natural history in the new nation, wrote about bird migrations through Pennsylvania, taking happy note of "those beautiful creatures, which annually people and harmonise our forests and groves, in the spring and summer seasons."[10] From his professorial post at the University of Pennsylvania, Benjamin Smith Barton likewise studied the multitudes of birds that passed within eighty miles of Philadelphia, from the rarely seen "Occasional Visitants," such as the Great White Owl (or Snowy Owl), to the more regular arrivals of "Passeres," none more numerous than the massive flocks of Passenger Pigeons. Barton attributed part of the attraction to the region's rural areas, where "the hand of man, by clearing and by cultivating the surface of the earth, contributes essentially to the greater uniformity in the temperature of climates," thus making the environment more inviting to migrants. In turn,

Figure 1. Mill Grove Farm, Perkiomen Creek, Pennsylvania, by Thomas Birch, ca. 1820. Oil on wood panel, 16 1/4 x 24 3/8 in. Object #1946.161. New-York Historical Society.

the annual migrations also influenced human behavior. He pointed, for instance, to the Pewee, "one of the earliest Spring birds of passage," typically arriving in the vicinity in mid-March: "We have seldom hard frosts after the arrival of this bird, which seems to give a pretty confident assurance to the farmer, that he may very soon begin to open the ground and plant."[11]

Mill Grove thus provided the perfect place for a bird-conscious boy like Audubon. He seemed little inclined to worry much about the seasonal call to "open the ground and plant," leaving most of the agricultural concerns to the farm's tenant, William Thomas, and his family. Instead, Audubon admitted that living in this new place liberated him and allowed him to indulge in all the engaging activities he had enjoyed back in France: "Hunting, fishing, drawing, and music occupied my every moment," he wrote; "cares I knew not, and cared naught about them."[12]

He did get to know the neighbors, and he came to care quite a bit about one of them, the young woman who would become his wife. Just across the road from Mill Grove stood another farm, Fatland Ford, which had lately come into the possession of another recently arrived immigrant, an Englishman named William Bakewell. At first, Bakewell's Englishness ran up against Audubon's Anglophobia—the English were, after all, the traditional enemy of France, and they had twice made his father a prisoner of war—but Audubon quickly came to realize that, except for being British, the Bakewells might be close to perfect neighbors. William Bakewell liked to hunt, was an expert marksman, and had a big house, beautiful dogs, and, perhaps best of all, an especially fetching seventeen-year-old daughter named Lucy.

Lucy immediately caught Audubon's eye, exuding "that certain '*je ne sais quoi*' which intimated that, at least, she was not indifferent to me." As soon as he sat down in the room with her, he just stared, "my gaze riveted," while she made polite chit-chat with her good-looking visitor. She looked good to him, too, and when she stood up to go help produce the family meal, Audubon noticed that "her form, to which I had previously paid but partial attention, showed both grace and beauty; and my heart followed every one of her steps." The rest of him, though, soon followed her father's steps: "The repast over, guns and dogs were made ready," and Audubon went out hunting with Mr. Bakewell and his boys.[13] But Audubon knew he had everything a nineteen-year-old could want, and right on the other side of the road. Mill Grove seemed a lucky spot to have landed, indeed.

To complement Lucy's je ne sais quoi, Audubon brought his own joie de vivre to the Bakewell household. Audubon let himself go before the Bakewells, giving them a one-man show of his exuberant spirit—a self-celebrating, scene-stealing penchant for performance that would stay with him even into his last years. Lucy's younger brother, William Gifford Bakewell, captured some of Audubon's eclectic talent in a quick survey of the sorts of prowess Audubon displayed before the family: "He was an admirable marksman, an expert swimmer, a clever rider . . . he was musical, a good fencer, danced well, and had some acquaintance with legeredemain tricks, worked in hair, and could plait willow baskets."[14]

Audubon himself was hardly the type to downplay his personal profile with false modesty, and he later described himself in those days as being "extremely extravagant." "I was ever fond of shooting, fishing, and riding on horseback," he wrote, and he took considerable pride in doing those things well and, equally important, well equipped. His guns, fishing tackle, and horses had to be the best. So did his clothes. He admitted that going hunting in satin breeches, silk stockings, and a ruffled shirt might have been a bit foppish, "but it was one of my many foibles, and I shall not conceal it."[15] (Two decades later, when first in London, he would do something of the reverse, going against cultural context by dressing in the garb of an American backwoodsman, but once again using a surprising sartorial display to draw attention to himself.)

On the other hand, he noted that he was "temperate to an *intemperate* degree," and his abstemious eating and drinking habits gave him an "uncommon, indeed iron, constitution." He existed, he said, "on milk, fruits, and vegetables, with the addition of game or fish at times, but never had I swallowed a single glass of wine or spirits until the day of my wedding."[16] Not many Frenchmen could make that claim, nor could many citizens of early nineteenth-century America, a bibulous country that one scholar has called the "Alcoholic Republic."[17] But Audubon had left one nation behind and was just getting his footing in the other, and he took personal pride in making the transition on his own terms. Still, boastful though he could be about the "iron constitution" of his youth, he could also later look back at the time and see himself as a work in progress: "I had no vices, it is true, neither had I any high aims."[18]

High aims or not, he still had birds. For all the time he spent hunting and fishing and riding and dancing and wooing Lucy, he always turned his attention back to his growing avian obsession. In the short time he first lived in Pennsylvania, he began a new project, "a series of drawings of the birds of America, and . . . a study of their habits." The term "study" seems especially apt, because it was in the account of his time at Mill Grove that Audubon described a now-famous but then-novel experiment in bird banding, which still stands as a much-respected contribution to American ornithological practice. In early April of 1804, finding a pair of Eastern Pewees in one of the many caves above Perkiomen Creek, Audubon visited the cave day after day until "the birds became more familiarized to me, and, before a week had elapsed, the Pewees and myself were quite on terms of intimacy." Audubon soon observed one of the most intimate moments of the pewees' relationship: the laying of six eggs and the feeding of the five young birds that survived. By that time, he said, "The old birds no longer looked upon me as an enemy," and they made no fuss when Audubon handled their young and "fixed a light silver thread to the leg of each, loose enough not to hurt the part, but so fastened that no exertions of theirs could remove it." Sure enough, the next year he caught some of the banded pewees in roughly the same area, and he thus determined that they migrated back to the neighborhood of their birth.[19] Audubon had not yet developed the systematic practices that might constitute a true scientific method, but with his persistent field observation and his innovative approach to bird identification, he was clearly taking some useful first steps toward becoming a serious naturalist.[20]

Audubon also wrote of using thread to make an innovative approach in his drawings of birds, an artistic technique that he would employ throughout his life. While again observing a pair of pewees and their "innocent attitudes," he wrote, "a thought struck my Mind like a flash of light"—the idea that the only way to capture nature on paper would be to represent the birds as they were in life, "alive and Moving!" If alive, most birds are indeed moving, and they don't tend to stay in place for long, so trying to draw them as they fly or flit from branch to branch could be frustrating, all but impossible, work for anyone. Audubon admitted he "could finish none of my Sketches." He knew how to draw dead birds, of course: "After procuring a specimen, I hung it up either by the head, wing, or foot, and copied it as closely as I possibly could."[21]

But dead birds just hung there, looking dead, and Audubon wanted to bring them back to life. He kept pondering the problem until early one morning, well before dawn, he had his "aha!" moment. He leapt out of bed, saddled his horse, and rode a fast five miles to Norristown, where, given the hour, "not a door was open." Having time to kill and being too agitated just to wait, he went on to the Schuylkill River, jumped into the water and took a chilly bath, and then retraced his ride back to Norristown. This time he found an open shop, and he bought what he needed: thin wire. Racing back to Mill Grove, he grabbed his gun, rushed down to Perkiomen Creek, and shot the first bird he could find, a Belted Kingfisher. Using the wire he had just purchased, he fixed the dead bird to a board and got its head and tail looking just right, and then he drew it on the spot. "Reader this was what I Shall ever call my first attempt at Drawing actually from Nature," he explained, "for then Even the eye of the Kings fisher was as if full of Life before me whenever I pressed its Lids aside with a finger."[22]

Drawing from nature—or at least recently deceased specimens put in a natural-seeming pose—became an important part of Audubon's artistic signature: "I have *never* drawn from a stuffed specimen."[23] In the end, that claim did not turn out to be altogether true, but it still served as a proud declaration of artistic independence.

Moving on from Mill Grove

Banding birds and arranging them in lifelike poses might have been fine ornithological and artistic activities, but neither one did much for the management of Mill Grove. Audubon's father already knew his teenaged son didn't yet have the talent and temperament to take on that sort of responsibility. When the boy was growing up in France, Jean Audubon had been mildly tolerant of his boy's fascination with nature—"so pleased to see my various collections," the younger Audubon wrote, "that he complimented me on my taste for such things"—but the father also wanted him to study something more practical, perhaps even more manly, like seamanship and engineering.[24] When young Audubon failed at those things, Jean Audubon had good reason to feel exasperated by his son's aversion to schooling and unprofitable-seeming predilections. Not surprisingly, when he sent the boy off to America for safekeeping, the elder Audubon also had good reason to feel that his son still needed adult supervision.

That came most immediately in the person of Francis Dacosta, one of Jean Audubon's allies from Nantes, whom the elder Audubon had enlisted to go to Pennsylvania to oversee both his lead mine and teenaged son. Young Audubon and Dacosta turned out to be a bad match, however. "This fellow was intended to teach me mineralogy and mining engineering," Audubon later wrote, "but, in fact, knew nothing of either." Indeed, Dacosta quickly found a place on Audubon's life list of much-despised enemies, becoming a useful villain in Audubon's narrative of his early life, representing the sort of treachery that could lead a young innocent astray. When Dacosta tried to curry a bit of favor by complimenting the young man's early work—"he assured me the time might come when I should be a great American naturalist . . . and I felt a certain degree of pride in these words even then"—the flattery faded fast. Instead, Audubon came to characterize Dacosta as a "covetous wretch, who did all he could to ruin my father, and indeed swindled both of us to a large amount."[25]

What Dacosta did most to ruin were Audubon's prospects for marrying Lucy, speaking "triflingly of her and her parents" and telling young Audubon it would be beneath him to marry into the Bakewell family.[26] In that regard Dacosta may well have been reflecting the feelings of the father, Jean Audubon, who also had doubts about the wisdom of his son's rushing into marriage. "My son speaks to me about his marriage," the elder Audubon wrote to Dacosta. "If you would have the kindness to inform me about his intended, as well as about her parents, their manners, their means, and why they are in that country, whether it was in consequence of misfortune that they left Europe, you will be doing me a signal service, and I beg you, moreover, to oppose this marriage until I may give my consent to it." Jean Audubon probably cared more about the Bakewells' means than their manners, and like all fathers trying to size up the prospects of the potential in-laws, he wanted to be sure they wouldn't be marrying into the family for the money: "Tell these good people," he concluded, "that my son is not at all rich, and that I can give him nothing if he marries in this condition."[27]

Young Audubon bristled at this intrusion into his love life, and he blamed it all on Dacosta. For a while, he fell into a "half bewildered, half mad" fury and thought about killing Dacosta, but an elderly lady "quieted me, spoke religiously of the cruel sin I thought of committing," and eventually talked him out of it. Thanks to those wise words, Dacosta stayed alive, and Audubon stayed out of jail and off the gallows. Instead, Audubon decided to head back to France, where he would make his case to his father. After a storm-tossed Atlantic crossing in the spring of 1805, Audubon arrived at La Gerbetière, his father's home near Nantes, where he happily fell into "the arms of my beloved parents."[28] Wasting no time, he spilled out his accusations against Dacosta— who, as it turned out, had already lost credibility in his relationship with Jean Audubon anyway—and the much-despised supervisor essentially ceased to be an issue.

Then, like any young man coming back home on what amounted to an extended vacation, Audubon took full advantage of the family largesse: "In the very lap of comfort my time was happily spent," he wrote. "I went out shooting and hunting, drew every bird I procured, as well as many other objects of natural history and zoology."[29] He also studied taxidermy with a family friend and physician, Charles Marie D'Orbigny, developing a skill that would become professionally useful in the decades to come.

But in addition to his further bird research, Audubon's trip to France in 1805–1806 produced two important results that would shape his return to the United States. First, he got his father's tentative permission to marry Lucy Bakewell. Despite Jean Audubon's initial doubts about the possible gold-digging ambitions of the Bakewells, he yielded on the marriage question, insisting only that young Audubon find some means to support a wife. Second, the elder Audubon arranged for young Ferdinand Rozier, the son of a family friend, to form a business partnership with his own son, whereby the two would go to the United States, try to make the Mill Grove lead mine a viable venture, deal with Dacosta, then seek out whatever other sort of business they could find to turn a profit.[30]

All the two young Frenchmen had to do, then, was to get out of France before Napoleon conscripted them. Rather than let the young men get snatched away into the military, the elder Audubon used his pull to help them get passports (albeit somewhat bogus-looking ones) and book passage on an American ship, the *Polly*, bound for New York. After a handful of harrowing oceanic adventures—enduring a ransacking at the hands of a British privateer, then making a close escape from a pair of British frigates, and finally running aground during a violent storm in Long Island Sound—the *Polly* managed to make it safely to New York Harbor in late May 1806. And there the two young Frenchmen disembarked and set out to begin what Audubon would later call "a partnership to stand good for nine years in America."[31]

Audubon in Business

As in many partnerships, the first few years were the most uncertain, but in some ways also the best. Audubon and Rozier initially had ambitious intentions of taking up residence at Mill Grove, making a go of the mining operation and the farm as well, and perhaps ousting Dacosta in the process. Unfortunately, neither had any experience in operating a lead mine, and neither had any interest at all in doing the difficult field labor that the farm required. They did, however, still have the obstacle of Dacosta, who, as a result of his earlier arrangement with Audubon's father, also held title to a large portion of the whole Mill Grove estate. He would be hard to move. Realizing they would probably not be able to beat Dacosta, then, and certainly not wanting to join him, the two young Frenchmen decided to sell him the remainder of Mill Grove.

That decision also led them in slightly different directions for a while: Rozier, who was not at all fluent in English, became a clerk in a French-owned importing business in Philadelphia, while Audubon went to New York to work for another wholesale import house, this one owned by Benjamin Bakewell, the uncle of Lucy Bakewell, the young beauty Audubon already knew he wanted to marry. Audubon stayed with the Bakewell business for about a year, from the fall of 1806 until the late summer of 1807, and he did a good-enough job to stay in Bakewell's good graces, a useful boon to Audubon's matrimonial aspirations. Since the elder Audubon needed some assurance that his son could support himself and a wife, the job with Bakewell certainly helped. It also helped Audubon pursue his other passion. During his time in New York he spent what free time he could away from the countinghouse, scouring the shoreline and wooded areas of the city for birds to draw. He also developed a friendship with Dr. Samuel L. Mitchill, a New York naturalist who allowed Audubon to practice taxidermy on his specimen collection, stuffing birds and mammals. On the whole, Audubon's brief stint in New York seemed reasonably well spent. Bakewell provided both income and indulgence, helping Audubon learn something about business, but also letting him roam the streets in search of birds.[32] Indeed, trying to balance business and birds became the main theme of Audubon's early days in the United States.

In August 1807, he and Rozier decided to go back into partnership again, this time getting away from the main East Coast cities and setting up a retail shop in the distant river town of Louisville, Kentucky. Having arranged for a starting stock of store goods, bought from Benjamin Bakewell on generous terms, they headed west by stage on August 31. By the latter part of September, they had started in business—but, as luck would have it, just before Thomas Jefferson's Embargo Act of 1807 disrupted trade everywhere, even for small-time operators like Audubon and Rozier.

Audubon couldn't do much about the bigger picture, and he had other business on his mind anyway: He wanted to marry Lucy. In March 1808, after he had been in Louisville for just over six months, he took the trek back to Pennsylvania to ask Mr. Bakewell for Lucy's hand, and on April 8, 1808, they were married at Fatland Ford. With that, the young couple headed off for Louisville, following the same rough route Audubon and Rozier had taken the previous year. After enduring almost two weeks of stagecoach bumps and flatboat exposure, they arrived in Louisville and settled in an extended-stay hotel, the Indian Queen. Surveying her new situation, Lucy sent her English cousin some hopeful-seeming words about the Louisville environment ("The country round is very flat, but the land is very fertile"), the inhabitants ("very accommodating"), and the houses ("some of them are very prettily laid out indeed"). Still, she confessed to being "very sorry there is no library here or book store of any kind for I have very few of my own and as Mr. Audubon is constantly at the store, I should often enjoy a book very much whilst I am alone."[33]

Being alone became a big part of the story of Lucy's life. As she would soon find out, her husband's habit of being "constantly at the store" would quickly dissipate, and he would spend as much time thinking about birds as business. Over the coming years, in fact, birds would become his business, and in pursuit of that business he would leave her alone for long stretches of time, most notably a three-year stint in England, 1826–1829, to begin producing *The Birds of America*. In the early years of their marriage, though, Audubon and Lucy would be together enough to have four children: two boys, Victor Gifford (born 1809) and John Woodhouse (born 1812), both of whom would turn out to be important assistants in their father's work; and two girls, Lucy (born 1815) and Rose (born 1819), both of whom died in infancy. Beyond being a mother, though, Lucy's main role in life eventually meant being a quiet contributor to the Great Work that defined Audubon's career and truly became a family business. Before any of that would happen, however, she had to adjust, many times over, to being married to a man whose head so often seemed to be in the clouds, his eyes typically turned to the tops of trees. Like Audubon's father, Lucy probably never expected her husband to become a bird artist.

She probably also never suspected that his life would become so affected by his first meeting with another—really, *the* other—bird artist in America.

Louisville Encounter

Every field has its famous, even defining moments, and there's a much-told Audubon story that merits a place in the apocrypha of early American art and science. The incident in question seems so implausible and yet so perfect that no one would dare have the gall to invent it completely—not even Audubon himself. And because the story comes from two sources, not just Audubon but the other main character as well, we might well assume it actually happened, even if not exactly as it has been told by either of them. Perhaps the best thing to say is that the story is *close* to being true, and it points to larger truths beyond the specific narrative details.

Audubon tells it this way: "One fair morning" in March 1810, he writes, he happened to be working behind the counter of the Audubon-Rozier store in Louisville, when "I was surprised by the sudden entrance into our counting-room of Mr. Alexander Wilson, the celebrated author of the 'American Ornithology.'" From the beginning, the story sets up a much-repeated contrast, with Audubon trying to take care of business but being distracted by birds or, in this case, pictures of birds. When Audubon took a look at Wilson's work, he saw something that filled his artist's eye with admiration, perhaps even envy— two large, leather-bound books with a total of eighteen engraved, hand-colored plates and over three hundred pages of accompanying text, volumes that were physically impressive in heft and visually striking in appearance.[34]

In those first few lines of his narrative, though, Audubon took quite a leap ahead, looking considerably beyond the time frame of the encounter at hand. "Celebrated author" would have been a bit of a stretch for Wilson, at least at the time he showed up in Audubon's store. A gruff and grumbling Scotsman who had worked as a weaver and then as a political activist and largely unsuccessful poet in his native Scotland, Wilson had failed at essentially everything he had attempted. Feeling financially frustrated and politically persecuted in Scotland, he left for the United States in 1794, but the new land of opportunity didn't seem to help. He faced yet another string of occupational failures, from weaving to day labor, the final straw being teaching in a school in Gray's Ferry, Pennsylvania, just outside Philadelphia. Wilson had worked as a schoolteacher before, and he approached this new post with something less than pedagogical enthusiasm: "I shall recommence that painful profession once more with the same gloomy, sullen resignation that a prisoner re-enters his dungeon or a malefactor mounts the scaffold," he wrote to a friend, offering a prediction that this new position would not turn out to be a success.[35] It didn't.

While in the Philadelphia area, however, Wilson had the good fortune to make the acquaintance of several of the city's prominent men—most usefully the naturalist William Bartram and the engraver Alexander Lawson—who became allies, both personally and financially, in encouraging him to try his hand at art. Even sympathetic friends might reasonably have surmised that, for a forlorn loser like Wilson, there seemed precious little left. Still, with some helpful lessons under Bartram's guidance, Wilson finally decided upon his life's true (and final) calling: "I am most earnestly bent on pursuing my plan of making a collection of all the birds in this part of North America," he declared. That collection would eventually become the nine-volume compendium of bird drawings and written descriptions, *American Ornithology* (1808–1814).[36] On the first page of the first volume, Wilson made the high-minded declaration that he wanted only to "draw the attention of my fellow-citizens . . . to a contemplation of the grandeur, harmony, and wonderful

Figure 2. Portrait of Alexander Wilson, probably painted by Thomas Sully, 1809–1813. Oil on wood, 23 1/4 x 22 inches. American Philosophical Society, gift of Dr. Nathaniel Chapman, 1822.

variety of Nature," adding that "*lucrative* views have nothing to do in the business."[37]

Wilson certainly turned out to be right about the "lucrative" part. Putting behind him the comparative comfort and support he enjoyed in Philadelphia, he started stumping the country off and on for a couple of years, lugging around examples of his work to show potential customers in order to secure subscriptions to his larger but still incomplete project. (At the time Wilson happened to come to Louisville, only the first volumes had begun to appear in print, in a limited edition of two hundred copies.) A subscriber to *American Ornithology* would be expected to come up with $120 for the full work, a hefty sum that only prosperous individuals and institutions could afford, and even many of them seemed disinclined to pay the price. In one instance, Wilson complained, a potential purchaser of *American Ornithology* "turned over a few leaves very carelessly; asked some trifling questions; and then threw the book down, saying—*I don't intend to give an hundred and twenty dollars for the knowledge of birds!*" After picking up his sample volumes and heading for the door, Wilson later grumbled that if "science depended on such *animals* as these, the very name would ere now have been extinct."[38] After suffering a depressing share of such insults and indifference, Wilson came to Louisville with no celebrity billing, just four letters of introduction that he hoped would open the doors of "all the characters likely to subscribe" and, most important, with the two volumes of his bird drawings, which he wearily lay on the counter in Audubon's store.[39]

As Audubon later related the opening moments of their meeting, Wilson's work seemed almost too good to be true and too good to pass up—at least at first viewing. Even though he could hardly afford to do so, Audubon was about to sign his name to the subscription list when, just then, his partner, Rozier, stopped him. After looking at Wilson's bird images, Rozier whispered to Audubon (in French, their common tongue) that Audubon's own drawings were "certainly far better" and that Audubon himself was just as good as Wilson: "You must know as much of the habits of American birds as this gentleman." A bit awkwardly and perhaps a bit too suddenly, Audubon backed away from his decision to subscribe to Wilson's work. Wilson seemed annoyed, Audubon continued, but he asked to see Audubon's own images, and Audubon laid a portfolio of his own drawings before his visitor. Wilson found it hard to believe that anyone else had also begun the same sort of work on birds, and, like any suddenly insecure author, he asked Audubon if he planned to publish his images. When Audubon said no, Wilson then asked if he could borrow some of Audubon's drawings, and Audubon said yes. Audubon also suggested the possibility of an artistic collaboration, or at least collegial attribution. Even though he remained convinced of his decision not to subscribe to Wilson's work—"for, even at that time, my collection was greater than his"—he did think that Wilson might be willing to print some of his images: "I offered them to him, merely on condition that what I had drawn, or might afterwards draw

and send to him, should be mentioned in his work, as coming from my pencil."
Wilson never took him up on the offer, however, and the Scotsman soon left
Louisville, "little knowing how much his talents were appreciated in our little
town, at least by myself and my friends." That, at least, was Audubon's version
of the encounter, which he published in the first volume of *Ornithological
Biography* in 1831.[40]

For his own part, Wilson never wrote much about his visit to Louisville,
and he never mentioned Audubon by name in *American Ornithology*. He did,
however, offer a few details in his diary. On March 19, he wrote that he examined Audubon's drawings, which he pronounced "very good," and he noted
that Audubon had two new birds, "both Motacillae," or warblers. The following two days he went out shooting, once with Audubon, but he also complained
he had "no naturalist to keep me company," thus excluding Audubon from that
category. Finally, he departed the town, giving a very different ending to the
story: "I bade adieu to Louisville . . . but neither received one act of civility . . .
one subscriber, nor one new bird," he grumped. "Science or literature has not
one friend in this place."[41] Neither, apparently, did Wilson, who left town with
a bitter taste in his mouth.

And that, after a week, was apparently that. Audubon wrote about seeing
Wilson only one more time, during a brief visit Audubon made to Philadelphia
in 1811, a polite but chilly-seeming meeting that left Audubon feeling that "my
company was not agreeable." No matter how that second encounter went—or
if it happened at all—that was the end of the personal relationship between
them.[42] Two years later, Wilson was dead, brought down by dysentery and the
general debilitations of too much time spent tramping around outdoors, trying
to find birds to paint and, perhaps even more difficult, customers to buy his
bird paintings.

So the story is told, from both Audubon's perspective and Wilson's, neither
of which may be absolutely accurate. Wilson's parting shot at Louisville and
its inhabitants might have been sharpened in print by Wilson's posthumous
promoter, George Ord, whose hostility to Audubon, as we shall see, grew to
be all but boundless. In turn, Audubon's more upbeat telling of Wilson's warm
reception in Louisville clearly came in response to Ord's published account
of Wilson's unhappy departure, thus putting himself in a more positive and
hospitable light. (Like Ernest Hemingway's treatment of F. Scott Fitzgerald in
A Moveable Feast, Audubon's description of Wilson in *Ornithological Biography*
underscores an important point: Whoever lives longer gets the last and most

self-serving word.) Still, for all the questions and caveats surrounding the competing narratives, the Audubon-Wilson encounter stands as the most famous human sighting in the history of American ornithology, and it invites speculation about the meaning of this remarkable meeting.

First, this unlikely encounter raises a simple but significant question: Was Rozier right? Was Audubon's work actually better than Wilson's? It would take a bird-by-bird analysis of the images both men had at the time even to begin to answer that question conclusively, and even then, it would probably be impossible to reach any all-encompassing artistic or ornithological judgment. Still, a one-bird comparison—in this case, of the Belted Kingfisher—can provide a good idea of the talents of the two artists at the time of their Kentucky encounter (see Plates 2 and 3). Wilson's kingfisher perches in profile in the midst of four other birds, three warblers and a thrush, dominating the picture in both size and prominence.[43] The bird's distinctive markings, particularly the reddish band that identifies it as a female, are sharp and well defined—but perhaps too much so, certainly more so than would be the case on a real bird. Audubon's kingfisher—drawn in 1808, well before Wilson came into his store, much less into his life—was also a female, but rendered with a better representation of the subtlety and irregularity of the color patterns and the texture of the feathers, especially in the bird's crest. One might acknowledge, of course, that the materials Audubon used in rendering his kingfisher in 1808—pastel, graphite, and ink—allowed for more subtlety than the final engraved version of Wilson's image. One might also complain—as some critics indeed have— that both Wilson's and Audubon's portrayals of the Belted Kingfisher seem a bit stiff, still adhering to the standard conventions of avian art by presenting the bird in profile and certainly not displaying the often dramatic vivacity that would later come to characterize Audubon's art.[44] Still, if Wilson had even a glance at Audubon's kingfisher—and it seems highly likely that he did—he would have had good reason to be impressed, perhaps even worried. Audubon had already become a remarkably avid observer and gifted illustrator of birds, and anyone who saw his work would have to take note of his skill. Alexander Wilson certainly must have.[45]

The artistic comparison, even competition, leads to a second question: What might have happened if the Audubon-Wilson story had had a happier ending? How might the course of American art and science have been affected if Wilson had lived longer and the two had indeed become collaborators, as Audubon said he had suggested? The two men clearly shared a common passion, and the work it required might also have been shared. The attempt to depict every bird in America defined an enormous, almost impossible-seeming agenda, certainly the sort of undertaking that might invite goodwill and a mutually respectful effort, especially in a society that did not yet have well-established science departments in research universities or substantial government funding to provide employment or ensure support. Neither Wilson nor Audubon had the personal financial resources to establish himself as an independent gentleman-naturalist, but both did have seemingly unlimited ambition, unflagging energy, and unmistakable artistic skill.

In the end, Audubon surpassed Wilson enough to claim a degree of celebrity and success that no other American naturalist had ever known—or, arguably, would know since—but he could scarcely see or talk about his own work without taking its measure next to that of Wilson. Throughout his long quest to illustrate all the species that would eventually grace the pages of *The Birds of America*, Audubon frequently relied on Wilson's *American Ornithology* as a ready reference. Even more often, however, he took it as a point of competitive departure, taking posthumous potshots at its author for decades to come.

Wilson's allies shot back, pursuing a professional struggle that became very personal, attacking Audubon with an intensity that may well have even exceeded Wilson's own, had he lived. Thanks to the work of Wilson's Philadelphia friends and defenders, Audubon would confront challenges to his ornithological accuracy and even integrity, with accusations (apparently true in a few cases) of having copied some of Wilson's images and published them as his own.[46] Wilson would forever be the "Father of American Ornithology" in the minds of his admirers, the original standard against which all subsequent work must be measured. No matter how much Audubon believed himself to have far exceeded that standard, he would always have his Scottish predecessor hovering like a grim ghost over his life. The edgy 1810 encounter between the two remained a problematic issue in Audubon's personal history and in the history of *The Birds of America*, later creating artistic and scientific comparisons that Audubon could never escape.[47]

In 1842, an American author writing under the nom de plume Christopher North celebrated the prospect of Wilson and Audubon in an ornithological pantheon: "'Alexander Wilson and John James Audubon!' We call on them—and they appear and answer to their names . . . they grasp each other's hand. . . . They are brothers, and their names will go down together . . . in all the highest haunts of ornithological science."[48] By that time, however, any notion of ornithological solidarity between the two had to be a fantasy. Instead of figuratively grasping each other's hand, they might more likely have been portrayed clutching each other's throats. Any notion of scientific collaboration had long since taken on a much more menacing aspect, leaving a legacy of unseemly enmity and competition that lasted throughout their lives, and even beyond.

First Taste of Financial Failure

In the meantime, at least for the decade after 1810, Audubon still had to take account of the other pressing imperative in his life: making a living in business. In writing about his professional trajectory, he almost always claimed to have pursed his passion for birds while letting pretty much everything else go. He was, he admitted, neither assiduous nor successful in his various business ventures, always having his eye more on the skies than on the bottom line: "Birds were birds then as now," he wrote, "and my thoughts were ever and anon turning toward them as the objects of my greatest delight. I shot, I drew, I looked on nature only; my days were happy beyond human conception, and beyond this I really cared not."[49] But he had to care about his life beyond birds, of course, because he had obligations to a wife and children and, for a while, to his friend and business partner, Rozier—but only for a while.

While Audubon's family always remained a source of emotional, sometimes even sentimental, attachment, Rozier soon fell from grace in Audubon's narrative of his Kentucky days. Rozier never took on quite the treacherous aspect of the despised Dacosta, but he did serve as a money-focused foil to Audubon's emerging self-portrayal as a man with a higher-seeming focus on art and science, too busy with the birds to spend time with business.

Soon after the in-store encounter with Alexander Wilson, Audubon wrote, he and Rozier "became discouraged at Louisville, and I longed to have a wider range." He would readily admit the fundamental flaw in the management of the original mercantile operation: "Louisville did not give us up, but we gave up Louisville. I could not bear to give the attention required by my business, and which, indeed, every business calls for, and, therefore, my business abandoned me."[50] He made only a tepid confession about his lack of attention, however, pointing instead to his fascination with pursuing his larger calling, all the while using the more practical-seeming Rozier to make the comparative point: "I seldom passed a day without drawing a bird, or noting something respecting its habits, Rozier meantime attending the counter," Audubon wrote. "I could relate many curious anecdotes about him, but never mind them; he made out to grow rich, and what more could *he* wish for?"[51] All Audubon seemed to wish for was being out of the store.

Instead, they both got out of the town and into another store, this one 125 miles farther down the Ohio River, in an even newer and less developed town, Henderson, Kentucky. One early observer unkindly described Henderson as a town of "about twenty houses, and inhabited by a people whose doom is fixed."[52] To be sure, Henderson had a population of only 159 in 1810, but doom did not seem to be the town's destiny, at least in Audubon's eyes at the time. The village was "quite small," Audubon admitted, "but our neighbours were friendly.... The woods were amply stocked with game, the river with fish; and now and then the hoarded sweets of the industrious bees were brought from some hollow tree to our little table."[53] Into that scene of Arcadian simplicity and contentment, Audubon added only a brief, half-sentence reference to Rozier: "I had then a partner, a 'man of business.'" The quotation marks around "man of business" served to separate Rozier's commercial inclinations from the "sports of the forest and river" preferred by Audubon, who "thought chiefly of procuring supplies of fish and fowl."[54]

Audubon did not belabor the difference any further in *Ornithological Biography*—he never mentioned Rozier again, certainly not by name—but in the more private space of the autobiographical "Myself," he offered an extended anecdote about their different approaches to business and perhaps to life in general. Soon after arriving in Henderson, Audubon writes, he and Rozier took a flatboat-load of whiskey and other goods down the Ohio River, headed to the Mississippi River to sell Kentucky's best beverage to settlers in the Missouri Territory. Starting out in a snowstorm in late December 1810, they had traveled only three days, to a few miles above the confluence of the Ohio and Mississippi, before learning that the Mississippi was covered with ice, meaning that their journey would have to wait for a thaw. In the meantime, they camped with some Shawnee Indians, and Audubon quickly made the most of the cross-cultural encounter: "I understood their habits and a few words of their language, and as many of them spoke French passably, I easily joined with their 'talks' and their avocations." The Shawnee people also joined Audubon in his own avocation, "and as soon as they learned of my anxiety for curiosities of natural history, they discovered the most gratifying anxiety to procure them for me."

While Audubon became the center of attention in this lively exchange of information, his partner sulked on the sidelines: "My friend Ferdinand Rozier, neither hunter nor naturalist, sat in the boat all day, brooding in gloomy silence over the loss of time, &c. entailed by our detention."[55] Again, the contrast with Rozier and his fretting over the loss of time and, presumably, money seems critical to the narrative.

Audubon goes on to describe a six-week icebound stay with native people— first Shawnees, then Osages—during which he happily hunted birds and bears with his newfound "Indian friends," played the flute for their amusement, sketched a "tolerable portrait of one of them in red chalk," and always carried on his study of birds and mammals, recording the results every night by the light of the fire: "I wrote the day's occurrences in my journal, just as I do now," he recalled almost two decades later, "and well I remember that I gained more information that evening about the roosting of the prairie hen than I had ever done before."[56]

Once the weather warmed and the ice on the Mississippi broke up enough to get their flatboat through, Audubon and Rozier eventually made their way up the river to Ste. Genevieve, where they quickly disposed of their cargo at a handsome rate of return: "Our whiskey was especially welcome, and what we had paid twenty-five cents a gallon for, brought us two dollars." But Ste. Genevieve, "an old French town, small and dirty," seemed not such a promising business site, at least not for a restless yet homesick man like Audubon. He quickly decided "it was not the place for me; its population was then composed of low French Canadians, uneducated and uncouth."[57]

Rozier apparently liked the people there just fine, and he decided to stay, so he and Audubon dissolved their partnership in April 1811, with Rozier buying Audubon out for a combination of cash and bills of credit. Rozier went on to do quite well in Ste. Genevieve, marrying a young woman of the town, fathering ten children, and eventually, in 1864, dying a prosperous pillar of the community at the age of eighty-seven.[58]

Audubon opted to take the money, such as it was, and run. Actually, he bought a "beauty of a horse" and rode back home to Henderson, back to Lucy and their infant son, Victor Gifford, and back to business. For a while, he would occasionally badger Rozier for more money, even going back to Ste. Genevieve a couple of times to try to collect—once walking all the way, he claimed, 165 miles in just over three days, "much of the time nearly ankle deep in mud and water," but never with much success.[59] Instead, he turned his attention toward new enterprises—but, again, never with much success.

The subsequent history he gives of his various financial ups and downs—mostly downs—in his Henderson years can best be compressed into a fairly (and perhaps mercifully) short narrative. But even a brief summary suggests that, for all his self-avowed aversion to the world of commerce, Audubon had an active, if not always canny, business sense. He may have sniffed disparagingly at Rozier as a "man of business," but his own entrepreneurial turn toward new ventures put him squarely amid the economic innovation—and turbulence—of his era.

Soon after parting with Rozier, for instance, Audubon formed a new business partnership with Lucy's brother, Thomas Bakewell, a young man of seemingly unwavering ambition. In 1811, Bakewell hatched a promising-seeming plan for opening an import-export operation in New Orleans, and he wanted to bring Audubon into it, seeing the value of both his Francophone-language abilities and his financial resources. Unfortunately, neither Bakewell nor Audubon could do much to control the context of larger global politics. Just as their new venture stood on the verge of opening its doors, the increasingly apparent prospect of war with Great Britain threatened to cut off the Atlantic trade, and the Audubon-Bakewell business went bust before it was ever really born. Audubon lost money—"My pecuniary means were now much reduced"—but he still kept faith in his brother-in-law, who moved to Henderson and joined Audubon in business there.[60]

Business prospects seemed promising for a while in Henderson, which, despite the economic uncertainties of the War of 1812, appeared to be heading into a business boom along with the rest of the region.[61] Audubon and Bakewell shared in the excitement as small-town storekeepers, expanding their business to a few downriver towns and engaging in land speculation on the side. They "prospered at a round rate for a while," and the Audubon family settled into comfortable-seeming circumstances, living in a well-appointed house, with handsome furniture, a piano, silver candlesticks, a substantial number of books indoors, and a newly dug pond outdoors, where Audubon could keep turtles for making turtle soup. (The pond was dug by Audubon's slaves. By 1813–1814, he had done well enough to buy nine slaves for just over ten thousand dollars, and even though he never said much about them or the larger institution of slavery in his writings, the people of color in his possession represented yet another indication of his financial standing in the early Henderson years.) Audubon seemed financially set and well satisfied: "The pleasures which I have felt at Henderson . . . can never be effaced from my heart until after death."[62]

Figure 3. Audubon's mill, Henderson, Kentucky. From Maria Audubon, *Audubon and His Journals* (New York, 1899).

But the pleasures would be effaced soon enough by other means. Audubon's young brother-in-law had ideas of going beyond mere storekeeping, and he took Audubon with him—down what eventually proved to be the path toward failure. First, young Bakewell got inspiration to embrace the cutting-edge technology of the era and "took it into his brain to persuade me to erect a steam-mill." Henderson had nothing like it, a combination grist mill and sawmill, both parts powered by a steam engine, and Audubon and Bakewell built a considerable structure on the banks of the Ohio River, just two-tenths of a mile from their store. When it opened in 1817, the mill was, according to the town's historian, "a great convenience" for the region, not to mention a showcase for Audubon's art as well: "The walls of his mill presented the appearance of a picture gallery, every smooth space presenting to the view the painting of some one or more birds."[63]

This huge project quickly became a huge headache, however: a slowly built, badly built, and ultimately overbuilt six-story structure that, even had it worked well, offered more capacity than the community needed. While other towns in the region enjoyed a better boom-time experience and became more important mercantile centers, Henderson remained a disappointment to its boosters.[64] Two of them, Audubon and Bakewell, soon had to realize that the region had too few customers to supply adequate demand for such capital-intensive technology; they had overbuilt and overspent, and they had even bet on the wrong materials. Soon after their mill opened, so the local story goes, good clay for brickmaking was discovered nearby, and the ensuing "building boom" favored brick structures over wooden ones, causing the demand for milled lumber to fall sharply. Moreover, the steam-powered mill could not compete with water-powered gristmills, and the Audubon-Bakewell mill business collapsed.[65] (The building lasted far longer than the business, and the town's local historian called it "perhaps the strongest frame in the city."[66] Today, the stone steps still stand.) In general, Audubon and Bakewell faced a perfect storm of adverse circumstances, and even though Audubon would admit that "the great fault was ours," he would also lament that "the building of that accursed steam-mill was, of all the follies of man, one of the greatest, and . . . the worst of all our pecuniary misfortunes."[67]

The first-person plural reference to "our pecuniary misfortunes" soon became first-person singular, and Audubon's alone. Soon after getting married in Henderson, Thomas Bakewell and his new bride decided that the town was not suitable for their social and financial aspirations, and so they moved away, leaving Audubon holding the almost empty bag of their joint business ventures. Glad as he may have been to see Bakewell go, taking his big ideas with him, Audubon stayed and suffered the economic consequences.

In addition to the steam-powered mill, Bakewell had involved Audubon in yet another steam-based misadventure, a partnership in a new steamboat named, with a striking lack of imagination, the *Henderson*. "This also proved an entire failure," Audubon later wrote with considerable understatement.[68] Thanks to Bakewell's quick exit from the scene, Audubon became embroiled in a complicated, shaky-seeming financial arrangement with another *Henderson* investor, a man named Samuel Bowen. The question of who owned what and who owed what to whom soon became all but moot when Bowen absconded with the *Henderson* and headed for New Orleans, with the intention of selling it. Audubon quickly took pursuit all the way down the Ohio and Mississippi rivers, chasing after Bowen in his own small skiff, along with two slaves, to recover his boat and his money, but he never saw either again. He sold the skiff and the slaves in New Orleans and went home. Then, in June 1819, when he and the evasive Bowen next encountered each other back in Henderson, both men felt wronged, enraged, and ready to kill each other—which they nearly did, in a street fight in broad daylight, not far from the behemoth Audubon-Bakewell steam mill. Audubon soon found himself under arrest for assault and battery, and even though he was quickly acquitted for acting in self-defense—the judge agreed that Bowen was a "damned rascal" who deserved to die—the Henderson era of Audubon's life had taken on a decidedly unhappy appearance.[69]

Audubon avoided jail in the Bowen fracas, but he soon faced incarceration because of an even larger financial crisis, the Panic of 1819. At a time when specie had become scarce because of the contraction in the European economy, largely unregulated local banks in all parts of the United States issued paper currency almost at will, and free-flowing money and other forms of paper-based credit seduced investors to take a shot at seemingly anything.[70] Audubon had been one of them—but only one. Unrealistic expectations had led to unbridled expansion everywhere, until the chain of debt began to weaken and eventually snap, leading to the downfall of people all over the country, at all levels of economic life. When the Panic of 1819 hit the Henderson region, the town's bank—which had been built with lumber from the Audubon-Bakewell sawmill barely two years earlier—collapsed, at least financially.[71] The crisis also struck Audubon especially hard, delivering what seemed to be the final financial blow. By that time, his money troubles amounted to much more than had been at stake in the Bowen imbroglio, and he "had heavy bills to pay which I could not meet or take up," his creditors came after him with a vengeance, and he "was assailed with thousands of invectives."[72]

On one level, there was nothing altogether disgraceful about financial failure in early nineteenth-century America, perhaps least of all in the volatile ups and downs of the 1810s, when businesses big and small hit the wall. (Audubon may have taken note of the fact that the Philadelphia firm of Bradford and Inskeep, Alexander Wilson's publisher, had declared bankruptcy in 1814–1815, soon after it printed the last volume of *American Ornithology*.[73]) On the other hand, looking at the larger financial situation still offered little solace when the pain became personal. Being a fellow sufferer in a national financial calamity couldn't pay an individual's bills, nor could it keep a struggling businessman out of jail.

As a consequence of this financial collapse, Audubon had to sell almost all the family possessions—his share of the mill, the house and its now-numerous furnishings, his musical instruments and much of his artistic equipment, Lucy's books, some farm animals, and, not to be overlooked, the remaining seven slaves—to his more successful and certainly supportive brother-in-law Nicholas Berthoud.[74] Even that was not enough. When Audubon went from Henderson to Louisville to try to clear up his financial situation, his creditors still hounded him, he was arrested for debt and put in jail, and he got out only by declaring bankruptcy. He left jail, he said, "keeping only the clothes I wore on that day, my original drawings, and my gun."[75] Those last two possessions proved critical, soon becoming essential keys to his future.

They had perhaps been a bit underutilized in the recent past. Over the years he spent in his "never-to-be-forgotten residence at Henderson, on the banks of the fair Ohio," Audubon always kept his good eye for birds—his ability both to shoot them in the bush and to render them on paper—and the region proved to provide the avian abundance Audubon needed.[76] Some thirty of his later written descriptions of birds in *Ornithological Biography* have a connection to his Henderson days, as did some of the images drawn during that period, which later made their way into the finished version of *The Birds of America.* Unfortunately, though, not many of the latter survived. In fact, he had probably suffered a net loss in the number of drawings, most famously when, as he recounted in the introduction to *Ornithological Biography*, a pair of Norway rats took up residence in a box of his work and gnawed the papers into nesting material. He was understandably distraught at the loss—"reader, feel for me"—but then, in classic Audubon fashion, he turned near-tragedy into triumph: "I took up my gun, my note-book, and my pencils, and went forth to the woods as gaily as if nothing had happened . . . and, ere a period not exceeding three years had elapsed, I had my portfolio filled again." He may indeed have done so, but very few images from the Henderson era now survive. It might even seem that instead of neglecting his business for the birds, as Audubon so often liked to tell it, he had actually done just the opposite.[77]

But again in classic Audubon fashion, he cast the collapse of his business and the attendant economic agonies of 1819 as an artistic epiphany, a realization that "nothing was left to me but my humble talents." "Were those talents to remain dormant under such exigencies? Was I to see my beloved Lucy and children suffer and want bread, in the abundant State of Kentucky? Was I to repine because I had acted like an honest man? Was I inclined to cut my throat in foolish despair?" To those rhetorical questions, Audubon had an emphatic answer: "No!! I *had* talents, and to them I instantly resorted."[78] And with them he moved on to the next step in his career.

Cincinnati Respite

After the debacle of bankruptcy, Audubon's immediate prospects had only one way to go, of course, and up they went—at least a bit, at least for a while. He did indeed resort to his artistic talents to support his family, taking on portrait commissions for around five dollars a head. Although he had little training (and even less interest) in portraiture, he gained a good reputation for doing the work quickly and effectively for his clients, drawing both the living and, when family members of a soon-to-be-lost loved one wanted one last likeness, the dying. He even did one posthumous portrait of the disinterred son of a Kentucky clergyman, which, he said, "I gave to the parents as if still alive, to their intense satisfaction." (As it happened, Audubon suffered the loss of a child of his own soon afterward: Lucy gave birth to Rose, named after Audubon's own half-sister in France, in 1819, soon after Audubon got out of jail in Louisville, but the infant girl died in early 1820, just seven months old.) Whatever his general disgruntlement over the less-than-agreeable artistic calling of producing quick portraits, Audubon kept himself sane with his unstoppable pursuit of birds. "In this particular there seemed to hover round me almost a mania," he later explained, "and I would even give up doing a head, the profits of which would have supplied our wants for a week or more, to represent a little citizen of the feathered tribe."[79]

Better still, he suddenly got a good, bird-related break. He heard that Dr. Daniel Drake, a prominent physician and president of the new Medical College of Ohio, in Cincinnati, needed an artist and taxidermist for the natural history collections he was helping assemble for another new institution in that city, the Western Museum. With a few letters of reference quickly sent to Cincinnati, Audubon soon found himself with a promising job offer—working on the museum's specimen collections for a decent-seeming salary of $125 per month. With a patron like Drake, Audubon might well begin to imagine a new future.

Drake was a man of lofty intellectual and civic ambitions.[80] Born in the same year as Audubon, 1785, he had had quite a bit more success in life, with a solid career in medicine and teaching already to his credit, along with several memberships in prominent learned societies, including, since 1818, Philadelphia's American Philosophical Society.

But the Philadelphia institution that Drake wanted most to emulate in Cincinnati was Charles Willson Peale's museum, or "Repository for Natural

Curiosities," the best-known and most successful exhibition space in the United States at the time. With a well-organized display of natural history specimens—plants, stuffed birds and mammals, even a mastadon skeleton—along with portraits of prominent Americans, Peale had created a site for both entertainment and instruction, a place where people could gaze about the gallery and behold some of the wonders of the American wilderness without even going outdoors. In the process, Peale gave his visitors a way to understand the connection between nature and nation, and he confidently expected that a museum of this sort could be a credit to its founder, to its city, and to the country as a whole.[81]

So did Drake, and he sought to create something similar in Cincinnati. Incorporated as a city only in 1819, Drake's Cincinnati seemed a pale reflection of Peale's Philadelphia, with just under 10,000 inhabitants in 1820, compared to almost 64,000 in the Quaker City. Like Philadelphia, though, Cincinnati had a hodgepodge population, with a combination of New England Yankees, Pennsylvanians, Virginians, and Kentuckians, not to mention a growing number of Germans and other European immigrants. Like Peale, Drake saw the new museum as a means of providing a civilizing influence over such a diverse society.[82] "As the arts and sciences have not hitherto been cultivated among us to any great extent," he observed, speaking the obvious truth, "the influence they are capable of exerting on our happiness and dignity is not generally perceived." But rooted in a foundation of scientific method, the Western Museum could be a place of celebration of the arts and sciences, where the eclectic collection of items "will rise from it in order and beauty, like those which start from the prepared canvass into imitative life, under the creative pencil of the painter."[83]

Audubon needed money more than he needed such lofty pronouncements about the rise of order and beauty, even with a positive word about the "creative pencil of the painter." Unhappily, he apparently didn't make as much as he had hoped in working for Dr. Drake, and he later complained that "the members of the College museum were splendid promisers and very bad paymasters." Anyway, his work for the Western Museum came to a fairly quick end, because he and his colleague Robert Best, the British-born curator of the collection, were "so industrious," he wrote, "that in about six months we had augmented, arranged, and finished all we could do for the museum."[84] Drake laid Audubon off at the end of April 1820, and even though he promised to pay Audubon for the work he had done, the money never materialized. Audubon and Lucy stayed on in Cincinnati, both teaching to make ends meet, both no doubt wondering what they might be able to do next.

Still, the time spent working at the Western Museum proved valuable in ways other than monetary. Audubon got additional experience in taxidermy, he found time to consult the museum's copy of Wilson's *American Ornithology* and other ornithological works, and he had the good fortune to display his bird drawings to prominent visitors to the museum—most notably Major Stephen Long, the expedition leader; Thomas Say, the naturalist; and Charles Willson Peale's son Titian, himself a budding artist-naturalist, all of whom "stared at my drawings of birds."[85] He also gained some valuable recognition for the useful work he had done. Elijah Stack, the president of Cincinnati College, wrote a letter of recommendation for Audubon, noting that "he has been engaged in our Museum for 3 or 4 Months & his performances do honor to his Pencil."[86]

Daniel Drake also put in a good word in a public venue. On the evening of June 10, 1820, in his formal address just before the opening of the Western Museum, Drake gave both Audubon and his profession a positive plug in his formal remarks before an audience of the museum's patrons. When speaking specifically about the field of ornithology, he made the obligatory bow to Alexander Wilson, acknowledging that to "this selftaught, indefatigable and ingenious man we are indebted for most of what we know concerning the natural history of our Birds." No sooner had he given Wilson this compliment, though, than he compromised it with an ornithological qualification based on his regional devotion to "that portion which we inhabit," the Ohio River Valley and the territory closer to the Mississippi River. While Wilson had "nearly completed" the study of birds of the Middle Atlantic states, Drake noted, he "must necessarily have left that of the Western imperfect." He went on to explain that birds don't typically migrate across mountains, but along rivers, such as the Mississippi, and from that geographical perspective he pointed out that "it is reasonable to conjecture, that many birds annually migrate over this country which do not visit the Atlantic states, and might, therefore, have escaped the notice of their greatest ornithologist in the single excursion which he made to the Ohio." That "single excursion" was, of course, the one that took Wilson to Louisville and Audubon's store in 1810. Without making any explicit reference to the touchy-seeming situation between Audubon and Wilson at the time, Drake quietly sided with his erstwhile employee Audubon, "one of the excellent artists attached to the Museum," as the more comprehensive of the two: Audubon "has drawn, from nature, in colored crayons, several hundred species of American birds, [and] has, in his port folio, a large number that are not figured in Mr. Wilson's work, and many which do not seem to have been recognized by any naturalist."[87]

Although Audubon was no longer in Drake's employ at the time of this talk, he could have taken two encouraging notes from Drake's remarks. First, he had to enjoy being called an excellent artist and, better still, receiving public recognition that he had a more extensive and comprehensive collection of avian images than Wilson had. Second, he could readily agree with Drake's argument that the limitations of Wilson's ornithological reach stemmed in large part from his spending almost all of his time in the East and only once venturing toward the West. Audubon knew that the millions of birds migrating in regions beyond the eastern mountain ridges represented fair game for dozens of new discoveries, and those could be his, not Wilson's.

With his prospects in Cincinnati seemingly at a dead end, with no interesting new work in sight, he decided to look downriver to find his future. From his current perch in the Ohio River Valley, no place seemed more immediately promising or proximate than the Mississippi Valley, which could in turn give him access to an even greater territory for exploration. On August 12, 1820, just two months after Drake's oration to the patrons of the Western Museum, Audubon bravely, perhaps brazenly, appealed to an even more prominent patron, Henry Clay of Kentucky, the Speaker of the U.S. House of Representatives. Audubon wrote that he had spent "the greater part of Fifteen Years in procuring and Drawing the Birds of the United States with a view of Publishing them," he wrote his fellow Kentuckian, but his collection of specimens were those that "usually resort to the Middle States only." Wrapping himself in the expansive nationalism he no doubt knew would appeal to an ambitious political leader like Clay, he spoke of his "desire to complete the Collection before I present it to My Country in perfect order." To do so, he continued, "I intend to Explore the Territories Southwest of the Mississippi . . . Visiting the Red River, Arkansas and the Countries adjacent." A few good words of introduction "from one on whom our Country looks up to with respectfull Admiration" could be enormously "Necessary to a Naturalist," and Audubon thus solicited the Speaker's support.[88] Within two weeks, Clay responded with a letter that recommended Audubon as "a Gentleman of Amiable and Excellent qualities, Well qualified, as I believe, to execute the object which he has under-taken."[89] Audubon now had just what he needed: a piece of paper that could open doors all along the Mississippi.

And that's where he headed next, now fully focused on making his living as an artist after all—and a bird artist at that.

Chapter 3

~~~

Making an Odyssey for Art and Ornithology

Without *any* Money My Talents are to be My Support and My anthusiasm My
Guide in My Dificulties, the whole of which I am ready to exert to keep, and to
surmount—

—John James Audubon, "Mississippi River Journal"

"The Watter is Low," Audubon observed as the boat moved into the Ohio
River's slow current, and so were his spirits. He knew what he wanted to do:
"to Acquire a true knowledge of the Birds of North America." He also knew
he had to get beyond Cincinnati to do it: "I Concluded that perhaps I Could
Not do better than to Travel, and finish My Collection or so nearly that it would
be a Valuable Acquisition."[1] He had big enough ambitions; he just hadn't fully
figured out a way to fulfill them.

In the meantime, the future must have seemed far overshadowed by the
recent past: Audubon's career had gone nowhere but down. The retail business
had eventually been a bust, bankruptcy had been an embarrassment, and even
his two interrelated loves, art and ornithology, held out precious little promise
for financial support, much less success. Making quick portraits of people for
five dollars a head seemed like hack work, and not very lucrative at that. Stuffing birds and animals as a museum taxidermist could hardly boost his imagination or reputation, and anyway, once the specimens went on display, that
was the end of that. But for all the disappointments he had had to face, Audubon also had to face the reality that he had obligations to his wife, Lucy (who,
with him, had suffered the recent loss of an infant daughter), and two young
sons, eleven-year-old Victor and eight-year-old John. Even though Audubon's
family would have to stay behind, they stayed very much on his mind. As he
made ready to head downriver, "the feelings of a Husband and a Father, were

my lot when I Kissd My Beloved Wife & Children with an expectation of being absent for Seven Months." Still, he stiffened his resolve and reminded himself that missing them would be a temporary loss in a longer-term process: "If God will grant us a safe return to our fammillies Our Wishes will be congenial to our present feelings Leaving Home with a Determined Mind to fulfill our Object."[2]

Those brave words came on the first page of a journal Audubon kept on his trip down the Ohio and Mississippi rivers to New Orleans, with fairly regular entries from the day he left Cincinnati, October 12, 1820, until the end of December 1821. He wrote primarily for his sons—"My Journal gives you a rough Idea of My Way of Spending the tedious Passage . . . to New Orleans"— but the preserved journal also serves the modern reader quite well. While some passages seem sketchy and uneven, others are almost eloquent in their descriptions, giving us an ornithologically detailed list of the hundreds of birds Audubon watched (and shot) on the trip, the landscape he saw along the riverbanks, and the ordinary (and sometimes extraordinary) people he encountered both on the boat and on shore. In the end, Audubon's 1820–1821 journal offers much more than a "rough Idea" of what he might have considered a "tedious Passage." At its best, it can take its place alongside the more professionally polished travel narratives of the early nineteenth century.[3]

At its heart, however, the journal takes us into the unfolding inner journey of this mid-thirties man on a mission. In all of Audubon's writings, his most significant subject is almost always himself, and even in his most successful written work, *Ornithological Biography*, he puts himself in the picture, right beside the birds. But in the case of the Mississippi River Journal, Audubon had not yet conceived, much less achieved, the sense of celebrity that would later shape his more self-conscious narrative of his life. Instead, he recorded the uncertain hopes and underlying vulnerabilities of a man whose commitment to a challenge far exceeded his confidence in its outcome.

The anxieties and frustrations that came to the surface in his Mississippi River Journal in 1820–1821 remained enduring concerns that would recur in Audubon's writings throughout the rest of his life. He fretted, first of all, about his financial situation, knowing that he never bore the burden of poverty alone, but also laid it heavily on Lucy and his sons, left behind with no assurance of support in his absence. Money worries in turn had an effect on his sense of self-identity as an artist-naturalist. Before he could make his collection of avian art big enough and good enough to become a "Valuable Acquisition," he would have to make a concession to other people's notions of what it meant to be an "artist," painting appealing portraits of whoever would pay the price and giving art lessons, largely to ladies and young girls.

And no matter what he did or where he went, Audubon always had an invisible companion in his mind: Alexander Wilson. By the time Audubon left on his Mississippi odyssey, Wilson had been dead seven years. In Audubon's mind, though, Wilson remained a competitor, the man who set the existing standards of ornithological art, the man whose work Audubon would frequently consult and almost as frequently criticize, the man whom Audubon felt the compelling need to surpass in order to define the measure of his own success. Even Henry Clay had brought up the specter of the late "American Ornithologist." In sending his generous letter of introduction for the Mississippi trip, Clay wondered if Audubon knew what he was getting into, given the possible expense of producing such an ambitious work: "Will it not be well for you . . . to ascertain the Success which attended a Similar undertaking of Mr. Wilson?"[4] Audubon could never escape the comparison thus imposed on him by others, and he would never cease imposing it on himself. In this as in many other ways, the pages of Audubon's Mississippi River Journal give us a preview of issues that would continue to dog him for years to come.

Flatboat Blues

If Audubon needed any reminder of the low state of his circumstances, all he had to do was to consider the boat he was taking downriver—a flatboat, several pegs down the scale of comfortable aquatic conveyance. In 1820, the year he decided to leave Cincinnati for the Mississippi River region, the steamboat was still a recently arrived marvel on the western rivers. The first steamboat to make the Pittsburgh–New Orleans trip did so in 1811 (owned and operated, coincidentally, by an acquaintance of Audubon's, Nicholas Roosevelt), and just a few years later, in 1817, steamboats began regular mail and passenger service from Cincinnati to New Orleans, usually making their way downriver in just over a week. The fortunate few who could pay for stateroom accommodations, about $125, could enjoy room and board all but equal to that found in the best hotels. Those who could only afford the Spartan conditions on deck had neither room nor board—they had to scrounge whatever sleeping space they could find amid the cargo and fix their own food—but they paid about a fifth of the fare, and they still got to New Orleans at the same time as their more prosperous fellow passengers.[5]

Either way, Audubon couldn't afford it. He couldn't even afford flatboat fare. He made an arrangement with a flatboat owner named Jacob Aumack, who offered him free passage in exchange for being the boat's hunter, shooting whatever he could to provide food for the others on board. One of those others was Audubon's young traveling companion, Joseph Mason, a twelve-year-old former art student from Cincinnati whom Audubon had engaged to be, like himself, an unpaid flatboat employee. Mason came along to help Audubon with the hunting and background painting and, in addition, to help Aumack with whatever unpleasant but necessary tasks he could assign the boy.[6] Traveling light, with only their guns and art supplies and a very few changes of clothing, Audubon and Mason joined Aumack and a handful of other men on the flatboat for what would turn out to be a long journey, both geographically and, for Audubon, psychologically.

The best thing to be said about a flatboat was that it made decent economic sense at the time—a cheap, simple, single-use, one-way vessel. Almost anyone could build this entry-level means of river transport for about fifty dollars or so, load it with upward of forty tons of goods, and then, once arrived at a downriver destination—typically after a month or more all the way to New Orleans—disassemble the boat and sell the scrap wood for whatever money it might fetch. In the first decades of the nineteenth century, thousands of flatboats floated down the Ohio and Mississippi rivers—almost 1,300 in 1816 alone—and flatboat transportation would remain a part of the river economy well into the steamboat era.[7]

A flatboat also defined an enclosed social space, sometimes even the site of a downriver rite of passage for young farm boys who made the trip.[8] One of the iconic genre paintings of the antebellum era, George Caleb Bingham's *The Jolly Flatboatmen* (1846), shows a group of eight young men taking their leisure on top of a flatboat, one of them dancing, one playing a fiddle, another keeping time on a metal pan, and the other five variously lounging around and enjoying the show, their oars horizontally at rest. With slow-flowing water below and clear, blue skies above, the painting offers an idyllic image of men at ease on the water, making the most of their riverine relaxation.

Behind this image of romanticized sociability, however, lay a much rougher reality. A modern history of the flatboat trade has described the boatmen of the Ohio and Mississippi rivers as being "as filthy as the dogs whose howls they

Figure 4. The Jolly Flatboatmen, by George Caleb Bingham, 1846. Oil on canvas, 38 1/8 x
48 1/2 in. Object #2015.18.1. Patrons' Permanent Fund. National Gallery of Art.

imitated," living on a daily diet of bacon and beans, and washing down their
meager meals with whatever beverage happened to be available on board—and
whiskey was always available. Once the boatmen got ashore at the end of the
downriver trip, they may have felt flush with a few dollars of wages to spend
for a few days on better food and more drink, but then they had to get back up
the Mississippi somehow, quite often having to make the journey on foot. Most
of them had no doubt become a good deal less jolly by that point.[9]

Audubon had no illusions about the romance of the river, and he would
certainly find little joy in this trip. He had been in the flatboat business before,
back when he and his Kentucky-based business partner, Ferdinand Rozier,
were moving goods from their store down the Ohio River to the Missouri
Territory in the winter of 1810–1811. Now, a decade later, he found himself
essentially bumming a ride on someone else's boat. His journal repeatedly
speaks of the "desagreable" discomforts of drifting downriver on a clumsy,
slow-moving wooden barge, being thrown together with men he didn't much

like, and living in squalid accommodations that gave him almost no protection from the elements, much less enough room to work. With only a small, claustrophobic cabin for shelter, Audubon and his boatmates remained constantly exposed to the weather, which turned out to be repeatedly rainy, windy, and surprisingly chilly for mid-fall. On the second day out, Audubon wrote, "The Wind Rose and brought us to Shore, it raind and blowed Violantly untill the Next Day," and for days more after that, morning frosts and temperatures below freezing seemed "desagreably Cold." As an artist he suffered under the circumstances of the cramped onboard environment, "drawing in a Boat Were a Man cannot stand erect." By November 1, another day with "weather drizly and windy," flatboat life had already left Audubon feeling flat himself: "Extremely tired of My Indolent Way of Living," he grumbled, "not having procured any thing to draw since Louisville." Even his dog, Dash, seemed to have her own case of the flatboat blues, looking to be "apparently good for Nothing for the Want of Employment."[10]

Ornithology on the River

For both man and dog, the best remedy for such ennui was to get off the boat and go on the hunt for food, which was Audubon's responsibility in his free-passage arrangement with Aumack. Almost anything counted in flatboat cuisine, and the search for fresh meat put him in pursuit of both mammals and, as always, birds. From the first day he boarded the flatboat, in fact, Audubon kept a regular record of the birds he saw and shot, and the entry for October 18 gives a good indication of the variety and abundance of the avian life along the Ohio River:

> Saw some fine Turkeys, killed a Common Crow Corvus Americanus Which I drew; Many Robins in the woods and thousands of Snow Buntings Emberiza Nivalis—several Rose Breasted Gros Beaks—We killed 2 Pheasants, 15 Partridges—1 Teal, 1 T. T. Godwit—1 Small Grebe all of these I have Seen precisely alike in all Parts—and one Bared Owl this is undoubtedly the Most Plentifull of his genus.[11]

(Audubon's occasional use of the Linnaean binomials in that passage also speaks to his self-conscious sense of identity as naturalist, and among his few flatboat possessions he carried a copy of William Turton's 1806 translation of Linnaeus's *Systema Naturae*. Turton's work served Audubon well as a ready reference for identification and classification, not to mention a model for drawing: One of the sketches in his journal is a detailed bird diagram copied closely from Turton.[12])

Despite all the discomfort and frustration of trying to draw in the cramped, uncomfortable quarters of the flatboat, Audubon kept at it, combining sketches of birds with close ornithological description. One "raw & Cloudy" Sunday morning in November, for instance, he got a "beautifull specimen" of a water bird he had never seen before, which he identified as an Imber Diver (and which is now identified as the Common Loon). He had to wait a while to get started on his drawing—"the Wind rendered our Cabin smoky I Could Not begin to Draw until after Dinner"—but he finished it over the next two days, while also recording field notes in his journal:

> It is with apparent Dificulty or a Sluggish disposition that these Birds rise out of the Watter & yet Will Not dive at the flash of a Gun—while on the wing are very Swift— . . . they frequently Dipp their Bill in the Watter, and I think have the power of Judging in that Way if the place Contains Fish = One I shot at; dove & raised again Imediatly as if to see Where I was or What Was the Matter.

Audubon took careful measurements of the bird's weight and dimensions, and he opened it up to see what it had been eating: "Contents of Gutt & Gizard Small Fish, Bones & Scales and Large Gravel—Body extremely fat & rancid."[13]

The day he finished his Imber Diver drawing, Audubon also "saw several Eagles, *Brown & White headed*," and a few days later he shot a "*beautifull White headed Eagle Falco Leucocephalus*," a male weighing eight and a half pounds and with a wingspan measuring over six and a half feet. Since his shot went through the eagle's gizzard, he couldn't determine its diet, but the dead specimen did give him the opportunity to make a close examination of the bird and become convinced that "the *Bald Eagle* and the *Brown Eagle* are Two Diferent Species." (In fact, they are not, the Brown Eagle being the immature version of the Bald Eagle.) Audubon continued to observe eagle behavior, especially the mating habits and other relationships between male and female. On the afternoon of December 1, he had the good ornithological fortune to watch two eagles copulate: "The femelle was on a Very high Limb of a Tree and Squated at the approach of the Male, who came Like a Torrent, alighted on her and quakled Shrill until he sailed off, the femelle following him and Zig zaging through the air."[14]

A week later, he made another observation of a male and female eagle pair, this one with a less happy outcome. "Mr Aumack Winged a *White headed Eagle*, [and] brought it a live on board," Audubon wrote, noting that "the Noble Fellow Looked at his Ennemies with a Contemptible Eye." Audubon then undertook a harsh-seeming scientific experiment with the captive eagle, tying a string to one of its legs, then making the wounded bird jump into the water. "My Surprise at Seeing it Swim well was very great, it used its Wing with great Effect and Would have made the Shore distant then about 200 yds Dragging a Pole Weighing at Least 15 lbs." When his assistant Joseph Mason went after it, the defiant eagle defended itself, and all the while its female partner hovered above and "shrieked for some time, exhibiting the *true Sorow* of the *Constant Mate*." This would not be the last time Audubon would have a close encounter with a captive eagle—thirteen years later, as we shall see, the story of his eye-to-eye showdown with a caged Golden Eagle would mark one of the most dramatic episodes in Audubon's artistic career—but it speaks to the ways that rough, even cruel treatment of live specimens could be a seemingly necessary, albeit unseemly, element of his scientific method.[15]

Longing for Lucy

Audubon's observations of male-female eagle relations may have been especially pressing on his mind, because he, too, experienced "the *true Sorow* of the *Constant Mate*." The farther he drifted downriver, the more his sense of separation from Lucy and his sons weighed on his mind, the more his slow-moving pursuit of an unpromising calling reminded him of his poverty—and theirs. Here he was, floating slowly toward New Orleans for a couple of months, shooting and drawing birds along the way, but really having no idea of what prospects lay before him in the longer run. In mid-November, after being afloat for five weeks, and as the boat finally left the Ohio River and turned into the Mississippi, he took sad note of the turn his life had taken. "Now I enter it *poor* in fact *Destitute* of all things . . . in a flat Boat a Passenger." In a brief reverie, he made the confluence of the rivers a metaphor for his own difficult history: "The Meeting of the Two Streams reminds me a Litle of the Gentle Youth who Comes into the World, Spotless he presents himself, he is gradually drawn in to Thousands of Dificulties that Makes him wish to keep Apart, but at last he is over done Mixed and Lost in the Vortex." Before being swept too much deeper into this emotional abyss, Audubon snapped out of it and turned his attention to a visual description of the way the "beautifull & Transparent Watter of the Ohio . . . Looks the More agreable to the Eye as it goes down Surrounded by the Muddy Current" of the Mississippi. Still, he could not suppress one last look at the river that had taken him away from Cincinnati, where he had left his family: "I bid My farewell to the Ohio at 2 o'clock P.M. and felt a Tear gushing involuntarily, every Moment draws me from all that is Dear to Me My Beloved Wife & Children."[16]

Audubon's longing for his family, and no doubt the guilt he felt for leaving them, continued to hang heavily on his heart. Sundays on the flatboat were a time for renewal of sorts, when Audubon would shave and wash—"anxious to See the day Come for Certainly a Shirt worn One week, hunting every day and Sleeping in Buffaloe Robes at night soon became soild and Desagreable"—but moon over Lucy: "On Sundays I Look at My Drawings and particularly at that of My Beloved Wife—& Like to spend about one hour in thoughts devoted to My familly."[17] Sometimes those thoughts made him imagine the worst: "While Looking at My Beloved Wife's Likeness this day I thought it was Altered and Looked sorrowfull, it produced an Imediate sensation of Dread of her being in Want."[18] He tried to write letters back home, but the river offered few opportunities for regular mail service, and letters could take six weeks or more to reach a recipient, no matter how beloved. Sometimes he could cheer up for a bit by remembering his mission, telling himself that "so *Strong* is my Anthusiast to Enlarge the Ornithological Knowledge of My Country that I felt as if I wish Myself *Rich again* and thereby able to Leave my family for a Couple of Years." Still, it was almost never that easy to take the long view, especially with his family so far off in the distance. By Christmas of 1820, when he had been away for two and a half months, he wrote of his "hope that My Familly wishes me as good a Christmas as I do them. . . . I hope to have Some tidings of them Tomorrow."[19]

As it happened, he did get some mail from Lucy the next day, a couple of letters posted in early November. Perhaps just as promising, he also happened upon, quite by surprise, his Kentucky friend Nicholas Berthoud, who was on a stopover while taking his own keelboat to New Orleans, Audubon's anticipated destination. Berthoud invited Audubon to join him on his keelboat for the rest of the trip downriver, giving him a welcome upgrade over Aumack's flatboat. But by New Year's Day, any encouraging effects of Lucy's correspondence and his improved accommodations had worn off, and Audubon could not avoid coming to terms with the dispiriting reality of his situation: "*I am on Board a Keel Boat going down to New Orleans the poorest Man on it.*"[20]

Ever the Observer

Poor as he was, Audubon had plenty of impoverished company in the Mississippi region, and he took note of the condition of the ordinary people he encountered, forming impressions that would later find a place on the pages of his published works. When he looked at riverside society, he frequently recoiled at the low state of the people's lives in their squalid communities. Landing at New Madrid, in Missouri Territory, one afternoon in November, he noted that "this allmost deserted Village is one the poorest that is seen on this River bearing a name," and the inhabitants looked shiftless and slovenly: "They are Clad in Bukskin pantaloons and a Sort of Shirt of the same, this is seldom put aside unless So ragged or so Blooded & Greased, that it will become desagreable even to the poor Wrecks that bear it on."[21] (Audubon neglected to note in his journal the possible economic aftershocks of the massive earthquakes that devastated New Madrid in 1811–1812, which may well have rendered it still a less desirable location for residence nine years afterward. Some years later, however, he did take note of the power of the earthquakes in the Mississippi region, when he wrote in *Ornithological Biography* about how "the earth was rent asunder in several places, one or two islands sunk for ever, and the inhabitants fled in dismay towards the eastern shores."[22]) A few days after passing New Madrid, he came upon two men and a woman in a skiff, "Too Lazy to Make themselves Comfortable, Lie on the Damp earth, near the Edge of the Watter, have *Racoons* to Eat and Muddy Watter to help that food down." Later still, he saw "two Women the remainder of a party of Wandering Vagabonds . . . these Two Wretches, Never Wash, Comb, or Scarcely clad themselves," barely surviving by doing a little sewing and washing and otherwise relying on the generosity of neighbors. Audubon painted, on the whole, a depressing picture of human jetsam washed up on the banks of the river, people who apparently headed westward "to proceed to the Promised Land" but wound up hopelessly stuck in the Mississippi mud. Still, he could hardly hold himself above them: "To Look on those people, and consider Coolly their Condition, then; compare it to Mine, they are certainly More Miserable to Common Eyes—but, it is all a Mistaken Idea, for poverty & Independance are the only friend that Will travel together through this World."[23]

Like many other American travelers of the time, Audubon made an exception for native people. "The Indian is More decent, better off, and a Thousand time More happy" than the wretched-seeming white people in the same region, he wrote, and he idealized their own "poverty & Independence" as a positive virtue: "Whenever I meet *Indians* I feel the greatness of our Creator in all its splendor, for there I see the Man Naked from his Hand and Yet free from Acquired Sorrow." But rather than truly finding a model for life in the native inhabitants, Audubon instead saw them, as he did almost everyone, as useful sources of ornithological information. He heard about an Indian chief on the Arkansas River who had shot three swans, one with a nine-foot wingspan, but "these Indians had Left when We arrived—a View of Such Noble Specimen would have been very agreeable."[24]

"New Orleans at Last"

When Audubon eventually reached the end of his Mississippi trip—"*New Orleans* at Last," he wrote on January 7, 1821—he found little that would immediately improve his mood. On his first day in the city, he received an invitation to a dinner party with some "good, well disposed, Gentlemen," but the loud talk and too much wine left him with a "bad head Hake." On the second day he walked around town "absolutely to Kill time, the whole City taken with the festivals of the day" in commemoration of the Battle of New Orleans, but someone picked his pocket, leading him to write acerbically that he would "remember... the 8th of January for ever." On the third day he made the rounds of a few acquaintances to begin looking for work, but when nothing turned up, he went back to Berthoud's keelboat and "remained on board... opposite the Market, the Dirtiest place in all the Cities of the United States." "My Spirits very Low," he wrote, and over time, Audubon's experience in New Orleans would take him lower and lower.[25]

The city itself shouldn't have been the problem. By the time Audubon got there, New Orleans was the fifth largest city in the United States—behind New York City, Philadelphia, Baltimore, and Boston—having come into the

United States in the same year Audubon had, 1803, when Thomas Jefferson's Louisiana Purchase added the whole Louisiana Territory to the new nation. Throughout the eighteenth century, since the city's founding by the French in 1718, New Orleans had developed a remarkably mixed population, with some of the region's original Native American inhabitants, primarily Caddos and Choctaws, who remained on the scene; Europeans, above all French and Spanish, but also immigrants from all over the continent, particularly southern Europe; people of African descent, both slave and free; and more recent arrivals from the West Indies, including several thousands from Saint-Domingue, slaveowners and slaves alike, refugees from the rebellions that had rocked the region at the turn of the century. The United States' acquisition of Louisiana created yet another influx of immigrants in the years before Audubon's arrival, making the city's 27,176 inhabitants the most diverse population of any other urban area in the nation.[26]

Some observers celebrated the city's mix. One of the best accounts of a newcomer to New Orleans comes from Benjamin Henry Latrobe, the British-born architect, who came to New Orleans in January 1819, exactly two years before Audubon arrived. Almost immediately, Latrobe's artistic eye quickly took in the sights of the exciting city, beginning with the main outdoor market, where he beheld a remarkable array of goods: "wretched meat & other butchers meat," but also fresher fare, "wild ducks, oysters, poultry of all kinds," along with a great variety of vegetables and fruits, including bananas, oranges, apples, sugar cane, potatoes, "& all sorts of other roots," and then "trinkets, tin ware, dry goods . . . more and odder things . . . than I can enumerate." But the goods were not the only things on display. Latrobe also described an energetic and cacophonous scene of five hundred or more people, "sellers & buyers, all of whom appeared to strain their voices, to exceed each other in loudness," all of whom reflected the rich ethnic and racial diversity of the city: "White men and women, & of all hues of brown, & of all classes of faces, from round Yankees, to grisly & lean Spaniards, black negroes & negresses, filthy Indians half naked, mulattoes curly & straight-haired, quarteroons of all shades, long haired & frizzled, the women dressed in the most flaring yellow & scarlet gowns, the men capped & hatted."[27]

To his credit, Latrobe took the time to look closely into other areas of New Orleans society, peering into all corners of the local culture and taking care to describe the different ethnic and racial groups that made the city so special. He liked the white women he saw at a fancy ball, for instance, writing appreciatively of their unpainted faces: "A few of them are perfect, and a great majority are far above the mere agreeable. . . . I could not see one face that had the slightest tinge of rouge."[28] In addition to looking into the polite entertainments of New Orleans's elite, Latrobe also took in the sights of the more vigorous outdoor dancing among the city's other major population, the people of color, hundreds of whom assembled each Sunday—the slaves' one day off—for a weekly festival of expressive celebration. Place Publique (or what was more commonly, albeit unofficially, called Congo Square) became the site of an open-air market and meeting place, where people of African descent rejoiced in their cultural heritage through dance and song. The city authorities often looked fearfully askance at such a large congregation of black people, and they occasionally tried to outlaw, or at least regulate, the jubilant gathering. Other white people came to the site as curious spectators to look on from the fringes, and Latrobe soon became one of them.

Like most white observers, Latrobe probably did not fully comprehend the cultural significance of everything he saw, but as a modern historian of slave culture in the city has noted, "His account is probably the best, most thorough observation available from the heyday at Congo Square."[29] Latrobe wrote that he happened to stumble on the gathering by accident while out for a walk one Sunday afternoon, when he "heard a most extraordinary noise" and discovered that it came from some "5 or 600 persons assembled in an open space or public square." Making a quick estimation of the racial identities involved, he noted that "all those who were engaged in the business seemed to be *blacks.* I did not observe a dozen yellow faces." He did observe the dancing men and women formed in circles, moving to music made by two drums and a stringed instrument, but he didn't completely appreciate what he saw: "A man sung an uncouth song . . . & the women screamed a detestable burthen on one single note. . . . I have never seen anything more brutally savage, and at the same time dull & stupid, than this whole exhibition." Latrobe might have noted that there was nothing more "brutally savage" than slavery itself, but he failed to make that connection. Still, he concluded, rather charitably and even credulously, that there "was not the least disorder among the crowd, nor do I learn on enquiry, that these weekly meetings of the negroes have ever produced any mischief."[30]

Latrobe took a dimmer and essentially dismissive view of New Orleans's Native American population, mostly Choctaws, whom he saw as "outcasts, the fag end of the tribe, the selvage, the intermediate existence between annihilation & savage vigor." After going through an exceedingly unflattering description of their appearance and behaviors—dirty and drunk, he said, "having strings of birds, squirrels, perhaps a raccoon or opossum, often ducks, which they either sell to the hucksters in the market or hawk about the streets"—he nonetheless found something positive to say about them: "They are most scrupulously honest. No theft of any kind has ever been charged to them, & their women are most scrupulously chaste."[31]

Latrobe was an architect, after all, not an anthropologist, and the point of considering his account here is not its accurate understanding, or even appreciation, of different cultures. Rather, his extensive and inquisitive exploration of New Orleans society, however biased or wrongheaded it may have been, still offers a standard of comparison for Audubon's own observations, which came just two years later. Audubon seemed considerably less interested and even less impressed; he certainly had less to say about the many textures of society in the city. Where Latrobe spent page after page on his perceptions of different cultures and customs in this remarkably diverse city, Audubon said what he had to say in less than two.

Like Latrobe, Audubon commented on the energetic scene he found in the Sunday morning market, "crowded by people of all Sorts as well as Colors, the Market, very aboundant, the Church Bells ringing the Billiard Balls knocking . . . the day was beautifull and the crowd Increased considerably." But immediately, in the next sentence, Audubon lost interest in the female part of the crowd, saying that "I saw however no handsome Woman and the Citron hue of allmost all is very disgusting to one who Likes the rosy Yankee or English Cheeks." Later in the day, he did see "some *White Ladies* and Good Looking ones," but he begged off going to the "quartroon Ball . . . as it cost 1$ Entrance I Merely Listened a Short time to the Noise."[32] With that almost offhand expression of disdain toward women of color and apparent indifference to the lively entertainments they could provide, he essentially ceased further discussion of the matter. His failure to look more deeply into New Orleans society in this written account of his initial 1821 visit seems striking, particularly given his later biographical association with the city and its surrounding region. For a man who would eventually even claim, less than fifteen years later, that his father had been "in the habit of visiting frequently . . . Louisiana" and had "married a lady of Spanish extraction" there, Audubon remained decidedly silent on the multicultural mix of New Orleans.[33] Whatever its energy and diversity, New Orleans never became an especially happy place for Audubon. Poor, separated from his family, facing the unhappy prospect of somehow supporting himself, and always preferring to spend his time and talent on his own art, he spent most of his time there in a funk.

In the first few weeks, he also spent most of his time looking for work, and he soon found that the world of art didn't offer much, certainly not in keeping with his artistic self-regard. When he had been in town for five days, he met "an Italian, painter at the Theatre," who seemed to like his work, but all Audubon could get from the theater management was an insulting offer to "paint with Mons. L'Italian" for a hundred dollars a month. "I believe really now that my talents must be poor or the Country," he grumbled. The following day he walked through the "Busling City where no one cares a fig for a Man in my Situation" to see John Wesley Jarvis, a local portraitist, but again the meeting amounted to nothing. Jarvis looked at some of Audubon's bird paintings "but never said they Were good or bad." When Audubon all but begged him for work as an artistic assistant, being willing to paint clothing and backgrounds and such, Jarvis proved at first evasive and then dismissive: "He very Simply told me he could not believe, that I might help him in the Least." Over two months later, Audubon finally got an audience with John Vanderlyn, the eminent "Historical Painter," who said some favorable words about Audubon's color and composition but ultimately offered only the faintest praise, saying that Audubon's works seemed "*handsomely done*"—hardly a forceful endorsement of Audubon's art. "Are all Men of Talent fools and Rude purposely or Naturally?" Audubon wondered.[34]

Audubon's fellow artists, "Men of Talent" or not, may have seemed nothing more than a source of discouragement, but women—white women in particular—became his artistic bread and butter. Throughout Audubon's two years in Louisiana, he made ends meet by painting portraits and giving art lessons, quite often for the wives and daughters of prominent men. Early on in his stay in New Orleans, he made a deal with Roman Pamar, a local merchant, to paint Pamar's three daughters. Audubon wanted twenty-five dollars apiece for head portraits, but Pamar wanted all three girls in one painting, so Audubon raised the rate to a hundred. To prove his skill, Audubon did a quick pencil sketch of one of the girls, Pamar liked what he saw, and he "*Civilly* told me that I Must do my Best for him and Left it to my self as to the Price." Several weeks later, Audubon's biggest and certainly most interesting commission came from a mysterious woman who accosted him on the street, asked him to do her portrait—full-length, and in the nude—and after more than a week of sociable posing, compensated him with the one form of payment Audubon might have preferred to cash, a top-of-the-line gun, worth $120. On another occasion, though, he did a portrait of the wife of a man "who Could Not spare Money" but offered only a woman's saddle in payment, "a thing I had not the Least use for." Still, a saddle seemed better than nothing, and with a rueful pun, Audubon concluded that "not to disappoint him I Sufered Myself to be Sadled." Working on one-off portraits had never been Audubon's idea of success, but it paid the rent and may even have contributed a bit to Audubon's personal self-esteem: "Seldom before My coming to New Orleans did *I* think that I was Looked on so favourably by the *fair* sex as I Have *Discovered* Lately."[35]

Making a Living on Lessons

Audubon's most promising opportunity came in the summer of 1821, when a wealthy woman, Mrs. Lucy Gray Pirrie, invited him to tutor her teenaged daughter, Eliza, in the necessary arts for a young woman—what Audubon would describe as "Drawing, Music, Dancing, Arithmetick, and Some trifling acquirements such as Working Hair &c"—at the family plantation, Oakley, over a hundred miles upriver from New Orleans, near Bayou Sarah. At first, Audubon figured he had "one hundred Diferent Plans . . . as Opposite as Could be to this," but the pay was decent—sixty dollars a month, along with a room in the plantation house for Audubon and his assistant, young Mason—so he took the deal and "found Myself bound for several Months on a Farm in Louisiana."[36]

Oakley was more than just a farm—a commodious house, full of family members and "constant Transient Visitors," surrounded by extensive grounds and cotton fields, worked by slaves—and Audubon did more than just teach drawing and such. He did his duty as artistic tutor for his "Aimiable Pupil Miss Eliza Pirrie," but he also spent as much time as he could on his own much-preferred project, "Hunting and Drawing My Cherished Birds of America." On his way to the plantation in June 1821, he realized how refreshed he felt to be out of New Orleans:

> The Aspect of the Country entirely New to us distracted My Mind from those objects that are the occupation of My Life—the Rich Magnolia covered with its Odiferous Blossoms, The Holy, the Beech, the Tall Yellow Poplar, the Hilly ground, even the Red Clay I Looked at with

Figure 5. Oakley Plantation House, Audubon State Historic Site, St. Francisville, Louisiana. Photo by Audubon State Historic Site staff.

amasement,—such entire Change in so Short a time, appears, often supernatural, and surrounded Once More by thousands of Warblers & thrushes, I enjoyd Nature.[37]

During his four months at Oakley, in fact, he wrote very little in his journal about working with Eliza, but page after page about birds: lists of species he had seen or hoped to see, extended descriptions of some of the ones he shot, and, best of all, a couple of accounts of first sightings.

The good times at Oakley came to an end, however, when Audubon was fired from his position. Eliza had been ill for a month, her doctor had warned against continuing her lessons with Audubon, and Mrs. Pirrie—never an altogether pleasant person, in Audubon's estimation—dismissed him on October 10, 1821. Seeking just a little more time to continue his ornithological work, Audubon appealed to Mr. Pirrie, a "Man of Strong Mind but extremely Weak of Habit," who drank too much and who, even in his sober moments, seemed to cower in the face of his wife's "Violent Passions"—the plantation was hers from a previous marriage, after all, and he was just the next husband, and a fairly feckless one at that. Still, Audubon managed to hang on for another five days, an awkward time when he felt "a remarkable Coolness . . . from the Ladies," but kept up a "close application to My Ornithology[,] Writting every day from Morning until Night, Correcting, arranging from My Scattered Notes all My Ideas." He and Mason finally "left this abode of unfortunate Opulence without a single Sigh of regret," but Audubon found it painful to leave the "sweet Woods around us . . . for in them We allways enjoyd Peace . . . [and] I often felt as if anxious to retain the fill of My lungs with the purer air that Circulate through them."[38]

Back they went to New Orleans, then, and back to the uncertain work of making a bare living through art—but not the sort of art Audubon wanted to do. From October 15, the day he arrived back in the city, through the end of 1821, the journal contains a series of dispirited entries about looking for work ("visited several Public Institutions where I cannot say that I Was very politely received"), enduring the jealousy of competitors in teaching art ("My Style of giving Lessons and the high rate I charge for My Tuition have procured Me the Ill will of Every other Artist in the City"), and actually having to give art lessons again and again ("Gave lessons at Mrs. Brand," "Gave a Lesson to Miss Pamar," "Gave My Lessons all round"). Finally, on December 18, Audubon recorded one much happier note: "My Wife & My Two sons arrived at 12 'o'clock all in good health." After fourteen disappointing and lonely months without his family and "all that renders Life agreeable to Me," Audubon mustered up his gratitude and "thanked My Maker for this Mark of Mercy."[39]

Louisiana Ornithology

All Audubon had ever really wanted in Louisiana was birds—birds and enough paying work to allow him to keep finding and drawing more birds. And draw birds he certainly did: Well over a third of the avian images that would later fill the 435 plates of *The Birds of America*, and at least 75 of the 100 images in the first volume, originated during his Mississippi-Louisiana period in the early 1820s; some of them—for instance, his near-iconic image of the now-extinct Carolina Parrot (or Carolina Parakeet), which he began at the Pirrie's Oakley Plantation in 1821 and completed in New Orleans in 1822—have become emblematic of his art.[40] By the same token, the pages of Audubon's journal that cover his time in Louisiana offer extensive lists and descriptions of the species he saw there, and references to the region recur throughout the five volumes of *Ornithological Biography*, such as his "having studied the habits" of the Purple Gallinule "under every advantage in Louisiana, and especially in the neighbourhood of New Orleans."[41] Audubon had chosen his destination well, and he made the most of the ornithological opportunity, getting down to work right after his arrival.

No sooner had he settled into New Orleans in early January 1821 than he "took My Gun, rowed out to the edge of the Eddy and killed a Fish Crow." Thus begins a series of ornithological entries in Audubon's journal, always searching for birds to draw, whether dead ones bought in the city's market or live ones shot in the surrounding environs. When he killed the first Fish Crow, for instance, "hundreds flew to him, and appeared as if about to Carry him off, but they soon found it to their Interest to let me have him." Audubon also bagged the birds common to coastal areas—pelicans, gulls, cormorants, ducks, geese—and welcomed the early arrival of migrants in the mild winter weather. "I had the pleasure of remarking thousands of purple martins travelling east-wardly," he wrote in the second week of February 1821, when the temperature sat at 68 degrees, and ten days later, he saw "Three Immense Flocks of *Bank Swallows* that past over Me with the rapidity of a Storm." Even though he seemed surprised at the birds' early arrival, he felt "pleased to see these arbingers of Spring," figuring that they would make it to Kentucky in about a month. Even if he stayed within the city, he could find that "the Market is regularly furnished with the *English Snipe* . . . Robins Blue Wingd Teals Common Teals, Spoon Bill Ducks, Malards, Snow Geese, Canada Geese, Many Cormorants, Coots, Water Hens, Tell Tale Godwits . . . Yellow Shank Snipes, some Sand Hills Cranes, Strings of Blew Warblers, Cardinal Grosbeaks, Common Turtle Doves, Golden Wingd Wood Peckers &c."[42]

As February turned to March, then April, Audubon marked the migrations that came and went during the Louisiana spring, noting that "to My Astonish-ment, the Many Species of Warblers, Thrushes &c that Were numerous during the Winter have all Moved on Eastwardly," but then, likewise to his surprise, he "heard the Voice of a Warbler new to Me, but could Not reach it."[43] That summer, when he and Joseph Mason moved to the Pirrie plantation at Bayou Sarah, Audubon took care of his tutor duties well enough, but he more happily spent hours, sometimes several days, ranging through the woods and relishing the profusion of big birds (ibises, woodpeckers, herons) and small (flycatchers, orioles, warblers of all sorts). After that job came to an abrupt end and he moved back to New Orleans, he still took pleasure in recording sightings of birds of all sorts—"Green Back Swallows, Gamboling over the City and the River the Whole day"—and sometimes making very detailed ornithological notes; his extensive description of the Brown Pelican covers two complete pages, and his journal ends with brief descriptions of over sixty "Water Birds of the United States."[44] Even now, Audubon's Mississippi River Journal contains the sort of careful and comprehensive field notes that a modern-day ornithologist might still find extremely useful.

"Mr Wilson Has Made an Error"

Audubon cared only for one ornithologist, one who had come earlier—the late author of *American Ornithology*. Alexander Wilson's work stood as the most comprehensive and authoritative work on American ornithology to date, and before Audubon could hope to surpass it, he had to see it. He frequently looked to Wilson's work for corroborating his own observations, but even more, he took quiet pleasure in correcting any errors or omissions he found in *American Ornithology*.

In the introduction to *American Ornithology*, Wilson noted the difficulty of building upon any scientific base in the United States, complaining that "from the writers of our own country the author has derived but little advantage." Thomas Jefferson had listed only 109 birds in his *Notes on the State of Virginia*, Wilson pointed out, William Bartram had extended the count to 215, and a few other gentlemen-scientists "have each enumerated a few of our birds." "But these," he concluded, "can be considered only as catalogues of names, without the detail of specific particulars, or the figured and coloured representations of the birds themselves."[45] In his own time, Wilson gained credit for ornithological supremacy over his predecessors by lengthening the list of American birds to 278, of which 56 had not been previously described.[46] Audubon would extend the number to over 500, quietly keeping score against Wilson in the process. At the beginning of 1836, when the final volume of *The Birds of America* had just gone to the engraver, he wrote a friend that "in Two Months from this date, I shall have at least exceeded Wilson in Numeric Species."[47] By that time, of course, Wilson had been dead almost a quarter century. He never died, however, in Audubon's mind.

On the very first day he set out from Cincinnati, Audubon wrote in his journal that one of the birds he and his boatmates shot, a Yellow-rumped Warbler, did not fit Wilson's description: "I feel perfectly Convinced that Mr Wilson has Made an Error in presenting the bird as a New Specie." Sometime later, on the Mississippi near the mouth of the Arkansas River, Audubon noted that "the *Prairie Hawk* that I see here is not the *Marsh Hawk* of Willson," and he provided the identifying marks that would distinguish it—"Lighter Colour, the Tip of the Wings Black and Only One Large Bend of Dark ending the Tail."[48] Still, with no extensive ornithological library at his disposal while still living on Aumack's flatboat, Audubon longed to see Wilson's *American Ornithology*, just to be sure about some of his own observations.

Audubon got lucky for a few days when he reached Natchez, when his friend Berthoud introduced him to a few useful Natchez acquaintances, one of whom "kindly procured Me Willson's Ornithology"—or as much of it as he had at the time. The Natchez subscriber had not yet received the ninth and final volume, which Audubon especially wanted to see, and "after many Inquiries . . . I was disappointed in my wish of examining it." James Wilkins, the Natchez resident who had the other volumes, raised Audubon's doubts when he "assured me that [Wilson's] Work was far from Compleat"—and far from complete in more ways than one. Wilkins said he himself had seen several species that Wilson had apparently missed, but as a businessman he didn't have the time to take careful note of them.[49] Audubon took note of Wilkins's observation, however, already keen to take every opportunity to record Wilson's omissions and mistakes.

On the other hand, Audubon was also keen to see what Wilson did get right, and he made good use of the time he had at his disposal. On the last Saturday in December 1820, with "the Weather very Cold, the Thermometer at 25," Audubon spent all day indoors, "writting the Name and Such Descriptions of the Watter Birds in Willson as would Enable Me to Judge Whenever a New Specimen" might come into his possession. He had Turton's recent edition of Linnaeus's *Systema Naturae* as a necessary stopgap, but in the end, only Wilson would do. Just before arriving in New Orleans, when he was starting to draw a tern he had shot, Audubon "ransacked Turtons but all without effect" to make a clear identification, and he needed better confirmation: "Yet I do not Consider this a New Specie untill I See Willson's 9th Volume."[50]

As soon as he got settled in New Orleans, he started seeking out any volume he could get, but only "in Vain," he lamented, "the high value set on that Work more Particularly Lately as rendered it extremely rare and the few who possess it will not Lend it." His luck was no better a couple of weeks later: "I walked a good deal about the City in search of Work & Willson's Ornithology but was not favoured with any success."[51] When he did get to consult Wilson, it sometimes made his own efforts seem redundant, as was the case with a "beautiful *Blue Crane*" (probably a Little Blue Heron) he purchased in a New Orleans market: "I Drew it and its Coresponding so well with A. Willson Description Stopd Me from writting it Myself."[52] The need to check himself against Wilson's work would continue throughout Audubon's Mississippi odyssey and, indeed, throughout his whole life.

On the other hand, nothing could be better than besting Wilson in a bird count. No sooner had Audubon begun to look into Wilson's work than he began to see some of its limitations. On February 17, 1821, he sent Lucy, his "Beloved Wife" back in Kentucky, a batch of drawings he had done during his Mississippi sojourn, and on the list of twenty birds he appended the notation "Not Described by Willson" to eight of them. He closed that day's journal entry with a contented sigh: "May I have the Satisfaction of Looking at these and Many More in good Order on My return the fruits of a long Journey."[53] A couple of months later, he wrote in his journal that "I am forced here to Complain of the bad figure that my friend Willson has given of the Warbler I drew yesterday," pointing out that in Wilson's painting the length of the bird's bill "exceed that of Nature 1/8 of an Inch—an enormous diference—and he has runed a broad White line round over the Eye that does not exist." Wilson had also apparently made the bill "Much too Large & Long" on his depiction of another bird, the Orchard Oriole.[54]

Part of the reason for Wilson's ornithological errors, Audubon thought, stemmed from flaws in his method: "Sorry I am to have to say that Mr Willson's Drawing could not have been Made from the Bird *fresh killed*," which, to Audubon, was the only way to work.[55] (If he made an occasional exception for his own use of bird specimens he bought in the New Orleans market, he never said so.) However much he might refer to "my friend Willson," Audubon did indeed see an enormous difference between himself and his ornithological rival, not just between some of the details in their work, but between the very ways the two men worked. If, as Audubon suspected of Wilson, a bird artist did not work from fresh-killed specimens, allowed himself to be in too much of a hurry, and, perhaps worst of all, relied on the hearsay of others, he could not be altogether trusted to be either complete or accurate.

The dubious details Audubon found in Wilson's work would only multiply—and be made public in print—over the years. That tendency to question and even challenge the accuracy and authority of Wilson's work opens up another aspect of Audubon's attitude toward his predecessor, a process of sowing doubt that became the dominant approach he took in producing his own massive work. And, whether justified or not, he certainly reaped what he sowed.

Philadelphia, City of Limited Love

As much as New Orleans afforded Audubon an opportunity to pursue his already-ambitious work on birds, he needed to leave. On the positive side, his wife and sons had joined him there in mid-December 1821 ("all in good health," he noted on the day of their arrival), but on the negative, he seemed doomed to having to make do with repetitive rounds of art instruction ("Gave 2 Lessons," he wrote day after day in late December, until finally being able to add "Paid our Rent this morning").[56] He remained in New Orleans through the early winter of 1822, adding to his collection of bird drawings, but in March, his fragile financial situation led him to move upriver to Natchez, where he took on more work in art instruction, first tutoring a teenage girl, then teaching at a nearby female academy. He even became a student himself, taking lessons in oil painting from a recently arrived portraitist, John Steen. He and Steen eventually went into business together as itinerant portrait painters, going around the region trying to make likenesses of plantation people, but the partnership didn't last very long: Audubon spent too much time chasing after birds, Steen thought, and the impatient painter wanted to move on more quickly, so he did.[57] Once again, Audubon was on his own and unemployed—and maybe on his way to becoming unemployable.

After that, he bounced around for a year, moving from one short-term job to the next—yet another art-teaching stint on yet another Louisiana plantation (from which he yet again got fired), more attempts at painting quick portraits and landscapes (for which he occasionally didn't get paid), and eventually some sign painting and steamboat decorating back in Kentucky (in which he could hardly engage his full artistic talent).[58] For someone whose aspirations seemed destined to reach the skies, this low-level labor could only lead to a dispiriting dead end.

Instead, Audubon headed toward the place in the United States that seemed to offer the best opportunity for putting himself in a better position to pursue his passion: Philadelphia. "I reached Philadelphia on the 5th April 1824," he would later write, "just as the sun was sinking beneath the horizon."[59] On a broader scale, he might well have seen the promise of a new dawn. By the 1820s Philadelphia had become, as one of the city's leading historians puts it, "a muscular commercial and industrializing city," with brains to match the muscle—no longer the nation's or even the state's capital, nor the nation's most notable emporium of trade or finance, but still "a center of law, medicine, science, and publishing." With a prominent university, several prestigious learned societies, and various other institutions devoted to the intellect, the city had become the "Athens of the Western World," the "Mecca of scientific men."[60] Boston had Harvard just across the river, of course, along with several prominent scientific societies—the American Academy of Arts and Sciences and the Society of Natural History—and was also home to the *American Journal of Science*. New York had its Lyceum of Natural History, where Audubon would eventually deliver his first scientific paper. But in the early decades of the nineteenth century, neither Boston nor New York nor any other American city could match Philadelphia for scientific significance. Audubon knew he needed to be there.

In Philadelphia, scientifically interested citizens, particularly middle- or upper-class urban dwellers, could find any number of opportunities to encounter the wonders of the natural world without even venturing beyond the city limits. Audubon himself had previously visited Charles Willson Peale's well-regarded "Repository for Natural Curiosities," the museum that for almost forty years had been the city's most accessible site for scientific inquiry, including the opportunity to draw from stuffed specimens (a practice that Audubon declined to admit to in his own case but, as we will see, frequently ascribed to Alexander Wilson).[61] Any member of the public able and willing to pay the entrance fee of twenty-five cents—which, admittedly, probably did not include the city's poorer residents—could get a ticket that offered an appealing promise: "NATURE. The Birds and Beasts will teach thee." They would teach, that is, according to Peale's notion of order. His idea of a museum of natural history had much more to it than being a mere repository for stuffed animals and other sorts of specimens: It could also be a source of social as well as scientific education for citizens of the new republic. By showcasing the systematic arrangement of nature according to taxonomic classification, the museum might well reinforce notions of rank and order in the minds of its viewers, reminding them that they, too, had an assigned place in the larger scheme of things. Science could thus be a means of supporting citizenship, or so Peale hoped. His famous 1822 painting *The Artist in His Museum* depicts Peale himself as scientific gate-keeper, standing prominently in the foreground and pulling back the curtain to invite the viewer into the well-ordered interior, with shelves of bird and animal specimens carefully arranged below rows of portraits of prominent American leaders. In his "Repository," Peale had created an unmistakable melding of natural history and national history, a didactic display for a demo-cratic society.[62]

Beyond the well-ordered interior of Peale's museum, Philadelphia offered other ways for people to satisfy their scientific interests, including those that might now be considered pseudoscience or mere quackery. Exhibitions of steam engines, hot-air balloons, and other such wonders of technology provided as much public entertainment as scientific instruction, and so did explorations of the inner workings of the mind, such as animal magnetism and phrenology. A Philadelphia academy for young men included phrenology in its curriculum, in fact, and local physicians formed America's first phrenology society in 1822. Two years later, in 1824, a Philadelphia-trained physician, Charles Caldwell, wrote *Elements of Phrenology*, the first American book on the subject.[63]

Audubon may have been impressed with the various attractions of Phila-delphia's scientific sideshow, but he was a man on a single-minded mission: to gain entrée into the city's artistic and scientific circles. At the time of his arrival, he had not yet perfected the manner of self-promotion he would later achieve, but he did have the good fortune to make himself known to the right people in the right places. He first reached out to an acquaintance from his earlier days at Mill Grove, Dr. James Mease, a well-connected physician with wide-ranging interests in science, from medicine to agriculture to horticulture to geology. Mease did not disappoint. He appreciated Audubon's work enough to offer him a series of introductions to some of the other prominent people he thought Audubon needed to know. Among the useful men in Mease's net-work was the celebrated artist Thomas Sully, who agreed to give Audubon lessons in his preferred medium, oil painting, in exchange for Audubon's giv-ing lessons in his, pastel, to Sully's daughter; Audubon would later call Sully a "kindred spirit." Mease also became Audubon's conduit to the recently arrived but immediately notable nephew of Napoleon, Charles Lucien Jules Laurent Bonaparte, the twenty-year-old Prince of Canino and Musignano, who had come to Philadelphia with his new wife to live for a while with the American wing of the Bonaparte family. Like Audubon himself, Bonaparte had intense ornithological interests, including an expanded version of Wilson's *American Ornithology*, but he never took Audubon to be an unwelcome competitor. Indeed, Bonaparte saw the possibility of an artistic and scientific collaboration, and even though he had already engaged one of the painter-named Peale sons, Titian Ramsey Peale, as the main artist for his project, he bought some of Audubon's drawings, one of which he incorporated into his own work.[64]

More important, Bonaparte provided Audubon with two apparently promising introductions—promising but ultimately unhappy, neither of them working out as well as Audubon might have hoped. First, Bonaparte took his new friend to a meeting of Philadelphia's Academy of Natural Sciences to show some of Audubon's drawings to the members, most of whom liked what they saw. One, however, did not—George Ord (1781–1866), Wilson's patron and champion, the man who guided *American Ornithology* to completion, promoted it in scientific circles, protected the posthumous reputation of its author, and, in 1824, planned to publish a new edition of Wilson's book on the not-so-distant horizon. Ord no doubt knew something of Audubon from Wilson himself while the "Father of American Ornithology" was still alive, and he may also have caught wind of some of Audubon's criticisms of Wilson's work. In any event, Ord needed precious little prodding to recognize Audubon as a threat to his stake in Wilson's legacy. Unlike the more generous (and certainly more self-confident) Bonaparte, Ord not only dismissed Audubon's birds—he took special exception to their occasionally odd postures and their elaborate botanical backgrounds—but also managed to have Audubon's application for membership to the Academy squelched.[65]

Thus thwarted at the Academy of Natural Sciences, Bonaparte took Audubon to meet the one man who might still give his birds the visibility they deserved: Alexander Lawson, Philadelphia's most celebrated engraver, a member of the critical craft that formed the city's infrastructure of art and science. Benjamin Franklin had been the most prominent printer in the eighteenth century, but by the second decade of the nineteenth century, Philadelphia's print culture had burgeoned into one of the city's central assets, running the gamut from newspapers and pamphlets to periodicals and books of all sorts, including illustrated works of art and science, none more notable at the time than Wilson's *American Ornithology*. Philadelphia also boasted some sixty engravers, none more notable than Lawson, Wilson's fellow Scotsman who

had put Wilson's birds so beautifully on the page.[66] Because Lawson was
Bonaparte's engraver as well, the young Frenchman had good reason to hope
that Lawson might also work with Audubon.

That hope died quickly in Lawson's shop, however, probably even before
Audubon arrived there. Lawson's commitment to Wilson's work had put him
in a long and close association with Ord, and the two formed a formidable
partnership that endured for years, even beyond the original publication dates
of *American Ornithology*. The Lawson-Ord arrangement had even become
something of a family affair: In addition to a son, Oscar, the only person he
trained in the craft as his successor, Lawson also had two daughters, Malvina
and Helen Elizabeth, who were colorists on the 1824 republication of Wilson's
work, for which Ord paid them quite well.[67] By the time Audubon arrived on
the scene, Lawson's ties to Ord had long been too strong to be undone, and as
soon as Lawson got wind of Audubon, he had no use for either him or his birds.
When presented with a specimen of Audubon's work, Lawson grumbled about
its excessive size and questioned its ornithological accuracy, huffily remarking
that "we in Philadelphia are accustomed to seeing very correct drawing." Even
when Bonaparte tried to reassure Lawson by saying that he would pay for some
of Audubon's drawings, Lawson snapped back, "You may buy them, but I will
not engrave them." Lawson would be a complete loss for Audubon, and he
never really had a chance of getting him as an engraver. When he later reflected
on the outcome of his 1824 visit to Philadelphia, he wrote, without naming
names, that "Wilson's principal engraver, amongst others, gave it as his opinion
to my friends, that my drawings would never be engraved."[68]

Notwithstanding the obstacles of Lawson and Ord, Philadelphia did not
turn out to be a complete bust for Audubon. In addition to Thomas Sully and
Charles Lucien Bonaparte, he made a number of acquaintances who would
prove to be valuable allies, some of these friendships lasting for the rest of
Audubon's life. Nobody did better by Audubon than Edward Harris, a young
gentleman-farmer from New Jersey, whose comfortable familial inheritance
allowed him to live the leisurely life of an amateur naturalist, collector, and
sportsman. His wealth also enabled him to encourage struggling artists, and
soon after he met Audubon, in July 1824, he generously bought all of Audubon's
unsold drawings, and at Audubon's price. (He would later do even more, becoming an early subscriber to *The Birds of America*.) The two became fast friends,
with Harris sometimes supplying Audubon with specimens and occasionally
going out in the field with him, even on extended journeys in the 1830s and

1840s, when Harris joined Audubon as a fellow explorer of the South and Upper Midwest. Audubon would give Harris the supreme scientific honor of naming a species of hawk after him, the Louisiana Hawk, or *Falco Harrisii*, describing Harris as "a gentleman who, independently of the aid which he has on many occasions afforded me[,] . . . merits the compliment of an enthusiastic Ornithologist." Throughout the pages of Audubon's *Ornithological Biography*, no one from Philadelphia appeared more frequently or more favorably than his "generous friend Edward Harris."[69]

More immediate assistance came from Gideon Fairman, a local printer who, unlike Lawson, admired Audubon's work enough to give him honest advice about getting it produced outside of Philadelphia. Lawson wouldn't print Audubon's birds, Fairman's own firm couldn't—it was too small and too specialized an operation for the big job Audubon already envisioned—and no one else in Philadelphia seemed able to take on the task either. Audubon also came to understand, from either Fairman or other Philadelphia friends, that New York would likewise not be an option: "I have reason to suspect," he wrote in his journal, "that unfriendly communications have been sent to the publishers from Philadelphia by parties interested in Wilson's volume, and who have represented that my drawings have not been wholly done by myself."[70] With engravers in the major eastern cities apparently out of the picture, the best prospect for high-quality engraving would have to be on the other side of the Atlantic. In the meantime, Fairman did offer Audubon a commission for drawing a bird for a New Jersey banknote—not the already-typical eagle perched majestically on a promontory in a patriotic pose, but a smaller and much more modest-seeming grouse, running along the ground.[71] As someone who had just five years before succumbed to bankruptcy in the Panic of 1819, the economic crisis created by the nation's flimsy banking system, Audubon might well have appreciated the irony that his first paying job for avian art would be for a picture on paper currency. If so, he never made the point in print but apparently just took the job and pocketed the fee. In any event, a one-off opportunity to draw a bird for a banknote did not offer any sort of financial future, and Audubon realized that Fairman and a few others were right: "The patronage which I so much needed, I soon found myself compelled to seek elsewhere."[72]

And elsewhere he went, first to New York City, then up the Hudson to Albany and across to Niagara Falls, then to Pittsburgh, Cincinnati, Louisville, and finally back down the Mississippi to Louisiana, where Lucy held a teaching job on the Beech Woods Plantation in Bayou Sarah (and where she had hung on after Audubon himself had been dismissed earlier in the year). As he looked back on his time in Philadelphia, he had decent reason to wallow in self-pity. He had essentially been sent packing, blackballed at the Academy of Natural Sciences and rebuffed by the one engraver who would have been up to the job Audubon wanted done. Soon after arriving in Louisiana, he wrote to a friend in the Philadelphia area about the necessary next steps: "With an almost despairing heart I shall leave America early this ensuing spring, and now bid you my farewell . . . for unless a success *scarce expected* should take place, I never will review this happy continent, will have to abandon my long acquired habits of *watching nature at work* and will droop moreso among the dreg of the world as it is called."[73]

Audubon never descended to the dregs, of course, and in his later published account of his decision to depart for the far side of the Atlantic, he would take a more upbeat sense of self-discovery: "For the first time," he wrote, "I communed with myself as to the possible event of my visiting Europe again; and I began to fancy my work under the multiplying efforts of the graver. Happy days, and nights of pleasing dreams!" No doubt reflecting his impatience to get on with the next chapter of his life, he then finished the account of his post-Philadelphia period with understandable narrative compression: "Eighteen months elapsed. I returned to my family, then in Louisiana, explored every portion of the vast woods around, and at last sailed towards the Old World."[74]

But even as he said his farewells to America, Audubon could not easily shake the sense of injury at the treatment he had received in Philadelphia. He had friends there, to be sure, but they could never overshadow his enemies in the city—and one in particular. In a last letter to Thomas Sully, he said, "I cannot help thanking Fairman, Peale, Neagle, Le Seuer and many others" for their various kindnesses, and he ended with one other conciliatory note: "Should you see Mr. Ord, tell him I never was his enemy."[75] Whatever pretense of goodwill that lay behind such a sentiment didn't last. In July 1826, soon after setting off on the ship that would take him across the Atlantic to England and, he hoped, the respect and success he so desperately sought, Audubon had a surprising sighting of a bird he had never before seen, and that sent him into an ornithological exaltation—but with a hostile personal edge: "The first mate called it a *Mure*!! Linnaeus never described this bird. Neither have I, not any of my predecessors—not even the very highly celebrated and most conspicuous Mr. George Ord of the city of Philadelphia, state of Pennsylvania, member of all the societies, &c., &c., &c., the perfect *academician* that laughed because [I said] a turkey could swim!"[76]

Audubon may have been close to drunk at the time. To deal with the tedium of the trip, he did what many impatient passengers might do: He drank quite a bit, and as a result, made more than a few tipsy-seeming entries in his journal. But inebriation can hardly be the whole explanation.[77] In his most sober moments, Audubon still had it in for Ord. For the next two decades, throughout his subsequent scramble for success with *The Birds of America,* Audubon would never forgive Ord for his early exclusion from the inner circles of science in Philadelphia, and the term "Philadelphians" became a spiteful synonym for Ord and the other men who he felt had done him wrong. He became almost as obsessed with his enemies as he had always been with his birds. But he did have his birds, and they would soon make him famous—eventually even in Philadelphia.

Chapter 4

Going into Business with *The Birds of America*

The purpose of this voyage is to visit not only England but all Europe, with the intention of publishing my work of the Birds of America. If not sadly disappointed, my return to these happy shores will be the brightest birthday I shall have ever enjoyed: Oh America, Wife, Children and acquaintances, Farewell!

—John James Audubon, *1826 Journal* (on board the *Delos*, at sea)

In Audubon's career, 1826 was the watershed year. By that time he had fully decided on his life's work, defined its nature and scope, determined how to get it done, and given it a name, *The Birds of America*. It would be a book—a very big book—and Audubon would come to call it his "Great Work." Great it certainly was—great in ambition, great in execution, and great in size and price. In its grandest and most complete form, the Double Elephant Folio edition, *The Birds* would be a huge, heavy, ungainly four-volume set of 435 plates, with each plate measuring 29½ x 39½ inches, each volume weighing more than forty pounds and requiring, as Audubon would put it, "two stout arms to raise it from the ground." It would also require a total commitment of around a thousand dollars to own.[1] *The Birds of America* would not be a book for ordinary people, even for the middle-class reading (or consuming) public. It was, almost by definition, a rare book from the beginning.

But great as it turned out to be, *The Birds of America* still invites one basic, maybe obvious question of its creator: How, let alone why, would anyone ever try to make a business, much less a living, with such a big, expensive book about birds? Audubon would continue to ponder that question, sometimes with worrisome self-doubt, throughout the time he worked on *The Birds of America*. But by the time he boarded the *Delos* in New Orleans in May 1826, he had cast his fate to the transatlantic winds, and his "intention of publishing my work of the Birds of America" gave him both destination and destiny. All he had to do was to get to terra firma in Great Britain and get down to business—and a business his Great Work would certainly be.

Entering the "Emporium of Commerce"

The world of art always reflects economic as well as aesthetic considerations, and the interplay between those two tells us a great deal about Audubon's role in making *The Birds of America* as well as marketing it. For years, despite repeatedly portraying himself as an artist and naturalist, too busy with birds to care for his business, he had been laying the groundwork for his high-stakes commercial venture in avian art. In addition to making dozens of drawings of birds, Audubon had made useful human connections as well, seeking support from prominent American allies who could help open the doors of opportunity, whether in the United States or elsewhere. When he embarked for Great Britain, he carried onboard the results of both—his well-packed portfolio of bird drawings along with several valuable letters of introduction, including ones from such political luminaries as Henry Clay of Kentucky and DeWitt Clinton of New York and a couple of well-placed friends, Charles Lucien Bonaparte of Philadelphia and Vincent Nolte of New Orleans. Clay's letter seemed supportive if not effusive, calling Audubon a "respectable citizen of the U. States [who] . . . has been, for some years, engaged in procuring drawings and preparing manuscripts in relation to the birds of America. He goes to England, with the purpose of completing his work, and having it published there."[2] Nolte, a businessman who had known Audubon for about fifteen years, got more to the point on the financial side of the enterprise. Writing to Richard Rathbone of Liverpool, who would soon become Audubon's frequent host and constant supporter in that city, Nolte prepared Audubon's way by praising his work, "a collection of upwards of 400 drawings, which far surpass anything of the kind I have yet seen." He also looked to the potential bottom line, concluding that Audubon's work "would prove a most valuable acquisition to any Museum, or any monied patron of the arts."[3] Nolte knew that Audubon's project would be a costly one and that the resources of a well-endowed institution or a "monied patron of the arts" could be critical.

The only thing that stood between Audubon and those potential patrons was the Atlantic Ocean, and he had to endure a tedious, two-month crossing—

"long and painful in the extreme," he would later complain to Lucy—before getting to Great Britain.[4] From the beginning of the boat trip, he became beset by impatience with the pace of going against the prevailing winds: "Never, if you can do otherwise, sail from New Orleans for Europe, in June, July, or August," he wrote in his shipboard journal, "as, if you do, you may calculate on delays incalculable in the Gulf, such as calms, powerful currents all contrary and worst of all the *Trade Winds* so prevalent during these months."[5] In his better moments, Audubon put his powers of observation to use, keeping his naturalist's eye out for porpoises, dolphins, flying fish, and other members of the "finny tribes," along with any birds that flew over the ship and sometimes landed on deck. A warbler of some sort came onto the ship and moved about for a few minutes "with great activity and sprightliness," but a Rice Bunting that stayed overnight wound up "exhausted, panted, and, I have no doubt, died of inanition."[6]

Dying of his own form of intellectual inanition, Audubon welcomed whatever onboard ornithology might come his way, and sketching specimens became a salvation of sorts, especially as the ship remained becalmed: "I should be dull indeed were it not for the fishes and birds, and my pen and pencil."[7] The farther one goes to sea, of course, the fewer the birds, and as he continued across the Atlantic, Audubon had no more serendipitous ornithological encounters. Instead, he spent time sketching members of the ship's crew and, more than occasionally, drinking.

Audubon regained his professional focus, however, as the *Delos* neared British shores. On July 18, 1826, his anticipation perked up as he sniffed the winds of "the emporium of commerce of England." Two nights later, when the ship at last dropped anchor in the Mersey River, he could look from the deck and see, no more than two hundred yards away, the lights of Liverpool. "I am in England," he wrote to Lucy in the pages of his journal. "With what success I shall go through my undertaking I shall be sure to inform Thee."[8]

At first, success seemed to come his way quite readily. He landed in Liverpool on July 21, a rainy Friday morning, and the smoke of the city immediately choked his lungs and stung his eyes. But over the next couple of days, once he cleared customs, lugged his baggage and portfolio of drawings to his rented lodgings, got his land legs under him, and got some food under his belt, he set out to put his letters of introduction to good use. The most valuable addressee of one of those letters turned out to be Richard Rathbone, a prosperous Quaker merchant and cotton importer, who introduced Audubon to a number of other

prominent men in Liverpool—most usefully his brother, William Rathbone
IV; the American consul, James Maury; and Edward Roscoe, himself an
accomplished botanical artist and patron of the arts. Everyone Audubon met
seemed to wish him only the best, and within a few days he could write to Lucy
that "all smile a welcome to me. Ah yes, I assure thee, a welcome." With the
help of his new friends Rathbone and Roscoe, he received an even more
emphatic sign of welcome, an invitation to put up his paintings in the city's
Royal Institution at the end of July. He had been in Liverpool just a little over
a week, but he had clearly arrived.[9]

The Naturalist as "Show Man"

When Audubon's work first went on display, on July 31, the prospect of exposing his paintings to "the public, the connoisseurs, and the critics" gave him
opening-day jitters, and he dreaded the "very casting askance of a single eye."[10]
He settled down, though, when he found a positive response from the Royal
Institution's viewers; within a week, he wrote with pride to his brother-in-law
Nicholas Berthoud that he had 250 drawings on display, and in one two-hour
period, over 400 people came in to see them, some from as far away as Manchester. Audubon could crow that "my fame as an ornithologist and artist has
flown from mouth to mouth with a rapidity that has quite astounded me," and
he finally felt "some hopes of success at last."[11]

Still, he expressed a nagging conflict between his new status as a celebrity
and his preferred status as an "ornithologist and artist." Having prominent
patrons and a measure of success was fine, but the prospect of becoming a
popular exhibitor before a paying audience seemed a bit of a problem, at least
for a while. When some members of Liverpool's Royal Institution had first
suggested that he charge admission to the show, Audubon recoiled with a sense
of injured honor: "My heart revolted at the thought," he wrote, "and although
I am poor enough, God knows, I could not think of doing such a thing consistently. . . . I could not, I repeat, think it consistent to become a mere *show man*
and give up the title of *J. J. Audubon, Naturalist*."[12]

Gradually, though, Audubon began to soften this high-minded stance, and
then he got over it completely. Like anyone playing to a paying crowd, he
counted the house ("Many persons peeped at my Birds as I peeped at *them*")
and even if he couldn't always tote up the daily take ("However, how many

English shillings were received, I cannot tell") he began to keep an eye on the income.[13] If people would pay to peep at his work, he would accept their shillings. *J. J. Audubon, Show Man* it would be, then.

A few weeks later, the popular response still seemed so positive that Audubon could write his son Victor that "it seems that I am considered unrivaled in the art of Drawing even by the most learned of this country. The newspapers have given so many flattering accounts of my productions and of my being a superior ornithologist that I dare no longer look into any *of them*."[14] But of course he looked, and he could only be delighted by what he saw. The Proprietors of the Royal Institution had given him a good promotional billing in the local press soon after the show opened, inviting the public to see Audubon's remarkable display: "MR. AUDUBON, of Louisiana, who has lately arrived in Liverpool with a large and splendid Collection of BIRDS, PLANTS, and other SUBJECTS of NATURAL HISTORY, . . . has been induced, at the request of the Committee, to exhibit his Collection more generally."[15] A few weeks later, the Liverpool *Mercury* carried a very glowing report on his show, noting that his drawings "display great science, accurate observation, and refined taste." Audubon's birds seemed much more than static specimens: "The great excellence of Monsieur Audubon appears in seizing the characteristic habit of each bird, and giving the semblance of life and motion." By the same token, Audubon's show transformed the staid exhibition space and gave the observer the exhilarating sense of being "in an aviary of skies, where birds of all plumage and of all dispositions are sporting around."[16]

The Liverpool show put Audubon well on his way to British acclaim, and exhibits in Manchester and Edinburgh would soon follow in the fall of 1826. "My plans are now fixed," he wrote Lucy the day the Liverpool *Mercury* review appeared in print. "I will continue to exhibit my drawings thro out the three Kingdoms and take especial care of the proceeds I assure thee for thy sake— I will keep myself at Drawing & Painting Birds & Quadrupeds for sale all the time, and I hope thus to do well." This approach seemed especially promising, Audubon added, because "my entry in England having been at once in the best Circles of Society will afford me a great length of continuation of the same favors."[17] With friends in high places in Liverpool, Audubon expected to find equally agreeable allies in other cities: "It is not the naturalist that I wish to please altogether, I assure thee," he wrote Lucy a few months later. "It is the wealthy part of the community, the first can only speak well or ill of me, but the latter will fill my pockets."[18]

Yet even as Audubon watched his pockets fill and promised to "take espe-cial care of the proceeds" for the sake of his family, he still resisted some of the more overtly commercial aspects of the exhibition process. Two men approached him at his show in Manchester, for instance, suggesting that they could help by providing music to accompany his exhibition. In a curious form of human field identification, Audubon observed that they were "Italians, by their noses and large mouths," and he thanked them but declined their services: He was showing birds, after all, not "Egyptian mummies or deathly-looking wax figures," and if "my songsters will not sing or be agreeable by themselves, other music would only diminish their worth."[19] Two days later, another man suggested that Audubon ought to paint a sign to hang outside his exhibition, and again Audubon declined. During his darker days in the United States, he had had to resort to sign painting himself to survive, and he could no doubt grasp the maxim that it pays to advertise. On the other hand, as he later explained to Lucy, if he painted a sign showing birds for free outside, people might "gaze at it so long that they would forget that 200 drawings are waiting to be examined for the mere trifle of one shilling."[20]

Above all, the most difficult part of exhibiting his work in public was listen-ing to the private conversations of his viewers, especially those who were critical of his work. From the beginning, he suspected that his drawings might be "seen by the shilling's worth, and criticized no doubt by the pound."[21] Some people didn't even pay the shilling. One day in Manchester, a little before opening time, he happened to overhear two men talking about his exhibit. "Pray, have you seen Mr. Audubon's collection of Birds?" one of them asked the other. "I am told it is well worth a shilling. Suppose we go now." But the other man had a decidedly different opinion. "Puh, it's all a hoax," he said. "Save your shilling for better use. I have seen them. Why the fellow ought to be drummed out of town." There's no accounting for taste, of course, and Audubon went on to receive considerable praise from other viewers later in the day. Still, the unkind words stung, and in recounting the incident to Lucy, Audubon told her he "turned pale and dared not raise my eyes lest I be known. But depend on it, I wished myself in America again."[22]

Some days turned out to be better than others, and some cities, too. "My time at Manchester has not been productive," he grumbled to Lucy, and he set out for a new exhibit in Edinburgh in late October 1826. There he had decent success, and as the exhibit neared its close in December, Audubon would declare the Edinburgh public to have "the eyes of the most discerning people in the world."[23]

In writing about his exhibition efforts a few years later, in the second volume of *Ornithological Biography*, Audubon took some of the sting out of the experience by portraying himself in circumstances similar to those of his now-deceased nemesis, Alexander Wilson. "I cannot say that the employment was a pleasant one to me, nor do I believe it was so to him," Audubon wrote, but if Wilson had had to stand "in a room crowded by visitors, holding at arm's length each of his large drawings, listening to the varied observations of the lookers on," Audubon could do so, too. Wilson had "acquired that fame, of which I was also desirous of obtaining a portion; and . . . I waged war against my feelings, and welcomed all, who, from love of science, from taste, or from generosity, manifested an interest in the 'American Woodsman.'"[24] (If Audubon's choice of the "A. W." moniker reflected some underlying connection, however inadvertent, to Alexander Wilson's own initials, he did not say. Some things about Audubon may never be known, but at best just suspected.)

The American Woodsman image became an image in a very literal sense. In addition to casting their eyes on Audubon's many paintings of birds, one thing the "most discerning people" of Edinburgh had the opportunity to examine was a painting of Audubon himself, done during his stay there by the Edinburgh artist John Syme.[25] At the time of his Edinburgh exhibition, Audubon took apparent pride in "my flowing curling locks that again loosely are about my shoulders," and Syme certainly captured them in his painting. It is a now-classic portrait of the dashing, forty-one-year-old Audubon, long-haired and clad in frontier garb, cradling in his arms a rifle, the always-critical tool of his trade, looking up at some distant bird, no doubt, and all in all conveying the very essence of the American Woodsman (see Plate 1). Audubon admitted that his portrait made him seem "a strange looking figure, with gun, strap and buckles, and eyes that to me are more those of an enraged eagle than mine."[26] But being "strange looking" made sense in the context of the time and place: Putting such a rustic-looking figure before a middle- to upper-class urban audience underscored the quasi-exotic nature of Audubon's self-consciously crafted identity, forcing people to see him as he liked to be seen, as a manly frontier naturalist carrying a gun, not a prettified painter in an artist's smock and holding a palette. Describing himself as an eagle perhaps also made sense: The painted Audubon became part of his own show of avian art, an arresting figure fixed on the wall among the birds rather than a man standing somewhat uneasily in the gallery among the viewers. "I preferred *it* to be gazed at [rather] than the *original* from whom it was taken."[27]

The "original" also began to return to his original decision for a different approach to displaying his bird paintings—in book form. He still had to exhibit his work for money to make ends meet, and he also supplemented his income by painting several one-off pictures of birds and animals and selling them as quickly as he could. But more and more, his thoughts turned to publication. On a day when only twenty people had come to see his Manchester exhibit— "Sad work, this," Audubon muttered in his journal—the American consul in that city, F. S. Brookes of Boston, came to visit and "advised me to have a subscription book for my work, &c., &c." In Audubon's journals, the "&c., &c." usually stood for something between boredom and annoyance, and at first he seemed skeptical about Brooke's suggestion: "It is easy to have advice, but to strike a good one is very difficult indeed."[28] Within a day, though, he had warmed to the idea and became committed to the new venture, even excited by it: "I concluded to-day to have a book of subscriptions, open to receive the names of all persons inclined to have the *best American illustrations of birds of that country ever yet transmitted to posterity.*"[29] The underscored emphasis Audubon gave to the second half of the sentence apparently revived and forti- fied his original sense of confidence—overshadowing Mark Catesby and all who came after him, and certainly surpassing Alexander Wilson—which had seemed clear enough before he left America, but had faltered occasionally, given the vagaries of the exhibition business in Great Britain. If he could just find a way to make enough money to avoid having to pander to the public and occasionally having to deal with the overheard insults of British philistines, he would be fine: "Then, my dear friend," he wrote to Lucy, "my exhibiting my work publicly will be laid aside for a while at least. *I* hope forever."[30]

By the time his Edinburgh exhibit was coming to an end in mid-December 1826, Audubon could indeed lay the exhibition business aside for a while and hope never to have to count the house again. He had done well enough finan- cially in the five-plus weeks his work had been on display, bringing in a little over £170, but he ended the experience with little enthusiasm. "Upwards of two hundred people were there," he wrote on the penultimate day, a bit bitter that so many had waited so long to attend. "The idea of its closing tomorrow had roused the dormant curiosity of the public."[31] Edinburgh's *Caledonian Mercury* had done its best to pique that public interest, calling Audubon's show "one of the most striking and beautiful illustrations of natural history that has ever been exhibited in this city." Noting that the show "has had numerous visitors, and has been universally admired by men of science and taste," the newspaper looked ahead to the prospect of his new project: "It is his intention, we understand, to have his sketches engraved and coloured, and to publish the whole in monthly numbers; which, when completed, will certainly form one of the most magnificent works on ornithology which has ever been given to the world."[32]

"Magnificent" the work would indeed be, but perhaps the most significant words in the sentence were the two set apart almost as an aside—"when completed." Publishing what Audubon would come to call his Great Work became the only work that mattered to him for more than a decade. At the time his Edinburgh exhibit came down, he had a clear idea of what he needed to do, but no real idea of how difficult the work would turn out to be.

From Exhibition to Publication

Audubon's decision to publish *The Birds of America* as a book immediately raises an imposing, even if obvious, question: Why the decision for such a big book? Or perhaps more to the point, what were the artistic and commercial implications of publishing such a big book? To be sure, Audubon was not the first artist-naturalist to publish a large, illustrated book about nature in America. Mark Catesby led off in the mid-eighteenth century with the two-volume, folio-sized *Natural History of Carolina, Florida and the Bahama Islands* (1731–1747); John Abbot and James Edward Smith followed with the two-volume *Natural History of the Rarer Lepidopterous Insects of Georgia* (1797); and Alexander Wilson, of course, set a high ornithological bar for Audubon with his nine-volume *American Ornithology* (1808–1814).[33] But Audubon's proposed book would be more ambitious and more extensive than any of those earlier works and, certainly in terms of its physical dimensions, the largest book ever before published.[34] At the outset, Audubon had already decided to depict his birds in the "size of life," as he put it, "presenting to the world those my favourite objects in nature, of the size which nature has given to them."[35] As one student of Audubon's work has aptly put it, "Audubon . . . made the book fit the birds; other naturalists made the birds fit the book."[36] Some of his friends sought to dissuade him from trying to publish such oversized images, arguing that a collection of such plates could never be affordable, maybe not even possible.[37] But Audubon was determined that drawing the birds "size of life" would be his defining contribution to both art and ornithology, and he never wavered from that position. "I must acknowledge it renders it rather bulky," he wrote with considerable understatement, "but my heart was always bent on it, and I cannot refrain from attempting it so."[38]

This particular sort of attempt, however, would be especially risky business. He admitted as much in the first month of his stay in Liverpool, writing to his brother-in-law Nicholas Berthoud that he understood the difficulties he faced, not to mention the price of failure: "Should I, through the stupendousness of the enterprise and publication of so large a work, be forced to abandon its being engraved, I will follow a general round of remunerating exhibitions and take the proceeds home."[39] Some three months later, when he should still have been riding high in the midst of his successful show in Edinburgh, he struck a grim note about the challenge he faced in "the publication of my enormously gigantic Work." He could well hope that "the work will be equal to anything in the world at present," but he also knew that undertaking such a multiyear challenge might result in overwhelming, even embarrassing, failure: "If I do not succeed I can return to my woods and there in peace and quiet die with the thought that I have done my utmost to be *agreeable* if not useful to the world at large."[40] Three weeks after that, however, he turned from the discouraged to the heroic, invoking the image of the Frenchman whose military reach had led Audubon to come to America in the first place: "Since Napoleon became, from the ranks, an Emperor, why should not Audubon be able to leave the woods of America for a while and publish and sell a *book*?"[41]

Ambitious though he had become, Audubon might still have asked an obvious follow-up question: Why should anyone buy his book, especially when it wasn't even finished yet? To be fair to Audubon's ambition, the question might invite an equally obvious answer: The avian images revealed a very rare artistic and scientific genius. Far from being the stiff-seeming profile portraits of birds that had been common in the past—even in some of Audubon's own earlier work—the works he exhibited in Great Britain seemed to take to the wing, as one of the appreciative newspaper reviews put it, "every feather appearing to be inspired with life."[42] He not only put the birds in motion, but also froze them in time and space, keeping the drama, even the occasional violence of the scene, safely contained on the page. By depicting his birds life-sized, he gave his viewers an image that seemed more real than reality itself, with the sort of detail that one could discern only by killing an actual bird and then having only a limp, bloody, and partly damaged dead specimen. Audubon brought the birds *back* to life, as it were, making them perpetually accessible, always available for viewing and close study. For naturalists and other bird enthusiasts, Audubon's images might thus seem a brilliant bargain, especially if some learned society were paying the price. Other potential purchasers might be less interested in ornithological detail than in simply collecting specimens, intrigued by the idea of acquiring the images primarily for the sake of having a complete set.[43]

At the time he decided on publication, though, Audubon knew that the most obvious consequence of trying to sell a book of such enormous proportions would be the cost to the consumer: The eventual product would be extremely expensive, much more than the vast majority of potential patrons could bear. Indeed, Audubon never intended to produce a best seller for a mass market. All he had to do was just sell a few hundred sets, he figured, and he would be financially set for life. Just as he had learned to count the house in the public exhibitions, he now learned to calculate the take on possible subscriptions: "If I can procure three hundred good substantial names of persons, or bodies, or institutions," he wrote his friend Richard Rathbone, "I cannot fail to do well for my family, although I must abandon my life to its success, and undergo many sad perplexities and perhaps never see again my own beloved America."[44] A little later he raised the target. He wrote Lucy that the Edinburgh engraver William Home Lizars estimated that if they could work together to produce and sell five hundred copies of *The Birds of America*, it would "make me and him Independent for the rest of our Days."[45]

Lizars, the man who made that estimate, no doubt knew what he was talking about. Just two years younger than Audubon, Lizars had inherited his father's Edinburgh engraving business when he was in his early twenties, and by the time Audubon came to Edinburgh, Lizars was engaged in producing two significant works on birds: Prideaux John Selby's *Illustrations of British Ornithology* and another book by Selby and Sir William Jardine, *Illustrations of Ornithology*.[46] Thanks to an introduction by a mutual friend, Audubon met Lizars at the end of October 1826, and the two took an immediate liking to each other, or they at least both saw the promising possibility the other offered. The first time Audubon opened his portfolio for Lizars, the engraver's eye lit up: "My God!" he exclaimed, "I never saw anything like this before." Three days later, he continued to marvel at Audubon's work, looking excitedly at

Audubon's drawings of turkeys, a hawk, and a crane, declaring them "wonderful productions." But when he came to a gory portrayal of the "Great-footed
Hawks . . . with bloody rags at their beaks' end, and cruel delight in the glance
of their daring eyes," he knew what he had to do: "*That* I will engrave and
publish."[47] Thus began the brief but remarkably valuable relationship between
Audubon and the man who promised to make him famous—and, better still,
wealthy.

J. J. Audubon, Entrepreneur

But before Audubon and Lizars could truly contemplate financial success, they
had to do the hard work of making and marketing Audubon's art, or what we
would now call the production and distribution processes in this remarkable
project—the biggest business venture Audubon had ever attempted, and the
one in which he would find his greatest success. However much Audubon may
have portrayed himself as having been lax in his previous professional efforts,
he had a very different story to tell and a very different persona to project
whenever he wrote about his work on *The Birds of America*. By finally merging
birds and business together, he found a way to direct his personal passion and
near-aggressive energy into an all-consuming commercial project. From the
beginning, the one constant in the story of *The Birds of America* was Audubon's
intense, sometimes obsessive involvement in every part of production: "I shall
superintend it myself," he declared at the outset, "both engraving and colouring and bringing up, and I hope my industry will be kept in good repair
thereby."[48] In making *The Birds of America* a successful enterprise, Audubon
made himself into an innovative entrepreneur and exacting supervisor.

The production process seemed straightforward at first. Audubon, of course,
initiated the work, producing the watercolor paintings of the birds, sometimes
cutting and pasting together several images done at different times, and then
delivering the picture to the engraver's shop. There the various employees
transformed the watercolor into an engraving in a complicated process that
depended on the workers' division of labor—and very skilled labor at that. The
basic outlines of each of Audubon's painted images had to be traced onto paper
and then printed in reverse on large, well-polished copper plates. Engravers,
working with sharp steel pens (or burins), etching needles, and other tools of
the trade then cut the lines into the copper, also using a concoction of nitrous

acid to add clarity, or "bite," to the image. Their artistry greatly impressed Audubon when he watched them work on "my Birds." "I was glad to see how faithfully copied they were done, and scarcely able to conceive the *adroit* required to form all the lines in a sense contrary to the model before them."[49] (The copper plates are indeed stunning still today, impressive in their detail and in some ways as visually striking and artistically remarkable as the finished images themselves.[50]) Once a copper plate had been engraved, printers inked it and pressed large, heavy sheets of high-quality paper into the grooves, producing an outline of the bird and the background that could then go to the colorists, who carefully hand-painted the picture, seeking to reproduce the colors as they appeared in the original drawing—quite a test of consistency from one colorist to the next, and one that they did not always pass. On the whole, the task of turning Audubon's original images into marketable engravings proved to be an extremely labor-intensive process that relied, most immediately, on the work of dozens of artisans, often working directly under Audubon's ever-critical eye.

But the work process went well beyond the engraver's shop. Unseen and unheralded others likewise made a critical contribution to the project: the papermakers who produced the huge, high-quality sheets Audubon required; the copper smelters who turned raw ore into clean ingots; the miners who extracted the copper ore from the earth in the first place; and so forth, back through all the prior steps of production. In that sense, *The Birds of America* was not just an extensive work of art, not just an example of the sole genius of the lone, struggling artist. It was, rather, an ambitious business venture that relied on a complex labor process and an extensive supply chain, an enterprise in which the artist became not just the designer of the work, but the administrative manager of dozens of people, many of whom could be called artists in their own right, and a marketer to prospective customers, many of whom he had to track down wherever he could find them, on both sides of a very wide ocean.

Audubon knew all about how to get birds, but not much at all about how to attract investors. In his 1827 "Prospectus" for *The Birds of America*—his initial public offering, as it were—Audubon laid out a highly leveraged project, promising that there would be three hundred images in all, comprising three volumes of a hundred engraved plates each. (The final product would eventually be four volumes with over four hundred plates, but that would come a decade later.) Once printed and painted, the images would be grouped together in sets of five, what Audubon called a "number," and distributed one number at a time, typically shipped in tin boxes, for the price of two guineas per number, at the anticipated rate of five numbers—or twenty-five plates—per year. "It would be advisable," Audubon told his potential customers, "for the subscribers to procure a portfolio, to keep the Numbers till a volume is completed."[51] Audubon might deserve some credit for thus preparing his subscribers for such delayed gratification, but he neglected to tell them that he did not yet have a complete collection of the images he had promised: There were still birds of America to be found, much less painted and engraved, and Audubon would spend almost a decade more procuring the avian specimens to make *The Birds of America* complete. In the meantime, he would dole the plates out slowly and trust his subscribers to trust him. *The Birds of America* would thus first emerge as a piecemeal publication, a book done in installments, rather like an ornithological equivalent of an extended Dickens novel.

But even within that piecemeal process, Audubon established a form of internal "rhythm" that had important implications for the total product. In the "Prospectus," he announced that each number of five plates would contain "one Plate from one of the largest Drawings, one from one of the second size, and three from the smaller Drawings."[52] This rhythm—big bird, middle-sized bird, three small birds—would then repeat in the next number, and so on. Audubon would later claim that "chance, and chance alone, had divided my drawings into three different classes, depending upon the magnitude of the objects which they represented," but in fact, chance had little to do with Audubon's approach to marketing his plates this way.[53] Instead, the reason for this rhythm seems clearly commercial, certainly not ornithological. Audubon no doubt understood that the big birds would probably be the big sell: People might generally put a higher value on the Great American Cock, or Wild Turkey, or maybe the Ivory-Billed Woodpecker than, say, a Chestnut-sided Warbler. Too wise to get rid of the big birds too soon, Audubon let them go slowly, keeping his customers waiting for the subsequent number. This sequence might not make for good science—it certainly had nothing to do with the taxonomic classification of birds—but it was very smart marketing.

Audubon also took a smart approach to inventory control. He started out by making only a few copies of the first few numbers, taking "good care to have only a very small stock of copies over the demand," never going much beyond the number of solid commitments he had from subscribers. More subscribers would enable Audubon to produce more copies at a lower cost per unit. He never used the term "economy of scale," but he clearly understood that once the copper plates had been engraved, the costliest part of the process was done. The plates could then be used to produce an increasing number of images, for which the main labor costs would be primarily printing and coloring. A hundred images of a single number, he explained to Lucy, could generate enough income to finance a hundred more, so that "when 200 subscribers will be had the amount will be . . . more than double *because* the plates are paid for by the first 100 copies sold and of course the printing coloring and paper is the only additional expense." The resulting profit on two hundred copies of five numbers per year would be, he assured her, "enough to maintain us . . . in a style of Elegance and Comfort that I hope to see thee enjoy."[54]

Some measure of comfort, if not true elegance, would eventually come, but not until Audubon had lengthened the list of subscribers. In the meantime, the process of marketing *The Birds of America* through subscription turned out to be a never-ending challenge. There was nothing new about publishing books through subscription, a practice that had been around at least since the seventeenth century to cover publication costs, particularly for illustrated works on natural history, which tended to be quite expensive. Unlike the modern-day author's advance from a single publisher, however, subscriptions had to come from a variety of sources, and authors publicly promoted a list of their subscribers in order to attract (and perhaps allay the fears of) others. The problem with the process was that it took near-constant effort to search out potential subscribers and, equally important, to get them to honor their commitments. Soon after starting production on the plates for *The Birds of America*, Audubon wrote Lucy that "my only present care is to procure plenty of Subscribers and manage to collect the money closely so as to enable Mr Lizars to keep his hands at work and *the work* in progress."[55] That would become an ever-present care, in fact, and both procuring subscriptions and collecting money bedeviled Audubon for over a decade.

He must have known what he was getting into. He certainly knew enough of the subscription-seeking exploits of Alexander Wilson to be prepared for the sometimes horrendous experience that could befall the artist/author, who typically had to endure much more rejection than admiration. Audubon reflected on Wilson's unhappy challenges in his own subscription seeking, noting "how often I thought during these visits of poor Alexander Wilson . . . [when] he as well as myself was received with rude coldness, and sometimes with that arrogance which belongs to *parvenus*."[56] Audubon's book would cost much more than the $120 Wilson wanted for *American Ornithology*, and he well understood the obstacle of cost. "I expect nothing in America in the way of encouragement," he wrote his son Victor. "700 Dollars is *rather high* for a Book on Birds in our Country."[57] (In the end, in fact, the full, four-volume Double Elephant Folio edition of *The Birds of America* would cost an American subscriber not $700, but around $1,000—an enormous amount for a book of any sort, roughly equivalent to the annual income for a middle-class household, and about double what a skilled male artisan could make in a year.[58]) Instead of trying to rely on cash-strapped individuals to purchase his work, he wrote Lucy, "I would greatly prefer having my Subscriptions confined to public institutions in America than to see added to it a number of the names of Men whose transient habits would only give us much trouble if not turn out to be bankrupts." If he did have to seek subscriptions from individuals, he wanted only those who "are *all* of good Standing and Wealthy—no Yankees here in Such matters."[59]

Early on, of course, Audubon's potential subscribers were not Yankees so much as individuals and institutions on the other side of the Atlantic, in Great Britain and France. He crowed to Lucy that King George IV had become both a patron and subscriber, and he planned to branch out by taking his work to "each of the scientific & public Societies . . . and by being present at each of the meetings will be able to exhibit and explain the nature of the work, and as the meetings are generally crowded by Noblemen & Gentlemen . . . I have some hope of reaping a tolerable harvest of names."[60]

"Trouble and Anxiety" in the Bird Business

Audubon's hopes of reaping that rich harvest with such prominent customers soon became compromised, however, by a series of frustrating experiences that caused him no end of tribulation. Booksellers who had agreed to serve as agents for Audubon—one of whom was Lizars's brother, Daniel Lizars—sometimes became slothful, simply failing to deliver the finished numbers to subscribers, and they did little to drum up additional business or collect the money already owed. That left Audubon to have to travel around the country to take care of business himself, usually finding exceedingly disappointing prospects. "One subscriber in a city of 150,000 souls, rich handsome and with much learning," Audubon wrote about Glasgow. "Think of 1400 pupils in one college!"[61]

He also had trouble with Lizars himself. The engraver sometimes let shoddy work leave the shop, an oversight Audubon could not abide. Finding plates that were not colored to his standards, Audubon said the sight of them "drew a sigh from my heart. Ah! Mr. Lizars! was this the way to use a man who paid you so amply and so punctually?"[62] The question of ample and punctual pay by Lizars to his colorists may have been the root of a bigger problem, in fact. While Audubon was on a trip to London in June 1827, he received a letter from Lizars telling him that the colorists had gone on strike and the work on Audubon's birds had come to a halt, news that sent "quite a shock to my nerves."[63]

Shocked or not, he looked around London and found a new engraving shop, this one operated by Robert Havell Sr. and his son, Robert Jr. In doing so, he also found a new and remarkably important partner in the younger Havell, who oversaw the production of Audubon's work from the beginning, as his father settled into retirement. Indeed, Audubon's long and often complicated artistic and business relationship with Robert Havell Jr. is crucial to the story, even the legend, of making Audubon's *Birds of America*.

Robert Havell Jr., one student of Audubon's art has declared, "was much more than the engraver of *Birds of America*. He was a genius . . . a fine artist in his own right, with a discerning eye for composition" that contributed enormously to enhancing the final look of Audubon's finished work.[64] Audubon himself would give Havell the ultimate ornithological compliment, naming a new species of shorebird after him—Havell's Tern or *Sterna Havelli*. "I consider him as one of the best ornithological engravers in England," Audubon wrote in *Ornithological Biography*. "I feel greatly indebted to him for the interest which he has always evinced in my publication . . . the engraving of which has cost him much trouble and anxiety."[65]

Much of that trouble and anxiety came from Audubon himself. Still smarting from the unhappy lesson in labor relations he had learned with Lizars and his striking workers, Audubon seemed not at all inclined to modify his approach to the dozens of artisans whose skilled work did so much to make his own Great Work so great. Even as he came to admire Havell's genius as an artist in the engraving business, Audubon never fully trusted him or anyone else engaged in the work to do it according to his very exacting standards. He always maintained extremely close supervision over all parts of production. When he first got settled in in London in 1827, he took up lodgings on Great Russell Street, across from the British Museum, and about a half a mile from Havell's shop on Oxford Street or, for a fast walker like Audubon, less than a ten-minute trip. Thus able to drop in on Havell easily and frequently, he became an almost constant (and, to the workers, probably overbearing) presence in the engraver's shop. "I pay my Engravers, colorers & printers as regularly as their work [h]as passed my last inspection."[66] The use of the "my" in that sentence stems from Audubon's financial responsibility for providing the money to keep the operation running, but it also suggests a sense of possessiveness and control that might well have caused confusion, even consternation, among Havell's employees, not to mention Havell himself, about who was ultimately boss. There was no confusion, however, on Audubon's part.

Nor was there any way for anyone to escape his oversight. Even when he had to be away from London at various times, he frequently received proofs sent from Havell's shop and thus kept close tabs on the quality of the work.[67] Writing from Liverpool in December 1827, he offered Havell praise for some of the plates he had seen—"it is with true & sincerest pleasure that I acknowledge to you my full approbation of them"—but he then followed the compliment with an admonition. He had received feedback from a few customers in Manchester that "the *Doves* were not all colored a like." Such inconsistency might not be at all surprising, since Havell at times employed upward of fifty colorists on the Audubon project, which made quality control from image to image a considerable challenge. Still, Audubon would accept no excuses, and he told Havell to speak to his workers about "the duty's of coloring carefully as on that also much of the credit of the Work depends."[68] Havell's letter engraver got off no better: He "must be dismissed or become considerably more careful and in fact must now correct his past errors." Audubon complained to Havell that some customers had been unhappily unimpressed with the quality of the plates they had received—one woman even said she "could not think of giving *house* room to any more such *trash*"—and he chided Havell to take better care to "redeem the caracter of a work of this Magnitude," reminding him of its "Importance to your own standing as a good Engraver and a Good Man."[69] Good as Havell might be in either regard, Audubon never forgot that the success of the work ultimately depended on his own taking care of the main role as ornithologist-in-chief: "While I am not a colorist and Havell is a very superior one," he said, "I *know* the birds."[70]

And know the birds he did. Audubon often fussed, sometimes even agonized, over every ornithological detail in the finished images, so that his critics in the scientific community (of whom he had more than a few) would not have additional ammunition. He was "principally desirous," he wrote in 1831, "that the Names of Birds plants &c should [be] *quite Correct* and in every Instance at the bottom of the Plate or on *the bottom side of it.*"[71] He took care to note a small but important detail on the Black-capped Titmouse (or Chickadee, as it is now more commonly known): "There must be *a white spot at the lower end of the black cap* next the shoulders," he wrote. "This white spot distinguishes our bird from the European of the same name, *which* is a different Species."[72] He wrote rather diplomatically to Havell about some waterbirds about to be engraved: "Allow me to ask you that the *Bills, legs & feet* are *carefully* copied from my Drawings—to Naturalists, these points are of the greatest importance." To make sure that Havell would not take offense, he added, "Do not think for a moment that I am *lecturing* you . . . [but] merely wish to enjoin you to keep a Masters eye over the Work in each of its departments." To make his point even more palatable yet important to Havell, he suggested that "who Knows but that your name if not your fortune is now connected with Mine and with my family?"[73]

The Audubon Family Business

A very important member of that family, Audubon's wife, Lucy, well understood the importance of her husband's struggles with the labor process. "This work of my husband," she wrote to her cousin, "is of such magnitude and such expense that it harasses him at times almost beyond his physical powers." She seemed also to have adopted her husband's uneasy attitude toward the workers involved in the project: "What with engravers, printers, and colourers, (all a disagreeable race) his mind and hands are full."[74]

In fact, *The Birds of America* became much more than "this work of my husband." Almost from the beginning, Audubon saw it as the family business—an enormous enterprise undertaken both *by* the family and *for* the family. He realized that if he were to have time to do all the work he needed to do to produce and promote his Great Work—creating more images, of course, but also overseeing the engraving process, seeking subscriptions, keeping track of payments, and so forth—he would need help, and who better to employ than his own wife and sons? "I am now better aware of the advantages of a family in unison than ever," he wrote Lucy in December 1826, "and I am quite satisfied that by acting conjointly and by my advice we can realize a handsome fortune for each of us."[75] Audubon's notion that his family members would

Figure 6. Lucy Bakewell Audubon, 1831, by Frank Cruikshank. Image #44214. New-York Historical Society.

work "by my advice" made clear his idea of who would be the boss, but he also knew that Lucy would have to be an important partner in the overall enterprise. Indeed, Lucy came to play an increasingly critical role that went far beyond that of wife.

During Audubon's first few years in Great Britain, Lucy stayed behind in America, still working as a tutor on a plantation in St. Francisville, Louisiana, to support herself and her sons, who were back in Kentucky. Separated from both her husband and her boys, she didn't much like her situation, and she certainly didn't like the people who employed her, whom she found "selfish beyond my calculations."[76] Still, Audubon saw the necessity of her being in Louisiana, and he wanted her to work for him as well, asking her to seek out subscribers in New Orleans and the surrounding area. "I wish thee to see if the Library of New Orleans and the College Library there would also subscribe," he wrote her in March 1827, but he added a warning that she remember that "I am not anxious to have subscribers that *will not pay well*."[77] He kept her up to date on his own business affairs in England, giving her periodic estimates of how much money he might make; complaining about how he had to postpone a subscriber-seeking trip because Lizars, his erstwhile engraver, continued to fail to provide the promised plates to subscribers; grousing about having to produce various small paintings for sale to survive in a city as expensive as London ("I do anything for money nowadays"); and wishing that he could find some public institution or perhaps "some Lord or others who may be sufficiently wealthy & possessed of taste to pay me well & suffer me to have my own way."[78]

He also wanted to have his own wife. Reading Audubon's correspondence with Lucy during his first trip to England seems a painful invasion of the couple's privacy, as his frequent entreaties about Lucy's coming to England reveal all the misunderstandings and frustrations of what was beginning to look like a failing marriage.[79] "It is now about time to know from thee what thy future intentions are," he wrote Lucy in late December 1826, near the end of his stay in Edinburgh. "Cannot we move together and feel and enjoy the natural need of each other?"[80] But a few months later, in the spring of 1827, he wrote in frustration that it was "alas quite useless for me to say that I would be *happier* if Thou wert with me." Since she was not, however, he went on to complain about his status as a married man living single in society: "I am married, every one knows it—and yet I have no Wife nor I am likely to possess one—. . . without my Wife and my Children, nay I am denied the privilege of every Father in the World, that of Judging what is best for them to do. . . . Such is the Situation of thy husband that after Years of Labours, in the midst of encouragement, the thought pangs of Sorrows fret my poor mind constantly"—and then he abruptly cut it off: "I will say no more."[81]

But, as always, he said quite a bit more. He tried to sweet-talk Lucy about the possibility of his coming back to America, where "I will still find my friend," and "walking together arms in arms we can see our Sons before us, and Lesson to the mellow sounding notes of the thrush so plenty in our woods of Magnolia."[82] In the meantime, though, he told her that he had to stay in England to "try to augment my subscribers," even though the experience of being there gave him "the blues with a vengeance."[83]

When the news of his blues failed to work, Audubon set his hurt feelings aside in favor of his simple business sense: "I think with my Industry and thy carefull good management we might do well," Audubon wrote her in August 1828. "Two heads are better than one and I think it would be of great mutual advantage, *the ultimate* success of my Publication, or in other words its completion."[84] But looking down the road to the time it might take for the end of the work to come, perhaps fourteen years in the future, Audubon warned Lucy ominously that "if it is thy Intention not to Join me before that time, I think will be best off both of us to Separate, thou to Marry in America and I to spend my Life most Miserably alone for the remainder of my days. . . . I am sick of being alone and from thee and how much longer I will be able to bear is a little doubtfull."[85]

However much Audubon might have wallowed in self-pity, not to mention self-interest, Lucy Audubon's side of the relationship offers a very different perspective. "If Audubon was trying to lure his wife to England," Lucy's biographer has written, "he surely took the wrong approach. As Lucy saw it, Audubon did not need a wife; he needed a trusted business partner, someone to supervise the work on the publication while he was off swaggering about Europe."[86] While Audubon was "swaggering about Europe," Lucy continued slogging away at her teaching job in Louisiana, barely making ends meet to support herself, much less help her sons in Kentucky and her husband in England. She wrote about her woes to her elder son, Victor, turning to him "to be my friend, my comforter and confidant" about the problems his father caused her. "From your Papa I am quite at a loss to grasp anything," she wrote, later adding that her husband seemed unable to understand anything of her situation: "Surely your father is blind to the real state of affairs, for these eight years I have relieved him of all expense but himself, paid him $500 . . . and three hundred since he went away." She even turned a bit grim about the toll the separation and her labors had taken on her, sighing almost suicidally that "I sometimes consider why I wish any longer to live, all my best days are over."[87]

But better days still lay ahead for Lucy, and therefore for Audubon as well. After all the epistolary back and forth, Audubon had finally decided to do the right thing for his wife—almost. In January 1829, he wrote her to announce that "I will sail for America (New York) on or about the 1st day of April next

. . . with a hope that I can persuade thee to come over here with me and under my care and charge." This generous-seeming gesture of meeting Lucy on her side of the Atlantic had its geographical limits, however, and he offered only to meet her halfway: "It is not my wish to go as far as Louisiana but as far as Louisville, Kentucky, where after my landing we will make arrangements to meet, never to part again!"[88] In fact, after he landed, on May 5, he spent almost six months searching for birds and painting new specimens, going from New York to New Jersey and Pennsylvania, before turning toward Lucy in Louisiana. Only in November 1829, after being gone for over three years, did he finally reunite with her, but when he arrived at St. Francisville, he wrote, all the emotional agony and mutual doubt about their relationship seemed to fall away immediately: "I went at once to my wife's apartment; her door was ajar, already she was dressed and sitting by her piano, on which a young lady was playing. I pronounced her name gently, she saw me, and the next moment I held her in my arms. Her emotion was so great I feared I had acted rashly, but tears relieved our hearts, once more we were together."[89] And together they stayed for the next several months, going up to Louisville to see their sons in mid-January 1830, then over to New York and down to Washington, DC, where they would dine with Andrew Jackson ("The President was kind and polite to us," Audubon reported) and eventually back to New York where, on April 1, they boarded a ship for the voyage across the ocean to England.[90]

Somewhere in this time of reconnecting and recommitting, Lucy apparently set aside her hurt feelings of the past few years—her self-sacrificing support for a seemingly ungrateful husband, her fear that he had lost romantic interest in her, her sense of growing old and tired and even toothless as she turned forty—and threw herself into *The Birds of America*. She became a very effective business partner, putting her head as well as her heart into her husband's work—so much so that it became her own. In the summer of 1831, she wrote her cousin again about the rigors of the work: "I am sure you will believe me when I tell you that our great Book demand all our funds, time, and attention, and since I came to England we have not indulged in any thing that did not appertain to the advancement and publication of the 'Birds of America.'"[91] The "our" and "we" in that sentence almost leap out of the letter. Like Abigail Adams, who began to refer to the family homestead in Quincy as "our" farm (and herself as "farmeriss") while her husband, John, was away serving as a member of the Continental Congress during the American Revolution, Lucy Audubon had also come to adopt the first-person plural possessive in speaking about their "great Book." Lucy became a partner in the production of *The Birds of America* for essentially all of her adult life, from the time she married Audubon in 1808 until the day she died in 1874.

Audubon also enlisted both of his sons in the work, hoping they could help him in producing the paintings. In late 1826, he urged the younger one, thirteen-year-old John, to "Draw a great deal and study music also, for men of talent are welcome all over the world."[92] In less than two years, however, he wrote Lucy with some impatience, saying that "I am glad to hear of John's improvement in Music but I regret that I do not hear of his drawing anything," telling her what an "*Immense Service* it would be to all of us that he should Draw as well as I do."[93] John did come to draw well, and he became an important artistic assistant within a few years. He also came to understand the uncertainty of the bottom line in the family business. In late 1833, while producing some drawings in Charleston, he wrote to his older brother, Victor, somewhat acerbically noting that "I am working that I may some day become a Second Audubon not to make a fortune." Still, he portrayed himself as a potential heir on a higher plane: "My wish is that I may some day publish some Birds or quadrupeds and that my name may stand as does my Fathers."[94] In turn, his father saw his younger son's promising potential, writing to Victor that "John has drawn a few Birds as good as any I ever made, and ere a few months I hope to give this department of my duty altogether to him."[95] Audubon had long since come to the conclusion that Victor's own artistic skills were comparatively lacking, and he could only wish that his elder son had followed the path of his younger one and taken the time to study "my style of Drawing & the Habits of the Birds of our Country a little more."[96]

But Victor had other useful attributes. When he later moved to England in the early 1830s to oversee the engraving process in his father's absence, he proved to be an excellent surrogate supervisor. Audubon wrote Victor to praise the work he and Havell were doing, noting that a recent batch of ten plates were "the best I ever saw of birds and that they do *Havell* and *yourself* my beloved Son great credit" and that he was therefore "delighted." Still, Audubon could seldom be fully pleased or at ease with the work coming out of Havell's shop, and he complained mightily to his son about the output of some of the workers: "*The letter engraver* is a miserable one. The horrible Mistakes which I have discovered in the plates of the 2d Volume are quite disgraceful to our Work— Mention this to Friend Havell and see that no errors of the same sort happen again."[97]

Even without his father's direct inspection of the work, Victor could be aggressive in keeping Havell and his workers up to the family standards. Once, when a couple of prominent subscribers withdrew their support after being disappointed with the images sent them, Victor chided Havell about the loss of business: "It is out of my power to say what could have produced this unfortunate event," he wrote, "but it is a strong hint for you to use every exertion towards making the work as excellent as possible." He went on to point the finger, as his father frequently had, at the colorists, noting that "I observe the colors on some of the birds *carelessly* daubed round the edges of the engravings, as well as being in some instances too bright or *gaudy* for the *true* color of the object." Victor knew, as his father knew, that in this sort of artistic and scientific enterprise, ornithological accuracy became critical to financial success: "So long as such faults are permitted to appear You will be losing subscribers daily."[98]

Audubon himself put the subscription problem in a larger familial perspective: Losing subscribers, whether through death or defection, "is a thing we can not help," he wrote Victor in late 1833, "but depend upon it that our *Industry*, our *truth*, and the regular manner in which we publish our Work—this will always prove to the World & to our Subscribers, that nothing more can be done than what we do." Indeed, he followed that statement with a celebration of the family's collective labor and its potential long-term legacy: "I doubt that if any other *Family* with our pecuniary means ever will raise for themselves such a *Monument* as 'the Birds of America' is, over their tomb!"[99] While they still lived, then, the members of Audubon's family helped make it possible for him to expand and eventually complete the remarkable enterprise he had undertaken for them—and, increasingly, with them.

Obstacles, Legal and Ornithological

Audubon's prospects for completing the enterprise, though, meant more than dealing with all the other troubling problems already facing him: drawing birds, overseeing their engraving and coloring, and tracking down and keeping track of subscribers. He faced two additional and seemingly difficult issues that loomed over the future of *The Birds of America*—one legal, one ornithological—both of which affected the very nature of the Great Work as a book.

To the legal obstacle, Audubon found a creative and surprisingly simple solution. His first engraver, William Lizars, had made him aware of a British copyright law of 1709, specifying that any book published in Great Britain had to be deposited, for free, in nine of the nation's libraries.[100] Given the significant expense of *Birds of America*, Audubon had no intention of giving so many copies of his work away gratis, and to avoid doing so, he took a very strict constructionist view of the definition of "book" in British law. If any publication that contained printed text qualified as a book, Audubon decided not to have any: He would publish *The Birds of America* as a collection of illustrated images, with just a title page for each volume. Even that caused him some concern. In 1831 he wrote to Havell asking, "Are you quite sure that the Tytle Page *Engraved as it will be* may not render the Work Liable?" He instructed Havell to destroy the proof sheets of the title page, if need be, and be careful *not* to include a table of contents, which would most certainly make *The Birds of America* begin to look awfully much like a book. After all, concluded Audubon, "I am far from having the wish to defraud the English Government."[101] But defraud the government or not, the point is that Audubon understood that, according to the letter of the law, words could be expensive; any written descriptions of his birds would have to come later. In the meantime, Audubon hoped to package *The Birds of America* in such a format that this big book should not be considered a book at all.

But the birds had to have words in the end. Knowing that his subscribers would want some measure of ornithological information to accompany the many images of birds, he began work on a separate and much less expensive publication of text only, what in time came to be the three-thousand-plus pages of *Ornithological Biography*, published in Edinburgh in five volumes between 1831 and 1839. At the outset, writing a book seemed not to be his strong suit. Audubon never doubted his ability with a paintbrush, but he struggled with the pen, especially if it meant writing in English, his second language. "I will now proceed with a firm resolution to attempt *being an author*," he wrote Lucy in October 1826. "It is a terrible thing for me, far better fitted to study and delineate in the forests than to arrange phrases with sensible grammarian skill."[102] Still, if he could arrange enough phrases to go along with his paintings, he would find a way to evade the British book laws and still satisfy his subscribers. To do that, Audubon knew he would need help, and in 1830, he once again admitted his weakness: "I can scarcely manage to scribble a tolerable English letter. . . . I know that I am not a scholar, but meantime I am aware that no man living knows better than I do the habits of our birds. . . . I cannot, however, give scientific descriptions, and here must have assistance."[103]

He found that assistance by hiring an Edinburgh-based naturalist and professor, William MacGillivray, whom Audubon described as "being possessed of a liberal education and a strong taste for the Natural Sciences." (MacGillivray had demonstrated much more than a "strong taste" for science some years earlier, when he walked from Edinburgh to London to visit the bird collections in the British Museum, covering some eight hundred miles in six weeks, subsisting largely on barley bread, and sleeping mostly out in the open.) As much as Audubon might have admired MacGillivray's talent and commitment, he still made clear his notion of their respective roles in producing *Ornithological Biography*: MacGillivray "has aided me," Audubon wrote at the outset, in the introduction to the first volume, "not in drawing the figures of my Illustrations, nor in writing the book . . . but in completing the scientific details, and smoothing down the asperities of my Ornithological Biographies."[104] Lucy also made her own contribution to the smoothing process, serving as a copyist for the words Audubon wrote and MacGillivray edited, getting the final manuscript prepared for print. While no doubt grateful for her unpaid labor, Audubon apparently took it for granted as wifely work, and unlike MacGillivray, Lucy received no mention for her copy-editing efforts in the published record.

In the end, Audubon and his assistants achieved a striking success, but it would take eight years to do so. Audubon knew the completion of his Great Work wouldn't be quick, and he addressed the issue directly in the first pages of *Ornithological Biography*: "As to the time necessary for finishing my Work, I have only to observe, that it will be less than the period frequently given by many persons to the maturation of certain wines placed in their cellars."[105] This oenological analogy may have seemed reassuring to some subscribers, but it barely masked the fact that Audubon did not even have all his work bottled, as it were. *The Birds of America* could not come to completion until Audubon could solve an even larger ornithological problem holding up the work: He simply needed more birds. Getting them meant going back and forth across the Atlantic three times in a matter of only a few years. In addition to his 1829–1830 trip back to America to fetch Lucy and bring her to London, he made two additional collecting trips to the United States—first to New York, South Carolina, Florida, Massachusetts, Maine, and Labrador, then back to New York and South Carolina between 1831 and 1833; and a few years later, southward again to South Carolina, Florida and the Keys, Louisiana, and Texas in 1836–1837.[106] As much as it pained him to surrender his careful, ever watch-

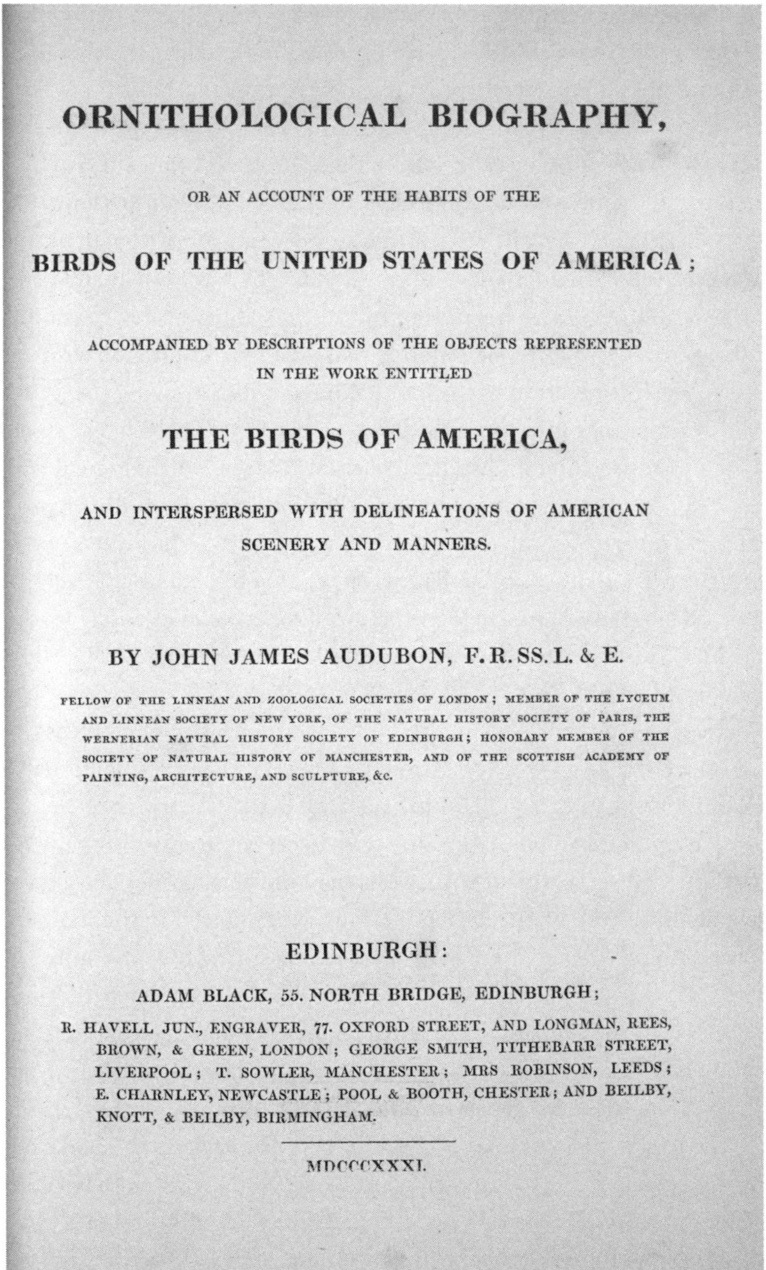

ORNITHOLOGICAL BIOGRAPHY,

OR AN ACCOUNT OF THE HABITS OF THE

BIRDS OF THE UNITED STATES OF AMERICA;

ACCOMPANIED BY DESCRIPTIONS OF THE OBJECTS REPRESENTED
IN THE WORK ENTITLED

THE BIRDS OF AMERICA,

AND INTERSPERSED WITH DELINEATIONS OF AMERICAN
SCENERY AND MANNERS.

BY JOHN JAMES AUDUBON, F. R. SS. L. & E.

FELLOW OF THE LINNEAN AND ZOOLOGICAL SOCIETIES OF LONDON; MEMBER OF THE LYCEUM
AND LINNEAN SOCIETY OF NEW YORK, OF THE NATURAL HISTORY SOCIETY OF PARIS, THE
WERNERIAN NATURAL HISTORY SOCIETY OF EDINBURGH; HONORARY MEMBER OF THE
SOCIETY OF NATURAL HISTORY OF MANCHESTER, AND OF THE SCOTTISH ACADEMY OF
PAINTING, ARCHITECTURE, AND SCULPTURE, &C.

EDINBURGH:

ADAM BLACK, 55. NORTH BRIDGE, EDINBURGH;

R. HAVELL JUN., ENGRAVER, 77. OXFORD STREET, AND LONGMAN, REES,
BROWN, & GREEN, LONDON; GEORGE SMITH, TITHEBARR STREET,
LIVERPOOL; T. SOWLER, MANCHESTER; MRS ROBINSON, LEEDS;
E. CHARNLEY, NEWCASTLE; POOL & BOOTH, CHESTER; AND BEILBY,
KNOTT, & BEILBY, BIRMINGHAM.

MDCCCXXXI.

Figure 7. Title page of Audubon's *Ornithological Biography*. Courtesy Rare Books Library, University of Pennsylvania.

ful oversight of his birds as they came off the engraver's plate, Audubon still knew he had to find more birds in the field. To make his Great Work as great as he had promised in his 1827 "Prospectus," the American Woodsman would have to venture back into the dark and danger of the American woods.

He would also have to venture back into the world of American science, an environment that could be just as menacing in its own way.

Chapter 5

Struggling for Status in Science

Among the greatest pleasures I have known, has been that derived from pursuing and faithfully describing such of our American birds as were previously unknown or but little observed. . . .

WILSON, who, it is acknowledged, made his figures from stuffed specimens in the Philadelphia Museum . . . had not seen the Anhinga alive or recently killed.
—John James Audubon, "Anhinga or Snake-Bird,"
in *Ornithological Biography*, IV

"Compare naturalists with any other sect," wrote an observer of American science in the early 1840s, "religious or irreligious, such as poets, philosophers, divines, admirals, generals, or worthies in general, civil or military, lay or clerical, and you will acknowledge that they are, peculiarly, a peculiar people, zealous in good works."[1] Peculiar some of them certainly were, and zealous as well, but not always as high-minded as one might expect: Good work did not necessarily breed goodwill. No matter how much early nineteenth-century naturalists might want to distinguish themselves as important participants in a lofty enterprise that served the larger interests of transatlantic science, and even the growing glory of the nation, they also paid heed to their own self-interest as well. The search for scientific knowledge, not to mention notoriety, has never been immune from less-than-generous behaviors, of course, but it seemed particularly so at a time when the study of natural history in the United States was still in what one historian has called its "discovery phase."[2] The prospect of identifying, describing, and perhaps even naming new species gave naturalists a remarkable opportunity both for making a significant contribution to science and for enhancing their own reputations. Sometimes, in fact, enhanc-

ing one's own reputation would mean diminishing that of someone else, and members of the naturalist "sect" could be as jealous as they were zealous.

Audubon came to know the nastier side of naturalists well enough, and in his struggle to gain credibility within the scientific community, he occasionally showed his own. He claimed, by all means, to be far above the petty bickering and backbiting that often accompanied the race to discover and describe new species. In the introduction to his first volume of *Ornithological Biography*, he wrote that "there seems to be a pride, a glory in doing this, that thrust aside every other consideration; and I really believe that the ties of friendship itself would not prevent some naturalists from even robbing an old acquaintance of the merit of first describing a previously unknown object." He confessed that he had "certainly felt very great pleasure" in finding a new species, but he assured his readers that he could take no pleasure in the competitive, even combative aspects of ornithology, and "I have never known the desire above alluded to."[3]

But he had, of course. However high-minded he considered his calling, Audubon came to be an active participant in the unfriendly underside of science. Certainly from the time he got on the wrong side of Alexander Wilson and his allies, he suffered attacks on his character and credibility, and he remained an embattled naturalist throughout his life. He did not suffer in silence, however. He struck back repeatedly, often taking pointed, even petty, shots at his suspected adversaries, most notably the long-dead Wilson or still-living naturalists, some of whom were indeed venomous, but others simply vulnerable.

Reeling in the "Odd Fish"

Of all the tall tales Audubon tells, none is more intentionally entertaining and, in the end, inadvertently revealing than the story of his encounter with the "Eccentric Naturalist." It dates from 1818, in Audubon's then-hometown of Henderson, Kentucky, and opens with the surprising arrival of a fellow student of science, who turns out to be received not so much as a respected colleague but more as the buffoonish butt of an extended joke. The joke, though, takes us into a more serious consideration of the nature of nineteenth-century science and, above all, the often edgy and competitive personal and professional relationships among American naturalists themselves.

"What an odd looking fellow!" Audubon said to himself as he saw the man get off the boat on the banks of the Ohio River. The traveler wore tattered, old-fashioned clothes that were stained with the juice of plants, and he carried what appeared to be a bundle of dried clover on his back. He happened to be looking for Audubon, and when the two met, the newcomer gave Audubon a note of introduction from a mutual friend: "My dear Audubon," it read, "I send you an odd fish, which you may prove to be undescribed." The note puzzled Audubon, and "with all the simplicity of a woodsman, I asked the bearer where the odd fish was." Expecting to see some sort of ichthyological specimen, Audubon seemed surprised when the man "smiled, rubbed his hands, and with the greatest good humour said, 'I am that odd fish I presume, Mr. Audubon.'"[4]

In the episode of "The Eccentric Naturalist," Audubon identified the man as "M. de T.," who proved to be an insistent visitor. He told Audubon he had come to Henderson to see Audubon's drawings, not so much for the birds as for the plants surrounding the birds, because he wanted to know if Audubon knew any form of flora he himself did not: "I observed some degree of impatience in his request to be allowed at once to see what I had." As he had done with Alexander Wilson several years earlier, Audubon opened his portfolio of drawings, and M. de T. quickly fixed his eye on a plant he thought did not exist. Audubon assured him that such a plant certainly did exist and offered to show him one the following day. "And why to-morrow, Mr. Audubon? let us go now." They went out immediately, and when they came upon the plant, Audubon said, "I thought M. de T. had gone mad. He plucked the plants one after another, danced, hugged me in his arms, and exultingly told me that he had got, not merely a new species, but a new genus."[5]

The day of scientific discovery continued well into the night. While the two men were sitting at the table that evening, the candlelight attracted a large insect, which Audubon identified as a Scarabeus, and which he assured M. de T. was strong enough to carry the candlestick on its back as it crawled across the table. "I should like to see the experiment made, Mr. Audubon," said the inquisitive naturalist, and so Audubon showed him that the insect could indeed carry the load. Later that night, after everyone else had bedded down, Audubon heard a loud commotion in M. de T.'s room and rushed in, only to find the man smashing at flying bats with Audubon's beloved violin. Again, M. de T. insisted that he had discovered a new species, and to restore peace and to save what remained of his musical instrument, Audubon smacked a few more bats and gave him the specimens. During the next few days, while M. de T. stayed with Audubon and his family, his strange-seeming behaviors continued, but in time the family became "perfectly reconciled with his oddities." Then, as suddenly and unexpectedly as he had arrived, the "Eccentric Naturalist" disappeared. "Whether he had perished in a swamp, or had been devoured by a bear or a gar-fish, or had taken to his heels, were matters of conjecture," Audubon concluded; a few weeks later, though, a thank-you letter from M. de T. assured Audubon that the man had arrived somewhere in safety.[6]

It was all a tall tale, a classic example of backwoods humor. In this story Audubon puts himself in the role of the rustic, in which the "the simplicity of a woodsman" stands in sharp and quietly mocking contrast to the "eccentricity" of the scientist. The whole visit, in fact, provides a parody of the scientific enterprise, complete with an "experiment" with a candle-carrying insect and the "discovery" of a familiar forest plant and the equally common bat. So intent is M. de T. on claiming a new species or genus that he lets his excitement give him a skewed view of his environment. Like James Fenimore Cooper's caricature of Dr. Battius in *The Prairie* (1827), Audubon's depiction of M. de T. in *Ornithological Biography* shows a "renowned naturalist" seemingly out of place in nature.[7]

But Audubon stops too soon, and he fails to tell us the rest of the story—a real incident with a real man that actually happened in the late summer of 1818. The naturalist identified as "M. de T." was in actuality Constantine Samuel Rafinesque (1773–1840), like Audubon the son of a French merchant living in a foreign land (in this case, Turkey) and a mother probably of mixed ethnicity (German and Greek). Rafinesque had traveled much of the world by the time he came to Kentucky—"He never held down a job for long," as one biographical profile puts it, "but preferred the life of a flamboyant traveler and collector"—and at age forty-five he had established a decent reputation as a naturalist on both sides of the Atlantic. In the United States he had been involved in the founding of the Lyceum of New York, and he had been elected a member of the Literary and Philosophical Society, which at the time amounted to far more scientific recognition than Audubon had gained in his thirty-three years. Rafinesque was, by most accounts, excitable and even eccentric, but he had also spent enough time collecting specimens of American plant and animal life to know more than a little about nature in the new nation.[8]

Audubon fooled him more than once and showed what Rafinesque didn't know. Audubon himself, however, apparently didn't know enough about

Rafinesque's scientific background at the time to take him seriously, and he told his visitor about various fictitious fishes and birds, even drawing pictures of some of them. Rafinesque, no doubt eager to take note of—and perhaps credit for—new discoveries, believed Audubon and, to make the matter worse, later published the erroneous information. Although he went on to secure a position as a professor at Kentucky's Transylvania University, then the most prestigious institution in the region, he had to live the rest of his life with the embarrassment of the false findings on his scholarly record. When Audubon published his account of "M. de T.," then, his description of Rafinesque as an "odd fish" provided an implicit (and perhaps impish) double entendre, a reference not just to the "Eccentric Naturalist," but to the false specimens that so easily deceived him.[9]

Taken as something more than a joke, the relationship between Audubon and Rafinesque, whatever its inherent humor, encapsulates several serious issues for early nineteenth-century naturalists. People supposedly devoted to the common pursuit of science could be as much given to competition as collaboration, and the race to identify and name new species in the seemingly endless and abundant American environment sometimes led them to report false, or certainly unverified, findings. In some cases, error even bred enmity, and jealous skeptics quickly cast accusations of scientific fraud. Personal jealousies sometimes overshadowed professional standards to the point that one naturalist's experiment did not have to be duplicated before being dismissed with disdain or ridicule by another. The existence of learned societies ostensibly offered a common source of authority, but not, as Rafinesque's case suggests, a guarantee against gullibility, much less mockery. Moreover, as Audubon would find out, admission to membership in such a society did not always depend on displaying one's own merit as much as on overcoming another member's malice. On the whole, whatever desire American naturalists had to see themselves as an emerging community with a collective national agenda, they could be as suspicious and hostile toward one another as they already were toward their European counterparts. The stakes seemed surprisingly high: Lacking a national tradition of established institutional support, naturalists in the new nation relied on their individual reputations to secure a respectable place in American science—and in American society as well. In that respect, Audubon's joke on Rafinesque was no laughing matter.

The "Real" M. de T. Surveys Science in America

Audubon published his account of "The Eccentric Naturalist" in the first volume of *Ornithological Biography*, which came into print in March of 1831, just a few weeks before another M. de T.—Alexis de Tocqueville—left France on a voyage to the United States. At the time, Tocqueville was a little-known twenty-five-year-old Frenchman who had become disaffected with his government and wanted to get away from France for a while. He and another young Frenchman, Gustave de Beaumont, got permission to leave under the ostensible premise of making a study of American prisons, which, to their credit, they dutifully did. But in the process of traveling extensively throughout the United States—from New York westward to Detroit and the Great Lakes; back east to Boston, Philadelphia, and Baltimore; down south to Nashville, Memphis, and New Orleans; then back across the South through Alabama, Georgia, and the Carolinas to Washington, DC, and New York—Tocqueville did much more than investigate incarceration. "I confess that in America I saw more than America," he famously wrote in the two-volume work he subsequently published, *Democracy in America* (1835, 1840). "I sought the image of democracy itself, with its inclinations, its character, its prejudices, and its passions."[10]

One of the things Tocqueville did not see in America was Audubon. Tocqueville's extensive journey in the United States between May 1831 and February 1832 coincided with a period of travel for Audubon as well. Audubon returned from England in September 1831 and headed south to South Carolina, Florida, and the Florida Keys for the remainder of 1831 and the first few months of 1832. Tocqueville crossed the Deep South in early January 1832, but he never went as far south as Florida and, most regrettably, never met Audubon. Equally regrettably, Audubon seems never to have read Tocqueville's *Democracy in America*, which ranged widely over all aspects of American society, from politics to sexual equality to slavery, and even as far as science. What the young Frenchman had to say about American science and "its character, its prejudices, and its passions" might have helped Audubon—as it now helps us—see his scientific experience in a larger cultural and intellectual context.

As much as Tocqueville admired certain aspects of American democracy, he admitted that the United States had not yet become the place to breed excellence in the life of the mind, at least in comparison to Europe. In a chapter whose title offered, at best, a tepid estimation rendered in an awkward-sounding double-negative disclaimer—"The Example of the Americans does not Prove that a Democratic People can have no Aptitude and no Taste for Science, Literature, or Art"—Tocqueville opened with an observation that the United States had little to tell Europe about the place of the arts and sciences in society: "It must be acknowledged that in few of the civilized nations of our time have the higher sciences made less progress than in the United States; and in few have great artists, distinguished poets, or celebrated writers been more rare." The reasons for that, he argued, stemmed not so much from the nature of democracy in general, but from the particular patterns of American history. The still-constraining hold of America's "strictly Puritan origin" and the expansive release inherent in the "new and unbound country" had come together to create a culture that instilled less interest in intellectual speculation than in economic opportunity. "In America everyone finds facilities unknown elsewhere for making or increasing his fortune," Tocqueville wrote. "The spirit of gain is always eager, and the human mind, constantly diverted from the pleasures of imagination and the labors of the intellect, is there swayed by no impulse but the pursuit of wealth." Yet this westward-seeking people remained tethered to their English origins across the Atlantic, seemingly still destined to be "that portion of the English people who are commissioned to explore the forests of the New World," thus leaving the rest of the British to "devote their energies to thought and enlarge in all directions the empire of mind." In that sense, the United States represented a quasi-colonial cultural outpost whose inhabitants could cling to just enough of the arts and sciences to avoid "relapsing into barbarism." He closed with a pale apologia for the United States as a demonstration project for democracy: "It is therefore not true to assert that men living in democratic times are naturally indifferent to science, literature, and the arts; only it must be acknowledged that they cultivate them after their own fashion and bring to the task their own peculiar qualifications and deficiencies."[11]

To the extent that Americans did pursue the arts and sciences according to "their own peculiar qualifications and deficiencies," Tocqueville continued, they were "More Addicted to Practical than to Theoretical Science."[12] They kept their eyes on the immediate prize, focusing on discernible details within their reach rather than grasping for "the essentially theoretical and abstract portion of human knowledge." Indeed, he argued, "They mistrust systems; they adhere closely to facts and the study of facts with their own senses." By "their own senses," Tocqueville meant those of each individual observer, a pragmatic skeptic who had little use for scientific precedent, schools of thought, scholarly language, or, for that matter, other scientific observers: "As they do not easily defer to the mere name of any fellow-man, they are never inclined to rest upon any man's authority; but, on the contrary, they are unremitting in their efforts to point out the weaker points of their neighbors' opinions." Favoring a fact-based practicality, plainspoken explanations, and intellectual independence, even isolationism, Tocqueville's America clearly seemed a place where science might "follow a freer and safer course, but a less lofty one."[13]

Declaring Independence for American Science

Tocqueville provided no detailed or compelling evidence in his overview of American science in the early republic, but he did hit the mark on several important issues that had long dogged members of the American scientific community. Above all, the allegedly ancillary (and implicitly inferior) role Americans played in the scientific pursuits of their European counterparts had become a point of injured pride during the middle of the eighteenth century. A few Americans—Philadelphia's John Bartram most notable among them— had engaged in regular correspondence with men of science in England and on the Continent, sending seeds, specimens, and other information about the American environment, and thereby gaining a good measure of scientific credibility in transatlantic circles. In turn, a few English allies—the London merchant Peter Collinson in particular—proved to be supportive promoters of American correspondents, helping them have their letters published in the *Philosophical Transactions* of Great Britain's Royal Society. Correspondence did not carry equal weight as actual articles, however, and Americans could not enjoy equal status as British members of the Royal Society; they remained, at best, colonial contributors to an Anglocentric scientific conversation, and their reports from the periphery had to be passed through the seemingly more important prism of British inquiry.[14]

Unable to be fully recognized participants in the most prestigious British organization, Americans began to create learned societies of their own, even before the increasingly anxious period preceding the American Revolution. The first of these American scientific societies, the American Philosophical Society (APS), seemed initially to accept its place in the provincial scheme of things. As Benjamin Franklin, one of the leading promoters of the new organization, dutifully explained, among the Society's goals would be to "produce

Discoveries to the Advantage of some or all of the British Plantations, or to the Benefit of Mankind in general." Almost as geographically separated from each other in America as they were from Great Britain, the "Virtuosi or ingenious Men residing in the several Colonies" would form their own network of correspondence, the nexus of which would be in Philadelphia, "the City nearest the Centre of the Continent-Colonies." The APS got off to a slow start, however—its members being "very idle Gentlemen," Franklin complained—and the society provided at best a nascent, albeit still shaky, source of American scientific identity.[15] Americans would never fully assert themselves until the increasingly frayed relations between the colonies and Parliament in the political realm caused the scientific connections across the Atlantic to begin to unravel, but in the run-up to the Revolution, scientific deference came more and more to be replaced by resentment and a growing sense of independence. In the years immediately following the Revolution, new American scientific societies would begin to blossom in the major cities of the new nation, and membership would become a sign of acceptance in the scientific community of the republic. But even then the move toward political independence did not bring complete intellectual independence. Although American naturalists might also consider themselves incipient nationalists, they could never negate their need to be taken seriously in transatlantic scientific circles. "They chafed to contribute to the various scientific debates that then raged abroad," one modern scholar explains. "Paradoxically, one was most certainly an *American* scientist when one succeeded in establishing a *European* reputation."[16]

Throughout the late eighteenth century, the most significant scientific debate facing American science came not from Great Britain, but from France. Indeed, any discussion of early American science almost certainly has to acknowledge the central, even essential, impact of one important publication—the fifth volume of *Histoire naturelle*, by Georges-Louis Leclerc, Comte de Buffon (1707–1788), which came into print in 1766.[17] Working with the rich resources of plant and animal life collected in the Jardin de Roi in Paris, which he supplemented by soliciting contributions from correspondents from distant parts, Buffon took on the task of producing a massive, multivolume work that would provide an encyclopedic survey of all of nature, in both the Old World and the New. Despite its imposing size and reach, Buffon's work had an accessible and engaging prose style that gained a wide readership among educated people on both sides of the ocean.[18] Building on the still-fresh foundations of the Enlightenment in France, Buffon took the notion of "history" in natural history seriously. That is, his view of nature did not depend on any notion of perfection at the time of Creation, but instead posited the possibility, even certainty, of change over time—and not always change for the better.[19]

The mutability of species thus played an important role in Buffon's fifth volume of *Histoire naturelle*, which put forth an argument for the "degeneracy" of nature in the New World. Building on assumptions about the basic environmental differences between Europe and the Americas, Buffon observed that the comparatively cool, humid climate of the latter had a profound effect on all forms of life, particularly in the animal world: "In America," he asserted, "animated Nature is weaker, less active, and more circumscribed in the variety of her productions . . . [and] the numbers of species is not only fewer, but . . . much smaller than those of the Old Continent."[20] Species endemic to the New World environment—quadrupeds, reptiles, even insects—were smaller to begin with, he argued, and those that were brought over from the Old World would eventually degenerate in size as well. Buffon likewise argued that human "savages" native to the New World were physically, intellectually, and sexually inferior to Europeans. He did, however, give a pass of sorts to Europeans who emigrated from the Old World, suggesting that they could eventually have a positive effect on other forms of life.

That possible exception notwithstanding, Buffon's theory of degeneracy quickly got the attention of scientifically inclined Americans, to whom size certainly seemed to matter. Particularly in the post-Revolutionary period, any notion of American inferiority would leave the new nation in the shadows of civilization, undermining national confidence right at a time when the new United States had just begun to bask in the light of independence. No less famous a champion than Thomas Jefferson mounted the most famous defense, in his *Notes on the State of Virginia* (1785), in which he took on the Frenchman's dismissive picture of American nature. He challenged Buffon's description of higher American humidity, arguing, first, that America was probably no more humid than any other region of the world and, second, that if it were, the damp climate would actually promote growth. Turning from the general environment to specific data about American flora and fauna, Jefferson counted, measured, calculated, extrapolated, made lists and tables—all to provide what he thought should be irrefutable proof of the healthy state of nature in the United States,

including its Euro-American and Native American human inhabitants. To
help make his case for the possibility of bigness, Jefferson noted the skeletal
remains of a huge yet still unidentified mammal, asserting that "it is certain
that such a one has existed in America, and that it has been the largest of all
terrestrial beings." Not having mastodon remains readily at hand, Jefferson
later sent Buffon the skin and skeleton of an American moose. Jefferson's
efforts seemed to have worked, and just before Buffon died, the Frenchman
was apparently on the verge of a partial public retraction in a forthcoming
volume of *Histoire naturelle*. The correction never came, however, and thus
the widespread resentment of his notion of American degeneracy did not
easily go away.[21]

The "Father of American Ornithology" Fights Back

In the early years of the nineteenth century, when American naturalists felt
increasingly compelled to declare their independence from Europe in general
and Buffon in particular, they needed to look no further than the field of
ornithology to find a model: Alexander Wilson, whose place in the pantheon
of American science would remain secure throughout the century. In 1878,
the eminent American ornithologist Elliott Coues asserted that "science would
lose little, but, on the contrary, would gain much, if every scrap of pre-
Wilsonian writing about United States birds could be annihilated." Those
were strong, strikingly uncharitable, and probably self-consciously provocative words, and Coues knew it would be an impossibly grave mistake to
diminish, much less annihilate, the earlier work of others. No matter how
much Wilson and other American naturalists of the early national era might
have tried to distance themselves from their predecessors, both European
and American, their very ability to work in the world of science rested on the
foundations that had been in place for at least two centuries. More to the
point, their attempt to do a distinctively American form of science remained
tied to a long tradition of transatlantic discourse that still continued to thrive
in their own era. Still, even though Coues could admit that Wilson had been
"no scholar" and had made his share of scientific mistakes in *American Ornithology*, he argued that "perhaps no other work on ornithology of equal extent
is equally free of errors." (A few pages later, Coues described Audubon's *Birds*

of America as "by far the most sumptuous ornithological work ever published"—a compliment, to be sure, but not carrying the same sort of scientific credential as that awarded Wilson.[22]) Just over a decade later, in 1889, Henry Adams likewise praised Wilson for having published "an ornithology more creditable than anything yet accomplished in art or literature."[23] American ornithology did not really begin with Wilson, of course, but it did derive much of its independence-seeking identity from him: For the "Father of American Ornithology," the term "American" became central to his own identity in his adopted country.

When it came to berating Buffon, Wilson wasted little time or energy. In the first volume of his *American Ornithology*, he noted the error Buffon had made in describing the Wood Thrush as a migrant from Europe whose song in America had degenerated into a harsh and unpleasant cry. Quick to defend the musical talents of the American species, Wilson countered that "the fanciful theory which this writer has formed to account for its want of song, vanishes into empty air." Writing later about the Blue Jay and its Canadian cousin, the Canada Jay, he again attacked the "theoretical reasoning of a celebrated French naturalist" about the allegedly debased condition of the American birds in the thousands of years since their migration from Europe, where, he added sarcastically "nothing like degeneracy or degradation ever takes place among any of God's creatures." In his defense of American woodpeckers, Wilson mocked the "eloquence and absurdity" of Buffon's overgeneralizations, labeling the famous French naturalist's "expressions improper, because untrue; and absurd, because contradictory." For his own part, Wilson contrasted his pragmatic reliance on "plain matters of fact" and the "plain realities of nature" against "all the narratives, conjectures, and fanciful theories of travelers, voyagers, compilers, &c . . . from Aristotle down to his admirer the Count de Buffon." Theory mattered for nothing, Wilson argued, when it remained "altogether unsupported by facts and contradicted by the constant and universal habits of the whole feathered race in their state of nature."[24]

Even after Wilson died, his patron, George Ord, continued in the same vein in the ninth and final volume of *American Ornithology*, lambasting "the labors of foreigners, who have interested themselves in our natural productions," only to reveal "how totally incompetent they were, through a deficiency of correct information." The only way to know American birds, in short, was to discard Buffon's fanciful theorizing for Wilson's field-based observing, through which, thanks to the late ornithologist, "we have as faithful, complete and interesting an account of *our* birds in the estimable volumes of American Ornithology, as the Europeans can at this moment boast of possessing in *theirs.*"[25] By the second decade of the nineteenth century, then, Wilson and Ord had helped transform the state of the argument, no longer rebutting Buffon from a defensive posture of trying to deny the degradation of American nature, but instead denying the value of European theory and asserting the scientific reliability of fact-based American independence—the perfect combination, in short, of natural history and national history.

Audubon Asserts Ornithological Authority

Audubon picked a different battle. By the 1830s, when he published his own ornithological one-two combination, *The Birds of America* and *Ornithological Biography*, he had comparatively little to say about Buffon, but much to say about Wilson and, more venomously, his posthumous supporter, Ord. The debate about Buffon's theory of degeneracy never seemed much of an issue for Audubon, but Wilson's ascendance to the title of the "Father of American Ornithology" certainly did: Throughout his life, Audubon never let go of his obsession with surpassing Wilson. He had come to know Wilson's work extremely well, and he had spent almost two decades pointing out Wilson's errors and omissions, both privately, in the pages of his personal journals and correspondence, and more publicly, in *Ornithological Biography*. Throughout the five volumes of his massive work, Audubon questioned and corrected Wilson dozens of times, much more than he did any other authority on ornithology. Alexis de Tocqueville did not have Audubon in mind when he wrote about men of science in America being "unremitting in their efforts to point out the weaker points of their neighbors' opinions," but when it came to Audubon's attentions to Wilson's opinions, the description surely fit.

In *Ornithological Biography*, Audubon sometimes took to talking out of both sides of his mouth. He occasionally gave Wilson almost gushing praise, commending his zeal, his "persevering industry," his fine eye as a "most excellent observer." He even gave Wilson credit for being "my predecessor," noting that "the Ornithology of the United States may be said to have commenced by Alexander Wilson," adding that "prior to the days of Wilson, very little indeed had been published respecting the habits of our birds."[26] In writing about a bird named for Wilson, Audubon could become almost rhapsodic: "Wilson's Plover! I love the name because of the respect I bear towards him to whose memory the bird has been dedicated." He even added an indirect reference to the time he and Wilson spent together in Louisville, before the memory of the visit turned sour: "How pleasing, I have thought, it would have been to me, to have met with him on such an excursion, and . . . to have listened to him as he would speak of a thousand interesting facts connected with his favourite science and my ever-pleasing pursuits." But then Audubon added his "alas!" qualifier: "Wilson was with me only a few times, and then *nothing* worthy of his attention was procured."[27]

In a number of subsequent references to Wilson, however, the narrative turned more toward the notion of "nothing worthy," or at least toward the errors in Wilson's work. To be sure, Audubon sometimes couched his criticism within a disclaimer of respect—"However highly I esteem the labours of Wilson . . . ," "Always unwilling to find faults in so ardent a student of nature as Wilson . . ."—but he then would be quite willing indeed to point out the faults he had found.[28] Some of the alleged errors might seem relatively minor, common enough among other ornithologists at the time (and certainly familiar to any ordinary birder today). Wilson sometimes failed to reproduce the bird's call accurately, to distinguish a male from a female, or to discern an immature bird from a full-grown adult.[29] In some cases, Audubon observed, the mistake came from Wilson's being "in all probability misinformed" by someone else, but in others the "error originated with Wilson, who has been followed by all our writers," thus compounding the problem by passing it along to those who accepted his ornithological authority.[30] Audubon's more serious challenge to that authority came from his skepticism about the geographical reach of Wilson's research, either his occasional failure to range far enough afield for a firsthand encounter with a particular species or, even worse, the possibility that he had made his drawing of the bird "from a stuffed European specimen in Peale's Museum in Philadelphia."[31]

Whatever the errors of his rival, Audubon nonetheless had the upper hand in the area that mattered most: his list. He wrote in the introduction to the final volume of *Ornithological Biography* that, in the time since Wilson's death, the list of species of birds that had been "discovered, figured, and described, is very great," even doubling the number in Wilson's work.[32] Indeed, the extent of Audubon's list stood as one of the main scientific selling points for *The Birds of America*, certainly a main element of his claim to fame as an ornithologist.

He not only made it a point of pride to note how much his list exceeded Wilson's, but he took as a direct challenge any notion that Wilson's work was still ornithologically adequate. He addressed that point very clearly in the second volume of *Ornithological Biography*, taking unmistakable aim at an unnamed ally of Wilson's who had apparently argued that Wilson had done enough. "Shortly after the death of WILSON," Audubon wrote, "one of the wise men of a certain city in the United States, assured the members of a Natural History Society there, that no more birds would be found in the country than had been described by that justly celebrated writer." Audubon added, with a bit of mockery, that "the orator has travelled much, having gone a few miles to the eastward of his own city, and even crossed the Mississippi; but, as he had predicted, he never discovered a bird in all his wanderings." But while the "wise man" remained in the city, assured that there would no more birds to discover, Audubon countered by noting that other, more industrious naturalists "have followed the track of Wilson, have extended their investigations, ransacked the deep recesses of the forests and the great western plains, visited the shores of the Atlantic, ascended our noble streams, and explored our broadest lakes;— and, reader, they have found more new birds than the learned academician probably knew of old ones."[33]

Audubon never mentioned George Ord by name here, but he hardly had to. Ord had a stake, both personal and financial, in promoting Wilson's work as "the" work on ornithology in the United States, and it made sense for Ord to declare the field essentially closed, all the birds discovered, all the good work done. For Audubon, going after Ord in this manner had precious little to do with bird identification anyway, but everything to do with Audubon's sense of his own scientific identity. Audubon's use of the term "academician" was one he frequently used to refer to Ord and his allegedly bookish approach to birds. Still, Audubon desperately wanted to be recognized as a legitimate man of science, particularly in Philadelphia, the city that remained the center of science in the United States, not just geographically, as Franklin had noted in the 1740s, but socially and intellectually. Audubon wanted in, and Ord had more than enough influence to keep him out—at least throughout the 1820s. By the time Audubon's sneering references to the "academician" and "wise man" appeared in print, in 1834, anyone who knew anything about Ord's role in Audubon's Philadelphia difficulties would have to know he had only one man in mind.

Figure 8. *George Ord,* by John Neagle, 1829. Academy of Natural Sciences, Drexel University.

At War with the "Philadelphians"

The long-standing and much-discussed struggle between Audubon and Ord could be cast in purely personal terms, as a case of persistent nastiness between two proud and prickly men, as both certainly were. It went well beyond them as individuals, though, spreading more widely throughout the scientific community on both sides of the Atlantic and certainly shaping scientific inquiry on the American side. Some of the pressing issues under dispute—Can turkeys swim? Can buzzards smell? Can rattlesnakes climb trees?—might appear almost laughable now, but they seemed very serious at the time, and not just to Audubon and Ord, but to other men of science, who had quite a few such questions to answer. (The nature of such scientific inquiry will indeed get its due in later pages. For now, one answer to all three questions may suffice: Yes, in some cases, but not especially well in any.) But the battle over the behaviors of birds, reptiles, and other forms of fauna needs first to be seen in light of the behaviors of those human students of science who became embroiled in the battle—Audubon and Ord, to be sure, but also their respective allies and even a few neutrals who hoped to remain above it all.

It would be hard to count most of Audubon's modern biographers among those neutrals. Ord has never fared especially well in the Audubon story, but he has been repeatedly portrayed in stark, unsympathetic contrast to the much more engaging image of Audubon. Francis Hobart Herrick, Audubon's first serious biographer, gave Ord credit for being a "quiet, persistent, and unassuming worker," but he also noted the earlier observation of Audubon's friend Richard Harlan that Ord had "no heart for friendship, having been denied that blessing by nature herself." More recent Audubon biographers have described Ord as a "pedantic, jealous aspirant to primacy," a "doughy, sharp-tongued man," and "a man known for his rudeness."[34] The Audubon-based perspective on Ord remains, on the whole, not at all flattering.

His apparent lack of personal charm, however, might well have been balanced by his success in both business and science. Ord was, like Audubon, the son of a ship captain who turned to land-based pursuits to make a good living for his family. His father, the senior George Ord, had first become wealthy by owning or insuring privateering ships during the era of the American Revolution, then by running a successful ship chandler and rope-making concern in Philadelphia. The younger George Ord, born in Philadelphia in 1781, moved into his father's business when he was nineteen and then, when his father died, in 1806, took over the operation, doing well enough to retire in 1829, still two years shy of fifty. Along the way, even though he apparently had little formal education, Ord developed a significant sideline in science that gave him considerable stature in Philadelphia's scientific community. He became a supporter and successor of Alexander Wilson, of course, completing the final two volumes of *American Ornithology*, but he also published his own works on zoology. In recognition of his scientific achievements, Ord gained election to Philadelphia's Academy of Natural Sciences in 1815 and then, two years later, the APS, eventually serving both institutions in several leadership posts.[35] One historian of early American science offers a somewhat disparaging assessment of Ord's scientific status in Philadelphia, arguing that by the time the city's "big men" of the early years of the century had died—Benjamin Smith Barton and John and William Bartram most prominent among them, but even Alexander Wilson as well—Ord and a few others "who remained to set the tone of art and learning were of the second order, with vested interests in what had gone before."[36] But whether of the second order or not, Ord certainly had the position to protect his vested interest in Alexander Wilson, and he proved to be a very effective gatekeeper for Philadelphia's scientific community, persistently doing his best to keep out one aspirant in particular.

Audubon desperately hated being kept outside the gate, even when he was far away from it. In 1826, soon after arriving in Great Britain, when he began to gain attention from many of the scientific societies there, he could still hardly appreciate his own achievement without taking a sarcastic swipe at Ord. When the Royal Academy of Arts in Edinburgh met in late 1826 to consider subscribing to Audubon's work, he wrote in his journal that "the meeting was very [large] and no doubt [a] very learned one. Thou knowest I cannot well say, but according to *Mr. Ord, who is learned and an academician,* I suppose each member here [to be] quite as much so, as least, as *Mr. George Ord.*"[37] But even as Audubon began to rack up a string of memberships in some of the most prestigious British societies—the Wernerian Society, the Antiquarian Society, the Royal Society of Edinburgh, and, even better, the Linnean Society and the Royal Society of London—he still kept an uneasy eye on Ord and his Philadelphia allies. When invited to write an essay for a new British publication, John C. Loudon's *Magazine of Natural History*, in 1828, he initially declined, "for I will never write anything to call down on me a second volley of abuse," he explained with a sense of defensiveness. "I can only write *facts*, and when I write these, the Philadelphians call me a liar."[38]

In time, though, Audubon's success in Great Britain could hardly be denied—his engravings for *The Birds of America* proudly bore the initials "F.R.S" and "F.L.S." at the bottom, underscoring his status as a fellow in the Royal Society and the Linnaean Society. In November 1830, he received additional good news from the American side of the Atlantic when he learned that he had been elected a member of the Society of Arts and Sciences in Boston. "Oh Philadelphia, Philadelphia," he asked rhetorically (and sarcastically), "is it true that I am not worthy of being one of thy Academicians?"[39] In September 1831, when he returned to the United States for a specimen-gathering and subscription-selling trip, he came back with enough renewed self-confidence to set aside his fear of the Philadelphians.

Eventually, even the Philadelphians had to yield, at least on formal grounds. Soon after arriving back on American shores, Audubon reported to Robert Havell back in England that "my enemies here are going down hill very fast, and fine reviews of the Work coming forth." To Lucy he later crowed that the "papers here have *blown me up* sky high. The Society of Natural Sciences of Philadelphia has *at last* elected me one of their members, the papers say *Unanimously.*" The APS had also made him a member that summer—Ord happened to be visiting England at the time and was thus temporarily unable to act against Audubon—and by the end of 1831, Audubon's status in the transatlantic scientific community seemed set, even in Philadelphia.[40]

"Sky High" in the American Press

Audubon wasn't just bragging. The papers did indeed blow him up, and some of them gave him more importance than even he might have expected. While Audubon had been pursuing production of *The Birds of America* in England, word began to circulate in the United States about his work, and by the first few years of the 1830s, he had become the subject of national—even nationalistic—acclaim. One American newspaper called Audubon's *Birds* "the most magnificent work ever ventured upon, we suspect, by individual enterprise." Another took its significance well beyond Audubon as an individual, calling it "an enterprize which will do much to advance the honor of our country" by vanquishing Europe in the field of ornithological art: "And it may certainly be regarded as a peaceful victory, not less than renowned than those of war." The only thing that tarnished the luster of the American triumph, some writers admitted, was that Audubon's own country had offered him "little of that encouragement which is derived from kindred taste in others."[41] He had had to find such support first, unfortunately, in Great Britain. "His native country was either unable to appreciate his genius, or unable to aid his efforts," said a South Carolina paper, "and England did both." A Philadelphia paper lamented that after so many years and so much personal sacrifice spent in bringing his work closer to completion, Audubon might have supposed that "this ardent labor and exertion should have been rewarded by the patronage of his country-men"—but no. He had to go to Great Britain, where he became the "Lion of the day," filling exhibit halls and gaining scientific honors, "which Philadelphia had refused." In the end, "had it not been for his own perseverance against the indifference of his countrymen, America would have no claim to so distinguished an artist. We are glad that he is patronized, even now—but it would have redounded more to the honor of our country, had not their aid come rather at the latest."[42]

Some writers even equated America's initial failure to embrace Audubon's work with a flaw in national values. Noting that very few Americans had subscribed to Audubon's work, one New York newspaper called it a "sad commentary on the patronage which the literary and scientific receive in this 'bank note' country." A Philadelphia paper likewise challenged the aesthetic taste and values of Americans, arguing that even though the cost of Audubon's work might be quite high, it was "rich, ornamental, instructive, and entertaining . . . much preferable to the merely personal gewgaws or transitory gratifications, upon which greater sums are so frequently expended."[43] Years before Oscar Wilde made his acerbic comment about those who know the price of everything and the value of nothing, Audubon's admirers in the press had begun to make that distinction clearly in his favor.

Still, if buying *The Birds of America* were to be a sign of national identity, even artistic maturity, one critical issue remained: Who would, or even could, afford to do so? That question opened up a larger conversation that encompassed considerations of the role of the private and public, from individuals and institutions to the state and federal governments, in the support of art and science. In Charleston, South Carolina, a local newspaper noted the high price of Audubon's work but still argued that "a copy should be secured in our city, at all hazards, though we resort to individual subscription for the purpose." A New York paper observed that while the price would be "indeed too costly, general speaking, for individuals," the city had among its wealthier inhabitants a few notable exceptions who could afford the work, and perhaps even some groups of people of more moderate means who could "associate together and present copies . . . [of] this admirable national work."[44]

The notion of *The Birds of America* as a "national work" had important implications not just for American pride, but even for American politics. In the early 1830s, several state governments entertained legislation to purchase Audubon's Great Work as a sign of support for the arts and sciences. Reporting on the introduction of a bill in the Louisiana legislature to purchase three copies of Audubon's work, a Virginia newspaper noted that "this is an honor but seldom paid to literature," adding that "Audubon's splendid work would be a valuable addition to every State, and other public Libraries in the United States."[45] Other works did not carry a price of upward of a thousand dollars, though, and cost-conscious legislators had to weigh Audubon's birds against the state's budget. In Massachusetts, for instance, the legislative proposal to purchase Audubon's work quickly developed into a debate about the role of the government in funding such an expensive project, particularly about the place of ornithological art in the greater scheme of the state's needs. One of Audubon's supporters heaped praise on him for his dogged commitment to his project, "pursuing the object of his researches, through the forests, with an assiduity, that no man ever did before." He also argued that it should be "the policy of the State, from time to time, to encourage science and art and works of public improvement"—perhaps especially the last of those. In addition to its beauty and scientific significance, Audubon's work had a very practical value "that would be of great use to agriculturists" by increasing common knowledge of birds and helping people understand better that "there are large classes of bird supposed to be pernicious, which turn out to be highly beneficial." Finally, Audubon's work could be considered a valuable financial asset: If Massachusetts ever fell into bankruptcy, "we can sell it for as much as we give."[46]

Not everyone bought the idea of buying Audubon, however. From Hampshire County in the western part of Massachusetts—a region that had long been suspicious of the machinations of legislators from Boston and its environs—came slyly humorous but still serious resistance to the use of state funds for pictures of birds. While admitting that the state did indeed have a constitutional responsibility to "cherish seminaries of learning, and to encourage societies for promoting agriculture, arts, manufactures, natural history, &c. in this state," the skeptical westerners argued that Audubon's work exceeded the legislature's ornithological jurisdiction: "Mr. Audubon is doubtless entitled to the praise which is bestowed upon him, but he does not belong to Massachusetts; his birds are the birds of America, very few of which are to be found in our fields or forests." Building on this strict-constructionist approach to avian art and science, the region's leading newspaper argued that if the members of the legislature still insisted on purchasing "the pictures which have dazzled our legislators," then the collection ought to be broken up and spread across the commonwealth, "so that all the people can have the pleasure of seeing at least a part of this state museum."[47]

The Massachusetts government did eventually agree to buy Audubon's work, as did Maryland, New York, South Carolina, and Michigan, not to mention the Library of Congress. So, too, did a number of wealthy individuals and learned institutions, so that by mid-1833, Audubon had over fifty subscribers in the United States. He never considered that enough, though, and throughout the 1830s, as he kept up his incessant search for more and more birds to include in the pages of *The Birds of America*, he likewise continued to stalk subscribers on both sides of the Atlantic.

Anglo-American Antagonism

For all the praise and puffery in the American press, Audubon could never rest on his success. He felt, with decent reason, that he continued to be stalked himself. Ord, his "academician" antagonist, fought a two-front war on both sides of the Atlantic, enlisting his most prominent, and certainly fiercest, British ally in Charles Waterton (1782–1865). Waterton was a notable naturalist of wide-ranging interests but with a professed preference for one in particular: "Mine is ornithology," Waterton wrote, "and when the vexations of the world have broken upon me, I mount it, and go away for an hour or two amongst the birds of the valley."[48] But the birds never fully pacified Waterton. He could visit no end of vexations on people whose opinions he questioned, even detested, whether about religion—he was defiantly Catholic—or ornithology. "Charles Waterton's religious anger was virile and majestic," a modern biographer has observed. "His ornithological anger had all the pride and fervor of religion but little of its humility."[49] When it came to Audubon, Waterton's humility seemed in especially short supply, and his anger fed on Ord's. Ord and Waterton had first met in 1824, when Waterton was touring the United States, and they became fast friends for life who, as a sympathetic contemporary put it, "kept up a regular correspondence on their common pursuit."[50]

That pursuit most often happened to be Audubon, and the transatlantic team of Ord and Waterton took their attacks far beyond Audubon's ornithology to his personality, engaging in an increasingly virulent form of character assassination. Waterton once dismissed Audubon's account of the love life of an owl as "fiddle faddle," and Ord echoed back that the story "may truly be called fiddle faddle."[51] But that was one of their more temperate terms. Ord soon went on to rail about "this impudent pretender, and his stupid book," looking forward to the time "for fully displaying all this man's incompetency and mendacity to the world." When Waterton published an article in London challenging Audubon's assertion about the olfactory limitations of the Turkey Vulture, Ord cheered from Philadelphia that it was a "double-cutter for the great Ornithologist, whose impudent impostures it is high time to expose." Not only did Audubon display "presumptuous vanity" that had no human parallel, Ord dismissively spat, but his work proved him to be a man of low breeding: "He is neither a scholar or a philosopher." Even worse, he considered Audubon a fraud and a thief who on at least one instance, and probably more, had "pilfered the whole of Wilson's supposed facts; and palmed them upon the reader, as the result of his own researches."[52] Egotist, imposter, liar, plagiarist—so went the accusations against Audubon, back and forth across the Atlantic for over a decade, never really ending even after Audubon's death.

But it wasn't Audubon alone. While the Ord-Waterton attack has understandably become the central scientific conflict in the telling of Audubon's life story, the immediate focus on Audubon as victim fails to take account of the collateral damage inflicted on other members of the scientific community. Ord and Waterton gave no quarter to anyone they identified as Audubon's ally. In April 1835, when Ord suspected that one of Audubon's Philadelphia friends, Dr. Richard Harlan, had organized a conspiracy "which deprived me of my chair of the Vice President" of the Academy of Natural Sciences, he spewed to Waterton that Harlan was "a fellow whose moral character is so infamous . . . that he is excluded from the society of gentlemen."[53] A few months later he turned his vituperation on Thomas Nuttall—who was, by then, one of Audubon's scientific acquaintances and an indirect contributor to Audubon's work—calling Nuttall "a presumptuous ass, who deserves the lash." For his own part, Waterton never suggested corporal punishment for Nuttall, but he did label him "a most superlative ass and blockhead."[54]

No one came to pay the price of an alleged alliance with Audubon more than William Swainson (1789–1855), who wound up taking unfriendly fire from both sides. Like Audubon and Ord, Swainson had little formal education, but he possessed a passion for natural history that consumed his life. In the 1810s he had spent time doing zoological and botanical study in both Sicily and Brazil, and by 1820 he was back in England, where he enjoyed membership in the Linnaean Society and the Royal Society and published an extensive list of books and articles. One of those publications, in 1828, gave a glowing review about some of Audubon's recent writings published in Edinburgh, noting that there was "a freshness and an originality about these essays, which can only be compared to the animated biographies of Wilson." Audubon might have groaned to see the Wilson comparison, but he had to be happier with Swainson's praise for his ability to evoke the "passions and feelings of birds" and, above all, his skill as an artist.[55] Audubon was so happy, in fact, that he soon visited Swainson at his London lodgings and then, in August 1828, invited Swainson and his wife to go to the Continent: "Would you go to *Paris* with me? . . . I will remain there as long and no longer than may suit your callings—I will go with you to Rome or anywhere, where something may be done for either of our advantage." Swainson's going gave him a chance to do more botanical and ornithological research in the museum at the Jardin des Plantes, while Audubon's trip became consumed with subscription selling.[56] Swainson's name opened doors, and Audubon lost no time in pushing his way through them.

After their joint journey to France, the two naturalists continued to keep up a friendly correspondence, which included Swainson's sympathetic support in the face of the unfriendly aspersions of Audubon's enemies. Waterton, Swainson observed, "is mad—stark, staring, mad."[57] He found Ord and some of his fellow Philadelphians to be utterly untrustworthy and essentially beneath contempt: "If these are samples of American Naturalists, defend me from ever coming into contact with any of their whole race." Still, when it came to defending Audubon against what he sarcastically called his many "*Ornithological friends*," Swainson admitted that the whole experience had left him feeling existentially discouraged: "I am sick of the world and of mankind, and but for my family would end my days in my beloved forests of Brazil."[58]

Swainson stayed on in England, however—for another decade, anyway—but his relationship with Audubon soon began to sour considerably. In August 1830, Audubon asked Swainson to work with him—or work for him, really—in writing what would become *Ornithological Biography*, wondering if Swainson might "*bear a hand* in the text of my work.—by my furnishing you with the ideas and observations which I have and you to add *the science which I have not!*" Audubon even suggested that their collaboration would work quite well if he and Lucy (who had recently joined him in England) could come and live with the Swainsons for a few months as boarders—on a BYOB basis, to be sure, as "we will furnish our own wines, porter, or ale."[59] Swainson took exception to the whole proposal, responding first that boarding "is never done in England, except as a matter of necessity or profession," then adding that the Audubons would undoubtedly become a bother to "our every-day domestic arrangements." But the larger issue for Swainson had to do with the division of intellectual labor—and the allocation of credit for it. "*You* have to speak of the birds as they are alive," he wrote, and "*I* to speak of their outward form, structure, and their place in the great System of their Creator." Audubon could write with all the flourish he wanted, that is, but Swainson would stick to distinct, and preferably separate, "Scientific Notes" to give the work good standing in the field of natural history. He also expected to be paid a suitable fee for writing and revising, and he insisted on a measure of credit as coauthor: "It would of course be understood that my name stands in the title pages as responsible for such portion as concerns me." The "of course," Swainson explained later, stemmed from principles of intellectual property: "I therefore must not suppose that you intended that I should give all the scientific information I have labored to acquire during twenty years on ornithology—conceal my name,—and transfer my fame to your pages & to your reputation."[60]

In fact, Audubon did intend that. He certainly had no notion of giving anyone else a place on the title page, and he had never expected to have to pay so much for scientific assistance. Again, his was a financially fragile, bare-bones business plan, which could bear precious little excess of expenditure. It was also his Great Work, an intensely personal project about which he had become as possessive as he was proud. Audubon's plans could not accommodate Swainson's demands, and the possibility of a partnership quietly died. "I have no wish to plume myself with others' feathers," he wrote to his friend and fellow ornithologist, Charles Lucien Bonaparte, "but neither am I willing to plume away the obvious benefit of my observations. Therefore, instead of W. Swainson at the rate of 200 pounds sterling per volume, I am employing Mr. MacGillivray for a little more than 50."[61] In addition to working for a lower wage, William MacGillivray would turn out to be a decidedly silent partner—the kind of collaborator Audubon always wanted.

Swainson's feelings may have been hurt and his respect for Audubon lessened, but to his credit, he never lambasted Audubon the way Ord and Waterton did. He confined his criticisms to quibbles over scientific classification, noting that Audubon's "nomenclature is altogether obsolete." In his discussion of American ornithology in his *Natural History of Birds*, he gave the requisite nod of respect to "the celebrated Wilson," whose work still seemed so thorough "that little has been left, comparatively, for those who have gone over the same ground." To Audubon, he gave a rather patronizing appraisal: "M. Audubon's two volumes of letter-press may be consulted to much advantage, but the scientific descriptions are destitute of that precision and detail which might have been expected in these days." The Episodes, those chapters about people and scenery that Audubon inserted to "relieve the tedium" of so much ornithological detail, Swainson dismissed as "not connected with the subject, . . . [but] particularly amusing."[62]

One might think Swainson had gone over to the other side. Audubon certainly did, complaining that Swainson had snubbed him in print and, even worse, that "he quotes Charles *Waterton* and speaks of him monstrous well!"[63] But Waterton still seethed with his own monstrous animosity toward Swainson, which he exposed to the scientific world in a short but scathing and decidedly ad hominem screed, *An Ornithological Letter to W. Swainson* (1837). "Sir, I have a crow to pluck with you," Waterton began. He first began to pluck away at Swainson's earlier support for Audubon, reaching almost a decade back to the 1828 review of Audubon's early essays, when Swainson had written that Audubon "drank of the pure stream of knowledge at its fountain head." To Waterton, Audubon had nothing to do with knowledge and everything to do with "an incorrect account of himself . . . [and] willful inaccuracies." But after taking his now-customary cuts at Audubon, Waterton plucked even more at Swainson himself, attacking his ornithological skills, ranging from Swainson's method of preserving bird skins ("wrong at every point") to his system of classification ("Your nomenclature has caused me jaw-ache"), finally concluding with "Believe me, Sir, you have a vast deal to learn, before you become an adept in ornithology." Waterton had been especially infuriated by Swainson's referring to him as an "amateur," and he turned the term back on him in a back-handed, hypocritical invitation to an imagined ornithological excursion:

> Climb then, with me, the loftiest trees;—range the dreary swamp;—pursue the wild beasts over hill and dale;—repair to alluvial mud flats;—follow the windings of creeks, and of the sea-shore;—and get yourself let down the tremendous precipice in quest of zoological knowledge. Worst come to worst,—this will at least, gain you the appellation of "Amateur" from the pen of supercilious theorists;—an honor, not to be sneezed at, in these our latter times.[64]

Swainson had no need of any assistance, much less any title, offered by Waterton, nor did he have any need for any more ornithological controversy. He soon moved as far away from England as he could, to New Zealand, where he lived from 1841 until his death in 1855. His final fifteen years in the Antipodes were reported not to have been altogether happy ones, but at least he never had to deal with Waterton or, for that matter, Audubon ever again.

"My Warfare & Welfare in This World"

The ornithological wars of the 1830s took a toll on everyone involved, and at some level, the combatants knew that. In his description of Wilson's Petrel—a bird named for his rival Alexander Wilson, after all—Audubon wrote almost wistfully about the way the birds coexisted so peacefully: "Social creatures! Would that all were as innocent as you! There are no bickerings, no jealousies among you."[65] Would that all were as innocent, indeed. As much as Audubon suffered from the bickerings and jealousies of his enemies, he expressed his own uncharitable sentiments toward other naturalists often enough, and not just about Ord and Waterton. Swainson's critical words about Audubon's "obsolete" nomenclature, for instance, must have hurt, particularly at a time when new approaches to systematic classification seemed to be multiplying with every succeeding decade.[66] Classification was not just a way to establish categories in natural history; it also offered a way to establish the authority, not to mention the prominence, of the natural historian. Put differently, classifying nature could be both a practice of science and a promotion of self. Looking at the classification landscape of the mid-1830s, Audubon lashed out at Swainson and other ornithologists who would put themselves forward in such a scientifically self-serving manner. "I feel so very sickened at all these puerile attempts," he lamented, "that I cannot reconcile myself to attempt anything of the kind."[67] He did reconcile himself to it, however, and in 1839 he published yet another classification system, *A Synopsis of the Birds of North America*, which offered his own way of organizing all the birds he had painted and described.[68] He would not let himself be left out of the natural history limelight.

Nor would he let himself rise altogether above the fray. For all his self-righteous sense of injury in what he claimed to be a one-sided attack by Ord and Waterton—"Selfishness has been the object of both, the charms of science in ornithology my only object"—Audubon never took complete account of his own role in the spiteful exchanges of the 1830s. The best he could do was to claim that "I never paid (personally) any attention to them through the press or otherwise," and that in his own work "not a word is there ... even in allusion to these beetles of darkness!" He could let sympathetic reviewers defend him in print, and he had valuable allies in several scientific societies who could present papers supporting some of his findings. In the meantime, he would pretend indifference to his enemies, hoping that they would be "greatly punished by my contemptuous silence."[69]

In 1834, for instance, Audubon sent a paper on vultures, written by his good friend John Bachman, to Richard Harlan, his loyal friend in Philadelphia, whom Audubon considered "a party concerned with my Warfare & welfare in this world." He asked Harlan to have it read before the Academy of Natural Sciences—and "have it read loudly, but without any flourish of trumpets or buggles—for after all Naked Truth is a Beautiful Demoiselle ... so very *respectable* that she will not even care about *The Happy & Most Learned Biographer of Alexr* Wilson."[70] Whether or not the members of the Academy wanted to choose sides in the Audubon-Ord "warfare," the paper about vultures came before them with Audubon's happy anticipation of leaving Ord's dead carcass on the field of combative ornithology. At the same time, Audubon could occasionally pretend that his enemies might be fading away. In 1834, he wrote of Waterton as a "dead letter," and a few years later he opined that "G. Ord is a poor devil after all, and perhaps not worth noticing."[71]

They never did fade away, however, at least not until Audubon himself did. In late 1852, almost two years after Audubon died, Ord could scarcely let his longtime nemesis rest in peace, much less on any laurels. He gave Audubon some credit for effort, writing to Waterton that he should be "fairly entitled to the merit of perseverance and industry. The elephant folio is a proof of that." But when it came to Audubon's talents as a naturalist and writer, Ord still gave no quarter: "Little reliance can be placed on his narratives, in consequence of an inveterate habit of mendacity, which should be Seen to have been the *primum mobile* of his intellect."[72]

By that time Audubon couldn't answer Ord's attack, but he hardly had to. He had already said essentially everything he could about his enemies in his correspondence and even in some of his ornithological writings, and the deep and unforgiving bitterness he felt toward them remains an inescapable part of his scientific and personal legacy.

But Audubon had another, and certainly better, legacy in science—and one very much of his own making. Stung and angered by the antipathy he had felt from Ord, Waterton, the "Philadelphians," and all the other "gentlemen of science" who had made his life so difficult, Audubon sought to separate himself from these despicable critics and their scientific snobbery. Instead, he embraced an alternative identity that stemmed directly from his personal involvement with the ordinary people of the new nation and his repeated exposure to the rugged regions of the American environment. Of all the prestigious titles he appended to his name—F.R.S., F.L.S., and so forth—the one he promoted most proudly was the one he gave himself, the "American Woodsman."

Chapter 6

~~~

Suffering for Science as the "American Woodsman"

Every student of nature must encounter some difficulties in obtaining the objects of his research, although these difficulties are little thought of when he has succeeded. So much is the case with me, that, could I renew the lease of my life, I would not desire to spend it in any other pursuit than that which has at last enabled me to lay before you an account of the habits of our birds.

—John James Audubon, "The Great Horned Owl,"
in *Ornithological Biography*

In June 1831, when Audubon left England on one of his specimen-gathering trips to the United States, a London newspaper reported on his departure with admiration for his ambition, but also with a quiet note of apprehension for his future: "This enthusiastic Naturalist is gone again to the woods. . . . It is his purpose to spend eighteen months or two years in exploring the western side of the Mississippi, up towards the Rocky Mountains. Should he survive, he intends returning to Edinburgh, and spending the rest of his days in arranging his collection, and publishing a continuation of his 'Ornithological Biography.'"[1] Audubon probably never had a chance to read this brief note, but if he had, he would have been pleased by seeing this sympathetic mention of the "enthusiastic Naturalist" setting off on a long and challenging trip to the West—not to mention the cautionary aside, "Should he survive." In Audubon's own unfolding narrative of himself as the American Woodsman, ornithology could be difficult and dangerous work, and only a courageous man of science would undertake doing it right. He had repeatedly faced the perils of his chosen profession, and he wanted the world to know it.

And the world was indeed coming to know it. By the early 1830s, Audubon was becoming a celebrity in the worlds of art and science, and newspapers on both sides of the Atlantic frequently commented on his comings and goings. When he passed through Philadelphia in September 1831, the Philadelphia *Gazette* reported, "In a few weeks he is off for the western and north-western forests, in whose pathless solitudes his nature appears to revel." A few weeks later, the Washington, DC, *Daily National Journal* praised Audubon, "whose mythological accomplishments have acquired him so high a reputation on both sides of the Atlantic," and it also took note of his proposed travels to Florida, the Gulf of Mexico, and then as far as the Rocky Mountains and perhaps even beyond. "This tedious, difficult, and perilous journey, Mr. Audubon has undertaken for objects purely scientific," the paper assured its readers. And so it went in newspaper after newspaper—not just updates on Audubon's many movements for the sake of science, but also admiration for the physical challenges he would face in pursuing his work.[2]

Throughout his life, he perfected his persona as the wilderness naturalist, the long-haired, buckskin-clad, gun-toting man of the woods, even on the far side of the Atlantic. Soon after he arrived in Great Britain for the first time, he became especially aware of his American identity, writing back to Lucy about his "sense of recollection" of his place in the world: "[Nothing] could make me relinquish the idea that in *my universe* of America, the deer runs free, and the Hunter as free forever."[3] As he became more widely known in artistic and scientific circles in the Atlantic world, he became increasingly committed to his embrace of America as his "universe" and to his self-styled image of himself as the "Hunter . . . free forever."

Audubon was not by any means the first American to adopt a woodsy-seeming identity when he went to the far side of the Atlantic—we think, of course, of the fur-hatted Benjamin Franklin in Paris in 1784—and Audubon's initial British image might be seen as only playing to type. To be sure, some people in Great Britain saw American excess in Audubon's guise and began to poke fun at it.

But Audubon clearly liked his own version of the look, and most of his subsequent portraits into the 1830s and 1840s depict him as a long-haired woodsman in the outdoors, almost always with a gun, never as a painter in the studio with an easel and brushes, never even with a dead bird ready to be drawn. For Audubon, costume spoke to character; the elements that became constants in his many portraits—wearing the garb of a man of the woods, displaying a gun as the critical tool of his trade, locating himself in the American

FANCY PORTRAIT, AUDUBON.

Figure 9. "Fancy Portrait: Audubon." From Thomas Hood, *Comic Annual* (London, 1836).

outdoors—defined the self-conscious core of his artistic and scientific identity (see Plate 4).

Manly Ornithology

Audubon may have posed, both in his paintings and in his own prose, as a rugged, outdoors naturalist embracing nature in its wildest forms, but it was not all pose alone. Audubon's approach to both art and science rested on one central premise: He drew the birds well because he knew the birds well.

Knowing the birds, of course, meant tracking them in the environment they inhabited, and that meant going into many dark and uncharted places:

> Many times, when I had laid myself down in the deepest recesses of the western forests, have I been suddenly awakened by the apparition of dismal prospects that have presented themselves to my mind. . . . At other times the Red Indian, erect and bold, tortured my ears with horrible yells, and threatened to put an end to my existence; or white-skinned murderers aimed their rifles at me. Snakes, loathsome and venomous, entwined my limbs, while vultures, lean and ravenous, looked on with impatience. Once, too, I dreamed, when asleep on a sand-bar on one of the Florida Keys, that a huge shark had me in his jaws, and was dragging me into the deep.[4]

In this one passage Audubon conjures up many of the menacing images that recurred in the long-standard literary descriptions of the American wilderness—dark forests, deadly quicksand, howling Indians, murderous backwoodsmen, loathsome snakes, ravenous vultures, with even a shark thrown in for special effect—that modern readers might now consider to be clichés. But he created this compendium of wilderness terrors not only to evoke a nightmarish vision of nature. Rather, he listed the rigors of the naturalist's life to impress upon the reader that there could be no other scientifically legitimate way to know nature. Even when taken at a discount for all their rhetorical excess, Audubon's many tales of the physical challenges of art and science in America contained an element of truth, and they certainly had a larger rhetorical point: The alternative to taking risks in the name of nature, he warned, would be to become that most sedentary of scientist, the "closet naturalist."

In Audubon's writings, "closet naturalist" recurred as a term of dismissive disrespect, typically applied to those gentlemen of science who gained most of their information from others, whether by acquiring a collection of specimens someone else had gathered or, even worse, by sitting comfortably in an armchair and reading from a printed text. Without actually getting out into the woods, Audubon wrote, "how difficult must it be for a 'closet naturalist' to ascertain the true distinctions of these birds, when, having no better samples of the species than some dried skins, perhaps mangled, and certainly distorted, with shriveled bills and withered feet."[5] Nothing could take the place of personal observation, of actually seeing the birds and counting the eggs, no matter what the personal difficulties. To be a true man of nature, then, meant taking a decidedly risky, even defiantly manly approach to the pursuit of science.

Audubon made masculinity part of the picture. Writing about his explorations of the woodlands to the west of Philadelphia, he described himself as "amazed that such a place as the great Pine Forest should be so little known to Philadelphians." He especially singled out for particular attention "the many young gentlemen who are there so much at a loss how to employ their leisure days," parading around town in their sartorial finery but not getting themselves out into nature. "How differently would they feel," he wrote, "if, instead of spending weeks in smoothing a useless bow, and walking out in full dress, intent on displaying the make of their legs, to some rendezvous where they may enjoy their wines, they were to occupy themselves in contemplating the rich profusion which nature has poured around them." Such feckless and implicitly feminine-seeming young men might partake of the specimens at Peale's museum, but they would never come close to encountering the fresh specimens an outdoor ornithologist could have at hand.[6] Audubon, by contrast, took care to assure his readers that he, Audubon the ornithologist, Audubon the artist, would pursue his calling with courage and face nature in its wildest form. On that point, Audubon himself had unmistakably made his choice, both for his primary activity as a naturalist and for his personal identity as the American Woodsman.

The Rigors of Research

Audubon was by no means the only American naturalist to get his hands dirty outdoors. Others may not have written so eloquently about the experience of studying birds and other aspects of nature, but they also engaged in the rugged, risk-taking manliness Audubon eventually came to celebrate in himself. Suffering for science became a badge of honor that separated the relative few who did rigorous field work from those who confined themselves to the indoor comforts of the "closet."

Consider, for instance, the case of Thomas Nuttall (1786–1859), a young, British-born naturalist who came to the United States in 1808 and settled in Philadelphia, where he became anything but the sort of dandified young gentleman Audubon later dismissed so scornfully. Indeed, Nuttall soon turned out to be a very effective field naturalist and eventually one of Audubon's valuable ornithological allies. He would work first, however, as a scientific apprentice for one of the leading lights in the Quaker City, Dr. Benjamin Smith Barton at the University of Pennsylvania. Barton had long been esteemed as one of the foremost men of science in America, but by the time Nuttall arrived in the United States, he no longer had the time or endurance to do extensive field work on his own. Nuttall, though, would do it for him.[7]

In the summer and fall of 1809, Barton sent the young naturalist on two short exploratory trips: one to Delaware (where Nuttall contracted malaria, or "ague," which would weaken him for subsequent expeditions) and another through New York, Niagara Falls, and Ontario. But in 1810, Barton had bigger plans for Nuttall, contracting with him to go all the way west into the Illinois country to collect plants and study all aspects of nature.[8] "You are to keep an exact Journal of your daily travels, and of the observations you may make," Barton instructed his young assistant. But then, in a clear declaration of intellectual property rights, Barton reminded Nuttall about who was really in charge and who would claim the scientific fame: "This journal, and all the observations you may make, are to be my exclusive property; and no part are to be communicated, without my consent, to any person." Barton did promise that, should he subsequently publish Nuttall's journal, he would "make a public acknowledgement, that the journey was performed by you, and . . . give you full credit for what services you may have rendered me."[9]

This sort of arrangement may seem strikingly familiar to many modern graduate students in the sciences who do the labor in the lab but still find themselves listed as distant coauthors on their professors' publications. What makes Barton's charge noteworthy, though, is the physical challenge he warned Nuttall would face on this long journey. To get from Detroit to Chicago and from there "to the Winnebagoes, and from thence to Superior," Barton told Nuttall he would no doubt have to get an Indian guide, advising him to "be cautious to obtain the best information, as to the character for integrity, sobriety, &c., of the guide." No guide, of course, could protect Nuttall from all the dangers of this trip so deep into the interior. That was a risk he would have to take in the name of science: "Always remember that, next to your personal safety, science, and not mere conveniency in travelling, is the great object of the journey. In pursuit of curious or important objects, it will be necessary to court difficulties, in traveling, etc."[10] With that, sounding a bit like a sterner professorial version of Audubon's American Woodsman, Barton signed off and sent Nuttall into the wilds.

And off Nuttall went, for what turned out to be an expedition that would take twenty months and cover thousands of miles. He left Philadelphia on April 12, 1810, taking heart in "the smiling aspect of *spring* . . . [when] the apathy of the vegetable is more than made up by the pleasing exertions of the feathered songsters." But the birdsongs soon left his heart, and soon his "eye wearied with the barren prospect of rocky mountains & black forests." Just ten days into the trip, when he came to Pittsburgh, the sight of this frontier town provided an already welcome alternative to the woods, allowing him to view "with satisfaction & delight the verdant meads & smoking cottages, where ere the fierce savages held their nocturnal dances, & the ancient forests planted by nature raised their towering heads into the skies."[11] Still, the darkness of the "nocturnal dances" of the former native inhabitants no doubt reflected the darkness of Nuttall's mood.

But he had his orders, and he pressed on, armed to the teeth with a fearsome array of nineteenth-century firepower—a double-barreled gun, a pistol, a knife, a pound of powder, a powder horn, and shot belt—none of which he ever learned to use to much effect. His lack of skill with the gun, that most necessary of scientific instruments in the field, sometimes left him guessing about the birds he saw. In early May, he saw "perched upon a tree in a swamp . . . a bird which to me seemed uncommon, by its long slender cylindrical bill, I should think it belonged to the *Grallae*. Its head & bill, were black, it had a conspicuous white circle round its neck, all the rest of the body which I could see was of a bluish ash-color. It had little or no tail. In flying it skimmed the surface of the water. *Alcedo Halcyon*?"[12] A Belted Kingfisher it no doubt was, and Nuttall got it—and he got it the way an amateur birder would today, by looking at its markings and observing its flight, an especially impressive sight identification in Nuttall's case, performed as it was without the benefit of binoculars and a pocket field guide. A week later, he saw a "fine songster about the size of the blue-bird, with a black head & red throat the rest of the feathers partly black & white." This time, though, he couldn't identify the bird—almost certainly a Rose-breasted Grosbeak—because he didn't shoot it and examine it closely as a specimen, as Audubon and most other nineteenth-century naturalists would have done.[13] Still, by the end of his expedition—which took him from Philadelphia to Pittsburgh to Detroit, around the Great Lakes and down the Mississippi to St. Louis, up the Missouri to the Mandan villages in the Yellowstone region, then back down the Mississippi to New Orleans— Nuttall did collect an extensive list of forty bird species, along with numerous insect, fish, reptile, and mammal species.[14]

After Barton's death in 1815, Nuttall undertook his own exploratory journeys, primarily to gather botanical specimens, going first to Virginia, Georgia, and the Carolinas in 1815–1816; then across Pennsylvania to Kentucky, Tennessee, and South Carolina in 1816–1817; and, after a little over a year and a half back in Philadelphia, down the Ohio and Mississippi rivers to the Arkansas and Red rivers in 1818–1820. On that latter expedition, in fact, he wanted to go as far as the Rocky Mountains, but the rigors of the trip took a toll on his health, leaving him fevered, famished, and faint.[15] All this suffering for the sake of plant and animal specimens seemed quite a sacrifice for science, especially on the part of a man who still could not swim, shoot well, or even build a decent campfire.[16] But Nuttall relished the experience, even though it nearly killed him.

Alexander Wilson also suffered serious physical challenges in doing his work for *American Ornithology*, and he, too, paid a price for the stressful challenges of field research. As Audubon did later, Wilson occasionally portrayed the American naturalist's work in strenuous-sounding terms, casting himself as a man of science struggling against the rigors required in research. He wrote about it to his brother in 1810: "Since February, I have slept for several weeks in the wilderness alone, in an Indian country, with my guns and my pistols in my bosom, and have found myself so reduced by sickness, as to be scarcely able to stand."[17] Wilson also described almost losing his life in the pursuit of a Pied Oyster-catcher (now called the American Oystercatcher) when he took to the water in pursuit of the bird at Cape May, New Jersey. He had wounded it with his gun, and as the bird tried to escape into the ocean, he plunged in after it—only to remember, too late, that he was still "encumbered with a gun and all my shooting apparatus." As the ebb tide started carrying him farther away from the shore, the overweighted Wilson had to choose between his own survival or his escaped specimen, and, almost reluctantly, he made his way back to the beach "with considerable mortification, and the total destruction of my powder-horn." As if to mock him, the wounded shorebird rose to the surface "and swam with great buoyancy out among the breakers."[18] Oystercatchers are not especially menacing birds, but for Wilson, it was the pursuit of the bird, the pursuit of science, that mattered. Living in the woods or diving into the ocean was the only way Wilson would have it.

In language Audubon could well appreciate, Wilson also complained of the tendency among naturalists to become too technical, to spend so much time separating birds into so many "Classes, Orders, Genera, Species, and

Varieties" that the resulting complexity and disagreement had "proved a source of great perplexity" to ordinary people. One of the main reasons for this confusion, Wilson continued, was the failure, or perhaps refusal, of other naturalists to observe personally "the manners of the living birds, in their unconfined state, and in their native countries."[19] The naturalist had to get out into nature, into the woods or even the ocean, to see the birds and do good work.

Good work turned into bad health for Wilson, who suffered from physical stress and sickness, finally dying of dysentery in 1813, at the age of forty-seven. In death, he received a scientifically supportive eulogy from his Philadelphia friend and ally, George Ord, who described Wilson as an outdoor ornithologist and certainly "no closet philosopher—exchanging the frock of activity for the night-gown and slippers." Wilson's knowledge, Ord explained, came not from reading books about nature, "which err," but from engaging nature itself, "which is infallible."[20] Ord's celebration of Wilson drew on some of the same masculine elements Audubon would later employ—in this case, the contrast between the woodsman's "frock of activity" and the more effeminate-sounding "night-gown and slippers" of the "closet philosopher," between knowledge derived from the "unwearied research amongst forests, swamps, and morasses" and the precious little that one could learn in the library. For the "Father of American Ornithology," as Wilson would come to be called, the true source of scientific authority could be found only in the field.

Guns in the Company of Men

No one pursued birds or science or, above all, the image of the manly naturalist more aggressively or effectively than Audubon himself. Never to be outdone by his ornithological rival, Wilson, he told his own tale about pursuing a Great Horned Owl so far into a swamp that he got "sunk in quicksand up to my armpits," only to be rescued at the last minute by his companions. The story may or may not have been exactly accurate, but it had a purpose: "I have related this occurrence to you, kind reader,—and it is only one out of many—to shew you that every student of nature must encounter some difficulties in obtaining the objects of his research, although these difficulties are little thought of when he has succeeded."[21]

Audubon defined ornithology as manly work, and he also liked to locate himself in the world of men, particularly the long-rifle hunters of the American frontier. Audubon certainly admired a good shot. In the chapter on "Kentucky Sports" in *Ornithological Biography,* he described skilled Kentucky sharpshooters who competed in a variety of gun-based games, using a bullet to drive a nail into a tree or snuff out a candle. Such outdoor entertainments gave "all the sportsmen" an opportunity to engage each other afterward, "to adjourn to some house, and spend an hour or two in friendly intercourse, appointing, before they part, a day for another trial." The combination of competition and camaraderie defined good fellowship among the frontiersmen of Kentucky, and it was the gun that gave them the measure of their male identity: "Every one in the State is accustomed to handle the rifle from the time he is first able to shoulder it until near the close of his career."[22] Audubon was a Kentuckian, too, at least for part of his life, and he certainly valued good shooting as a significant measure of male prowess.

He also valued a good story, and none served him better than the one he told about his old Kentucky backwoods hero, the "daring hunter, the renowned Daniel Boon." Boone proved to be especially adept at the frontier amusement called "barking off squirrels"—"a delightful sport," Audubon called it, especially when the sportsman doing the barking was Boone, and especially when Audubon claimed to go with the famous frontiersman into the woods, where "squirrels were seen gambolling on every tree around us." "My companion, a stout, hale, and athletic man, dressed in a homespun hunting-shirt, bare-legged and moccasined, carried a long and heavy rifle, which, as he was loading it, he said had proved efficient in all his former undertakings, and which he hoped would not fail on this occasion, as he felt proud to show me his skill." In Audubon's telling, the old hunter carefully wiped and loaded his weapon and then pointed to a squirrel on a distant tree and told Audubon to "mark well the spot where the ball should hit." Boone took careful aim and fired, the shot echoing through the surrounding woods and hills. "Judge of my surprise," Audubon wrote, "when I perceived that the ball had hit the piece of the bark immediately beneath the squirrel, and shivered it into splinters, the concussion produced by which had killed the animal . . . as if it had been blown up by the explosion of a powder magazine."[23] That was good shooting—Boone killed the squirrel without actually hitting it—and Audubon was duly impressed with such a dramatic display of marksmanship.

Audubon also liked to associate himself with such a celebrated marksman and icon of frontier masculinity, even when the hunting was done. He told a story about how he once "happened to spend a night . . . under the same roof"

Plate 1. *John James Audubon*, by John Syme, 1826. Courtesy of the White House Historical Association (White House Collection), Washington, DC.

Plate 2. *Belted Kingfisher,*
by Alexander Wilson. From
American Ornithology, 9 vols.
(Philadelphia, 1808–1814),
plate 23.

Plate 3. *Belted Kingfisher*
(*Megaceryle alcyon*), by John
James Audubon, 1808. Pastel,
graphite, and ink on paper, 22 x
41 centimeters. MS Am 21 (50),
Houghton Library, Harvard
University.

Plate 4. *John James Audubon,* by G. P. A. Healy, 1838. Courtesy of the Museum of Science, Boston.

Plate 5. *Golden Eagle (Aquila chrysactos)*, by John James Audubon, 1833. Watercolor, pastel, graphite, and selective glazing on paper, 38 x 25 1/2 inches. Object #1863.17.181. Courtesy of the New-York Historical Society.

Plate 6. *Golden Eagle (Aquila chrysactos)*, detail, by John James Audubon, 1833. Object #1863.17.181. Courtesy of the New-York Historical Society.

Plate 7. *Golden Eagle,* by John James Audubon. From *The Birds of America*, Royal Octavo First Edition (New York and Philadelphia, 1840–1844), no. 3, plate 12.

Plate 8. *Great Black-backed Gull (Larus marinus)*, by John James Audubon, 1832. Watercolor, graphite on paper, with mat, 39 x 53 inches. Havell plate 241. Object #1863.17.241. Courtesy of the New-York Historical Society.

Plate 9. *House Wren (Troglodytes aedon)*, by John James Audubon, ca. 1824. Watercolor, graphite on paper, with mat, 29 x 23 inches. Havell plate 83. Object #1863.17.83. Courtesy of the New-York Historical Society.

Plate 10. *Purple Martin (Progne subis)*, by John James Audubon, 1822. Watercolor, graphite on paper, with mat, 27 x 35 inches. Havell plate 22. Object #1863.17.22. Courtesy of the New-York Historical Society.

Plate 11. *Sora (Porzana carolina)*, by John James Audubon, 1821, 1833. Watercolor, graphite, collage, pastel, black chalk, black ink, and gouache on paper, laid on card, 23 x 29 inches. Havell plate 233. Object #1863.17.233. Courtesy of the New-York Historical Society.

with Boone. Like a starstruck fan of a famous athlete, Audubon gushed about his hero's skills with a gun and his physique: "The stature and general appearance of this wanderer of the western forests approached the gigantic. His chest was broad and prominent; his muscular powers displayed themselves in every limb; his countenance gave indication of his great courage, enterprise, and perseverance." When the day's "shooting excursion" had concluded and night fell, Boone declined to sleep in a bed, Audubon said, but "merely took off his hunting shirt, and arranged a few folds of blankets on the floor, choosing to lie there than on the softest bed."[24] And so the two men bedded down for the night—Audubon presumably in a bed, but Boone shirtless and on the floor—preparing themselves to resume the hunt the following day.

Audubon's Boone story was a good one, to be sure, but not a true one—a fantasy, perhaps, but also a complete fiction. Audubon did once write to Boone and ask to go hunting with him, in 1813, but Boone turned him down. Boone was almost eighty and almost blind at the time, an aging woodsman of smaller and weaker build than in his prime—hardly the sort of partner to join Audubon on a hunting excursion.[25] An 1834 review of Audubon's work in the *Western Monthly Magazine* questioned the possibility of the Audubon-Boone story, noting that Boone hardly met the standard of "gigantic" but "would have been called a fair example of the *middle-sized* man." The critic went on to observe that "this propensity for exaggeration is apparent throughout Mr. Audubon's book . . . and no western man can read his descriptions without a smile of incredulity."[26] Audubon had little fear of anyone's incredulity, though, and that underscores an important point: To read Audubon for the absolute truth is to miss his larger meaning. Audubon's agenda, rather, was to create an effect (or an affectation): the image of the naturalist as a "wanderer of the western forests," just as bold as Daniel Boone, just as good with a gun. If a tale about going shooting with Daniel Boone would make the point in print, then it would work well enough.

In real life, no one came to embody Audubon's love of gun-based male bonding better than John Bachman, whose skills as a naturalist and a shooter placed him in the pantheon of Audubon's ornithological associates. When Audubon first met Bachman in Charleston, in October 1831, the two hit it off immediately. In addition to being "an ornithologist, [and] a philosophical naturalist," Audubon wrote, Bachman possessed another important scientific skill: "He shoots well." Bachman took Audubon out to the countryside in search of birds, and he quickly became "everything a kind brother could be to

Figure 10. *Portrait of Daniel Boone by John James Audubon. Painted in Missouri a Short Time Before Boone's Death.* Audubon Museum, Audubon State Park, Henderson, Kentucky.

me."[27] Audubon eagerly told Lucy about the manly scientific activities he and
Bachman pursued during his stay in South Carolina: "Out shooting every
Day—Skinning, Drawing, Talking Ornithology The whole evening, noon,
and morning."[28] It seemed an ideal existence for Audubon, and he reveled in
the camaraderie of the occasion. The physical and social experience of being
an artist and scientist in company with a fellow naturalist made Audubon as
happy as any hunter out with his favorite companions. Audubon quickly
came to consider Bachman a close friend and equally close watcher of birds,
and the phrase "my friend JOHN BACHMAN" would recur repeatedly in
Ornithological Biography, almost always to cite the learned Lutheran minister as a source of ornithological authority and, on occasion, masculine
outdoor activity.[29]

Bachman's most extended appearance in Audubon's writings came in the
chapter "The Long-billed Curlew," when Audubon recounted an excursion in
November 1831, with "several friends," including Bachman, and some "servants,"
or slaves. At night, when the party retired to its shelter for supper, they had
plenty of good food with them—"Fish, fowl, and oysters . . . some steaks of
beef, and a sufficiency of good beverage"—but no one to serve as cook, Audubon added, "save your humble servant." Taking charge of the food preparation,
Audubon noted one other missing ingredient, salt, but he had a solution: "I
soon proved to my merry companions that hunters can find a good substitute
in their powder flasks. Our salt on this occasion was gunpowder, as it has been
with me many a time." Sprinkling gunpowder on the various meats, the hunters went at the meal with their hands, having no eating utensils (yet another
omission from the packing list), and soon their fingers and faces were smeared
with "villainous saltpetre," a mess that "only increased our mirth." After a
couple of days of shooting, the use of gunpowder as a substitute condiment
became almost a ritual of the hunters' feast, a way of taking into their bodies
the substance that had given their guns the power to kill. After that, they "spread
out our blankets on the log floor, extended ourselves on them with our feet
towards the fire, and our arms under our heads for pillows," and had a deep,
sound sleep.[30] In this narrative, bedding down with Bachman and his friends
echoes Audubon's alleged overnight encounter with Daniel Boone, but this
story has a much better chance of being true. If Audubon couldn't really go
hunting with Daniel Boone, John Bachman would certainly do.

This manly companionship with Bachman comes into sharper focus with
the appearance of another member of the Bachman household in Audubon's

writings, Maria Martin, Bachman's sister-in-law (and, later, wife). For all his undeniable self-promotion as an artist, Audubon could not claim complete credit for every brushstroke that made up his images, and he frequently employed background painters, especially for botanical detail. Maria Martin became one of his most valuable assistants in that capacity. When Audubon first met Martin at Bachman's, in 1831, she was already doing drawings for Bachman, a task she would continue for years. "She is my right hand still," Bachman wrote Audubon, "paints for me, keeps butterflies and even toads and snakes if I should wish them."[31] She also drew for Audubon for years, well into the early 1840s, and Bachman assured him she loved the work: "Drawing is an amusement to her and to gratify you will always afford her pleasure."[32] Audubon referred to her in print as "my amiable friend Miss Martin," but in his correspondence with Bachman he took a more playful and essentially patronizing tone: She was almost always "our sweetheart," a term that skirted any notion of professional collaboration, much less compensation. Audubon once complained, perhaps somewhat humorously, that "our sweetheart . . . may be becoming somewhat slack in ornithological delights or still worst on botanical ones," but he blithely added that she "may be forgiven, being one of the *slenderer sex*."[33] Slack or not, though, she still supplied the artwork Audubon needed, and sometimes he needed it immediately. "Tell our Sweetheart that *whatever* Drawings She May have on hand and ready for me, the Sooner She Sends them the better," he wrote Bachman in early 1836, when he was beginning his last volume of *The Birds of America*. "I will pay her in Kisses at least when I Meet you all."[34] Audubon's school-boyish snickering about kisses for his "sweetheart" may have concealed a deeper sexual attraction for Maria Martin—one modern novel imagines a long-distance love affair, asking, "Who is this woman who haunts him from south to north, she who waits in hot Charleston painting butterflies and flowers?"[35]—but Audubon seems to have stayed safely on the male side of the line with John Bachman, his good friend and, almost equally important, fellow good shot. That may well have been the ultimate point for Audubon: The proof of any man lay in his shooting—something Audubon himself did often and well.

Sighting Down a Scientific Instrument

Bewick's Wren is a smallish, mostly brown bird that, like other wrens, has an upward-pointing tail and a downward-curving bill and feeds primarily on insects. It lives year-round in the western parts of the United States, from Texas across to California and up the Pacific Coast to the southwestern corner of Canada, but its winter range also includes a little bit of Louisiana. It was there, "about five miles from St. Francisville, in the State of Louisiana," that Audubon first saw this bird, which he "shot on the 19th October 1821" (about eight months after painting the nude picture that got him his *souvenir* firearm in the first place). Thus begins his description of the Bewick's Wren in the first volume of his *Ornithological Biography*, noting the specific date and place of the bird's death and, as it happens, of its depiction: "My drawing of it was made on the spot." Audubon then goes on to recall that some eight years later, in November 1829, "I had the pleasure of meeting with another of the same species" in almost the same place. This time, though, he initially "refrained from killing it, in order to observe its habits."[36]

In the passage that followed, Audubon does indeed seem to take pleasure in looking at the little wren, putting himself in the picture while he watches the bird's quick movements hopping along a rail fence, flying to a nearby tree, catching insects, and sometimes "uttering a low twitter," then returning to the fence to "continue its avocations already described." In a sense, Audubon's passage makes a verbal circle of the bird's recurring "avocations," creating the impression that its hopping about from fence to ground to tree and back again could go on indefinitely. But then, in the next short sentence, without any word of transition, much less warning, Audubon pulls the trigger on both the wren and the reader: "I shot the bird, and have it preserved in spirits."[37]

That was the way he worked. For Audubon to depict the bird, to preserve it not just in spirits but in paint and in words, he had to kill it. In fact, this understated and certainly unapologetic account of the death of an individual wren represents just one of many such scenes Audubon offers his readers: His writings, especially his journals and published works, are full of shooting and killing birds. More to the point, this one wren is just a single specimen among the thousands, probably tens of thousands, of birds Audubon killed. In discussing his method of drawing birds, Audubon explained that "the birds, almost all of them, were killed by myself, after I had examined their motions and habits, as much as the case admitted, and were regularly drawn on or near the spot where I procured them."[38] Audubon's encounter with the Bewick's Wren offers a perfect illustration of this approach: Again, he "refrained from killing it, in order to observe its habits," but then once he had seen enough, he shot it and did his drawing "on the spot."

Sometimes, of course, he did more than observe and shoot a single bird for the sake of art and science. Audubon loved to go out shooting for the sheer enjoyment of the hunt and, in some cases, the shared experience with his fellow hunters. He wrote a provocative passage about a shooting expedition on the St. John's River in Florida, when he and some sailors went out to hunt Bald Eagles. When they approached a tree occupied by two adults and two immature eagles, "the old ones flew off silently, while the young did not seem to pay the least attention to us, this being a part of the woods where probably no white man had ever before put his foot, and the Eaglets having as yet had no experience of the barbarity of the race." Whatever form the "barbarity of the race" might take, Audubon did not let any qualms deter him from his desired endeavor, or from making note of the marksmanship of the men on the hunt. Visiting another nest, they bagged three eaglets in four shots, using a telescope to aid their long-range shooting.[39]

Audubon and his naval companions could congratulate themselves on being as good marksmen as any backwoodsman barking a squirrel. And congratulate themselves they did, by making a meal of the birds when they returned to their ship: "In a few hours these young birds were skinned, cooked, and eaten, by those who had been 'in at the death.'" (On a culinary note, Audubon observed that the young eagles "proved good eating, the flesh resembling veal in taste and tenderness.") But not all of the men "in at the death" remained in at the dinner: Audubon also notes that "one of us only did not taste of the dish, simply I believe from prejudice."[40] He thus separates himself and most of his companions from the sole other, the "one of us only" who did not partake of the ritual feast and who excluded himself—or perhaps was excluded by the others—because of "prejudice," perhaps even principle. This unnamed member of the group serves as the exception whose difference helped define the bond that brought the others together, both in the field and at the table: Those who had been "in at the death" shared not just a meal, but a male experience as fellow hunters, as fellow men.

In some cases, though, the thrill of the hunt took an excessive, almost obsessive turn. Many Audubon observers, both admirers and critics alike, have taken note of his unabashed boast about his seemingly insatiable appetite for shooting birds. Once again, when he was in Florida collecting specimens, he wrote that "I and four negro servants proceeded in search of birds and adventures," and Audubon, at least, got both. They rowed their boat into a narrow bay, where they "came in sight of several hundred pelicans, perched on the branches of mangrove trees, seated in comfortable harmony, as near each other as the strength of the boughs would allow." Seeing so many pelicans packed so close together gave Audubon an opportunity he could hardly pass up, and he began shooting as fast as he could, "bringing down two of the finest specimens I ever saw" with his first shot. "I really believe I would have shot one hundred of these reverend sirs, had not a mistake taken place in the reloading of my gun." Because of the mistake, though, the pelicans flew away, and Audubon complained of his disappointment with the rest of the day's shooting: "The birds, generally speaking, appeared wild and few," and he then explained that "I call birds few, when I shoot less than one hundred per day."[41]

In this instance and in several others, Audubon's excitement could lead him to apparent excess, and whatever artistic or scientific need he had for specimen gathering could turn into what might strike some modern observers as overly aggressive greed. This sort of story certainly might not always sit well with many people who become aware of Audubon's approach to ornithology for the first time: The image of the artist armed, the painter as predator, can be jarring and disconcerting to someone whose first Audubon encounter comes through the beauty of his avian images. But in Audubon's defense, there could be scientific method in the massacre. Dead birds make a crucial contribution to what can be known about live birds, helping disprove erroneous assumptions and provide new information. Modern ornithologists have at their disposal detailed bird guides, high-powered binoculars, spotting scopes, cameras, and mist nets to help them observe birds in the field. Sight records are sometimes inadequate, and even the most skilled scientist can make occasional identification mistakes. To be more certain, they still rely on dead specimens, birds in the hand, as it were, to give them a wide range of ornithological information: taxonomy, anatomy, plumage, diet, reproduction, migration, geographic variation, and a host of other issues. No matter what some nonscientists may think, ornithologists still cannot know everything there is to know about birds. Thus the collecting still continues, and birds sometimes have to die in the process.

No one could argue in Audubon's era that ornithologists already knew everything there was to know about birds. Though several hundred years old, the field was still in its scientific infancy, particularly regarding the study of the birds of North America. Especially for an artist-naturalist like Audubon, who wanted not just to study birds but to extend the list of all the avian species in America, the collection of fresh specimens became an absolutely essential part of a national task. Audubon knew that proof lay in a pile of carcasses. In answer to European naturalists who doubted the presence of night herons in the United States, for instance, Audubon said, "I wish these people had been with me and my friend Bachman. . . . How strange it would have appeared to such assertors of notions to have seen a boatload of Night Herons shot in the course of a few hours, and that too in the winter season."[42] Later, taking aim at terns near Key West, he noted that he and his crew "continued to shoot until we procured a very considerable number." When the shooting stopped and he began picking the dead birds out of the water, he suddenly realized that he discovered a new specimen among them: "I perceived from the yellow point of its bill that it was different from any that I had previously seen, and accordingly shouted 'A prize! a prize! a new bird to the American Fauna!' And so it was, good Reader, for no person before had found the Sandwich Tern on any part of our coast."[43] With a new bird not just for his own life list, but for that of the nation, Audubon could call that a good day of shooting, indeed.

Ornithology in Captivity

One particular specimen, however, puts the rest of Audubon's bird killing in a different light, offering an arresting exception to the rule. In the winter of 1832–1833, while he was working in Boston and waiting to go on a trip to Labrador, he depicted the Golden Eagle, a dramatic-looking image that turned out to be an enormous struggle, as much emotionally draining as it was artistically challenging. Audubon's several accounts of his encounter with the Golden Eagle, drawn from both *Ornithological Biography* and personal correspondence, give us a revealing insight into Audubon's coming to terms with the way he worked—or not, as the case may be.

As Audubon tells it, a Boston museum director named Ethan Allen Greenwood offered him a live specimen of a Golden Eagle, which had gotten itself caught in a fox trap in New Hampshire and had then been brought to Boston in a cage. When Greenwood showed the bird to Audubon, it was a case of ornithological desire at first sight: "I directed my eye towards its own deep, bold and stern one . . . and I determined to have possession of it." Audubon bought the bird, covered the cage with a blanket "to save him from the gaze of the people," he explained, and took the bird back to his quarters.[44] From there he wrote his son Victor about his "great gratification of purchasing this day the finest *Golden Eagle* alive which I ever have seen." Better still, the bird was a bargain: "I paid $14.75 a pretty good price."[45]

Keeping a large bird of prey in one's rented rooms might strike many people (certainly innkeepers, above all) as strange and dangerous behavior, but to Audubon and some of his fellow naturalists it seemed very much within the range of common practice. One of the most famous stories in the history of early American ornithology concerns Audubon's fellow artist-naturalist Alexander Wilson and an Ivory-billed Woodpecker. Wilson first encountered this magnificent (and now probably extinct) bird near Wilmington, North Carolina, and his initial response upon first sighting a new species was, as usual, to shoot it. He only wounded the woodpecker, though, and when he went to capture it, the bird began to emit, as Wilson tells it, "a loudly reiterated, and most piteous noise, exactly resembling the violent crying of a young child." Wilson covered it up and went on into Wilmington to get a room for the night, and since the woodpecker cried like a child, Wilson pretended that it was one: He asked a hotel landlord "whether he could furnish accommodations for myself and my baby." When he finally uncovered the "baby" and revealed it to be a huge woodpecker, "a general laugh took place" among Wilson and the hotel keeper and the other people on the scene. Neither Wilson nor the landlord would be laughing for long, however. As soon as Wilson let the woodpecker loose in his room, the bird began hammering at the walls, breaking off big chunks of plaster and knocking a hole in the lath. Wilson tried to restrain the big bird by tying a string around his leg, but the woodpecker then went to work on "the mahogany table to which he was fastened, and on which he ... wreaked his whole vengeance," and he even drew blood from Wilson himself. Injured and no doubt further in debt to the hotel keeper, Wilson nonetheless found himself impressed that the bird "displayed such a noble and unconquerable spirit that I was frequently tempted to restore him to his native wood." But he didn't. He kept the woodpecker, which "lived with me nearly three days, but refused all sustenance, and I witnessed his death with regret."[46]

Audubon tells several similar stories, and most of them have an equally unhappy ending. When he was living in Henderson, Kentucky, he kept a Wild Turkey and a Trumpeter Swan as pets for a while at his home, although the swan kept running away and finally succeeded in escaping for good. Audubon later wrote ruefully that "pet birds, good Reader, no matter of what species they are, seldom pass their lives in accordance with the wishes of their possessors." The same turned out to be true with injured birds. Audubon once took in a Swallow-tailed Hawk (or Swallow-tailed Kite) he had wounded in the wing, and even though he tried force-feeding it, the hawk "refused to eat, kept the feathers of the head and rump constantly erect, and vomited several times part of the contents of its stomach." After a few days, the defiant bird "died from inanition." Something similar happened to a smaller bird, a Bonaparte's Fly-Catcher (or Canada Warbler) that Audubon wounded in Louisiana. "I carried it home," Audubon wrote, "and had the pleasure of drawing it while alive and full of spirit." When the bird became unhappy about being cooped up in a cage, Audubon let it fly around the room. But indoor captivity was still too confining, and the next day Audubon found the bird "very weak and ruffled up, so I killed it and put it in spirits." Thus the small bird went from being "full of spirit" to being put in spirits in less than a day, and its fate underscored Audubon's observation that keeping birds in captivity seldom made for a happy ending.[47]

Ornithological Gothic

Audubon's indoor encounter with the Golden Eagle turned out to be even worse—an experience that proved especially disastrous for both of them. Even though Audubon wanted to keep his newly bought bird away "from the gaze of the people," he himself went eye-to-eye with the eagle again as soon as they arrived back in Audubon's rooms: "I placed the cage so as to afford me a good view of the captive, and I must acknowledge that as I watched his eye, and observed his looks of proud disdain, I felt toward him not so generously as I ought to have done."[48] This intense watching, this looking each other in the eye, had an unsettling effect on Audubon. As he wrote Victor soon after buying the bird, "The eye—aye—the Eye of Birds is like that of Man, each has its own peculiarities."[49] In this case the "peculiarities" of this eye-to-eye equality left the human captor feeling uneasy, even threatened, by the "proud disdain" in the eye of his aquiline captive.

At first, in fact, Audubon even considered letting the eagle go: "I several times thought how pleasing it would be to see him spread out his broad wings and sail away toward the rocks of his wild haunts" in New Hampshire. But this vision of seeing the eagle flying free soon was interrupted by an unspoken, disembodied, and apparently silent suggestion: "But then, reader," he says, "some one seemed to whisper that I ought to take the portrait of the magnificent bird"—and painting it meant killing it. Most notably, that sentence signals that Audubon did not take responsibility himself for deciding to paint—or to kill—the eagle: "Some one seemed to whisper," Audubon says. He explains that he did what he did for the sake of the "reader," and that sealed the bird's fate: "I abandoned the more generous design of setting him at liberty, for the express purpose of shewing you his semblance."[50] In a sense, Audubon portrays himself primarily as an intermediary between the whispered suggestion of this disembodied "some one" and the apparent need of the "reader" to have the eagle's image preserved in paint.

Even the initial approach Audubon took to killing the bird appears inadvertent and comparatively passive: After having decided to paint the bird, Audubon spent a whole day just watching its movements—again, always watching. He then spent another day deciding how he wanted to position the bird for the painting—this time seeing it in his mind's eye. And only on the third day did he decide how to kill it—and even here, he still failed to take full responsibility himself. Audubon consulted with his "most worthy and generous friend" Dr. George Parkman, and they talked about various approaches to death—electrocuting the bird or suffocating it with smoke—both ultimately concluding that death by suffocation "would probably be the easiest for ourselves, and the least painful to him." So Audubon tried to suffocate the eagle by putting its cage—or "his prison," as Audubon calls it—into a small room, covering the cage closely with a blanket, putting a pan of burning charcoal in the room, closing the door and windows tightly, and waiting for the eagle to die.[51]

It didn't work. After a few hours, Audubon writes that he "opened the door, raised the blankets, and peeped under them amidst a mass of suffocating fumes." There the eagle still stood, Audubon continues, "with his bright unflinching eye turned towards me, and as lively and vigorous as ever!" Audubon then shut the door and continued trying to smoke the eagle for the whole night, but all he succeeded in doing was to make the apartment all but insufferable for himself and his son John, who was helping him. The next morning, to make the fumes even more toxic, Audubon added some sulfur to the smoldering charcoal, but again the "stifling vapors" had a worse effect on both Audubons than on the bird: "The noble bird continued to stand erect, and to look defiance at us whenever we approached his post of martyrdom." Finally, to finish off the defiant bird and to make the martyrdom complete, Audubon "thrust a long pointed piece of steel through his heart, when my proud prisoner instantly fell dead, without even ruffling a feather." This final encounter with the eagle ruffled Audubon's own feathers: "I sat up nearly the whole of another night to outline him and worked so constantly at the drawing, that it nearly cost me my life. I was suddenly seized with a spasmodic affection, that much alarmed my family, and completely prostrated me for some days."[52] It might appear Audubon was embellishing a bit here, as he did so often in *Ornithological Biography*, playing up the intensity of his struggle with the Golden Eagle to add a little drama to please his readers. In fact, one student of this story draws an explicit parallel between Audubon's tale and Edgar Allan Poe's story "The Tell-Tale Heart," in which Poe's narrator feels the urge to kill an old man because the old man's eye refuses to let him go: "One of his eyes resembled that of a vulture," Poe writes, "a pale blue eye, with a film over it." The stare of a large, apparently menacing bird makes a chilling image in Poe's tale, and a nineteenth-century reader of Audubon's account might likewise detect an element of popular gothic fiction in the tale of the Golden Eagle. But such comparisons aside, Audubon did not crib his plot line from Poe: "The Tell-Tale Heart" came into print in 1843, several years after Audubon's own tale appeared in *Ornithological Biography*.[53]

Audubon's personal correspondence helps corroborate the truth of his agony. In early February 1833, Audubon wrote to Victor, who was in England overseeing the engraving of the paintings, and the exhausted artist had no real reason to play to an audience: "I am just now quite fatigued by the drawing of the Golden Eagle," he wearily admitted, "which although it will make a splendid plate, has cost me sixty hours of the severest labor I have experienced since I drew the Wild Turkey."[54] Six weeks later, when he had finally finished the painting and recovered a bit, he summoned up enough bluster to write his good friend Richard Harlan and tell him all about his "experiments" with the Golden Eagle and, never forgetting his nemesis, calling the experience a "fine tale this to relate to Monsieur G. Ord of Philadelphia!"[55] But he also wrote Victor to tell him that the fatigue he felt in doing the Golden Eagle had indeed been a bit more serious, and that he had suffered a "Severe fit of Ilness last Saturday"— a mild stroke, really—adding, with a note of hopeful reassurance, that "it was of only about one hour duration and at the exception of a general temporary weakness I am again *Myself!*"[56] In fact, he would not be himself for quite some time, and never fully so. In this case, we can take Audubon's suffering seriously, as he certainly did: The stroke frightened him and made him much more aware of growing old, even at age forty-eight.

Revision Through Inversion

To get an even better measure of the impact of the Golden Eagle episode, we have another artistic product to consider: the painted image, the original watercolor that apparently caused him such agony. In its various versions, this painting invites close (and perhaps conflicting) analysis.

Clearly, Audubon does not show us the eagle in the cage, covered with a blanket and locked in a room, choking with smoke. Instead, he portrays the eagle in its mountaintop environment, what Audubon called "the rocks of its wild haunt"—the place it would have returned to had Audubon set it free. And he shows the eagle rising high into the sky, seemingly in triumph, clutching its prey in its talons, shrieking out a call. A correspondent to the Philadelphia *Gazette*, who viewed the painting soon after it had been completed, wrote about the impressive power of Audubon's image: "The picture seems instinct with life, and one can almost see the tremor of the bird and hear the rustling of the wings. The *scenery* . . . is of the most faithful character, and shows his botanical as well as ornithological skill."[57] What this writer neglected to point out, however, was that the scenery contained an exceptional element for an Audubon painting: a small image of a man down in the lower left-hand corner, a buckskin-clad hunter carefully inching his way across a log over a deep crevice, using a hatchet to pull himself along; to make his situation even more precarious, the hunter has to maintain his balance with a gun and a large dead bird, most likely an eagle, strapped to his back (see Plates 5 and 6).

Taken together, those two images—the large, screaming eagle and the small, struggling man—offer not just a visual inversion of scale, but also a narrative inversion of the story Audubon tells us of his experience with the eagle. Above all, the painting transfers the central act of violence from the human to the nonhuman. It's the eagle, not Audubon, who does the killing in the painting. It's the eagle, not Audubon, that impales its victim—and right in the eye, putting out the organ that had been so central to Audubon's account of his encounter with the caged eagle in the smoke-filled room.

There's another significant inversion in this story, this one ornithological: Audubon notes on his painting that the eagle is a female, and therefore we must assume that the live specimen Audubon kept captive in his room and then killed and posed for his painting was also a female. But in telling the story of his encounter with the eagle, Audubon always uses the male pronoun: "I watched his eye, and observed his looks of proud disdain. . . . There stood the eagle on his perch, his bright unflinching eye turned towards me."[58] Why does Audubon make such a sex change for the eagle in the text? In this case, the inversion of male and female cannot be just inadvertent, not just an ornithological mistake, not just a slip of the paintbrush or pen. Is it that he can't bring himself to take a female eagle, keep her in captivity, and then murder her by poking a piece of steel through her heart? Is such a murder, even of a female bird, something perhaps too shocking for antebellum audiences, so that Audubon feels he has to change the bird's sex to make it male and therefore perhaps more menacing so that he can kill it? In this sense, the story seems to be not just about the eagle, but about Audubon, and about how he comes to terms with his own violence against the bird by making it a male adversary, his gender equal in this eye-to-eye confrontation.[59]

As if to underscore this transfer of violence to the eagle, Audubon offers an additional section right at the end of his chapter on the Golden Eagle, a story allegedly told by Dr. Benjamin Rush of Philadelphia about "the effects of fear on man." It seems that some Revolutionary-era soldiers found a Golden Eagle's nest on the cliffs over the Hudson River, and to get the eggs in the nest, they lowered one man down by a rope. As soon as he got to the nest, though, an adult eagle attacked him, and to defend himself, the soldier drew out a knife and started slashing at the bird. Unfortunately, he only succeeded in cutting the lifeline holding him, and he looked on in horror as the rope began to unravel. His companions pulled him up just in the nick of time and saved him from a deadly fall, but the experience so terrified him, Dr. Rush (and Audubon) tell us, "that ere three days had passed his hair became quite grey."[60] (One might even suggest that Audubon refers to this story indirectly in his painting, making a kind of visual pun of the soldier's suddenly gray hair and the white hare, as both are attacked by the eagle.) Clearly, a viewer/reader should be able to make the connection between the story of the soldier dangling over a cliff and the small portrait of the hunter—indeed, probably a self-portrait—struggling to cross a precipice, steadying himself and pulling himself along with a small hatchet and carrying a gun and an eagle on his back. Both appear on the verge of a disastrous, no doubt deadly fall.

What is Audubon trying to say here? Several students of Audubon's art have offered various approaches to this part of the painting, most of them describing it in terms of near-heroic struggle, or at least the challenges Audubon faced in being an artist-naturalist.[61] But the connection between Audubon and the big bird goes beyond the challenges of doing art and even science.

Another reading of this painting could take us further into the realm of the painter's psyche, to a deeper consideration of Audubon's "struggle with this composition." Where some commentators see Audubon's self-portrayal as an image of his rising to the challenge, we can also see the danger of his collapse. Here again is this small man, headed downward, bearing his burdens on his back (which, if Audubon were a more religious artist and more given to Christian symbolism, might well be interpreted as a cross), teetering above a vast abyss and about to lose his balance—and not just his physical balance, but his psychological balance as well. In short, we can read this image not so much as an expression of artistic effort, but of personal agony, of the "spasmodic affection, that . . . completely prostrated me for days," the remorse, perhaps guilt, over killing the captive bird in such a direct, personal manner, not shooting from a distance, as he usually did, but taking it in his hands and plunging a piece of steel through its body, through its heart. This was a different sort of death for Audubon, and it apparently shocked him, perhaps as it still might shock modern readers.

The story doesn't end, though, with Audubon's original painting, complete with the small self-portrait of the artist as an endangered species of man. In fact, in subsequent versions of this painting and the accompanying story of Audubon's killing the eagle in his room, both the image and the narrative were modified, even sanitized, to take Audubon himself out of the picture, as it were, and to transfer the violence wholly to the eagle. Somehow, in the process of turning Audubon's original watercolor into the engraving that eventually became part of the published version of *The Birds of America*, the image of the hunter/artist, the self-portrait, vanished. Unfortunately, there seems to be no letter or note, no "smoking gun" communication, as it were, from Audubon to his engraver back in England, Robert Havell, or from Havell to Audubon in which either one suggests taking the human figure out of the picture. To be sure, signs of human beings are very rare in Audubon's bird paintings, and those that do get in are usually background cityscapes, showing evidence of human habitation, but no human beings themselves.[62] The sole exception, the picture of the Snowy Egret, contains a South Carolina plantation in the background and, coming from it toward the egret, a small, gun-toting human figure that might be a miniature self-portrait of Audubon (and again, a self-portrait of the artist armed). Moreover, that last detail stayed in, surviving the transition from Audubon's original watercolor to Havell's engraving for the Double Elephant Folio version (see Plate 7). There may well have been nothing inherently "wrong," then, about showing Audubon armed in the engravings of the birds. But in the particular case of the Golden Eagle, the drama behind its demise seems to have caused the image to be altered—or, again, sanitized.

The same thing happened in the Royal Octavo edition of *The Birds of America*, the smaller, less expensive, and more widely accessible version published in the early 1840s. Not only was the image of the hunter/artist removed, but so was the whole story of Audubon's killing the eagle in the hotel room. What stayed in, though, was the story Audubon tells (or, really, the story Audubon tells us Benjamin Rush told) about the Revolutionary soldier's being attacked by the Golden Eagle and being so frightened that his hair turned gray.[63] The point, then, is that by the time the image and the story of the Golden Eagle had gone through subsequent transformations—from Audubon's original watercolor to Havell's engraving in the large Double Elephant Folio edition to the small lithograph in the Royal Octavo edition, and from the original narrative in *Ornithological Biography* to the abbreviated description in the smaller edition—Audubon the eagle killer had been removed, and only the eagle remained as the agent of violence. The blood was on the eagle's talons, not on Audubon's hands.

Back to Business as Usual

It would make a nice story to say that this exhausting encounter with the Golden Eagle became a transformative experience for Audubon, a scientific epiphany that changed the way he worked, that he never killed another bird—nice, perhaps, but ultimately naïve. Audubon kept on killing birds for the sake of art and science, and he kept on writing about doing so in his personal journal and in *Ornithological Biography*.

The next significant episode in Audubon's own story came soon, in the spring of 1833, after he had recovered from the Boston debacle enough to become "again *Myself*," when he set off on his already-planned trip to Labrador. He chartered the schooner *Ripley*, with Captain Henry Emery and crew, and outfitted it to his needs, with a "very good parlor, dining room, drawing room, library &c" built below deck, all at a considerable expense—$350 a month, plus supplies, which would turn out to be a total of around $2,000 in all. Taking his son John Woodhouse and four other young men along with him to help with the shooting and collecting, Audubon seemed eager to get back into the out-doors to take possession of even more species. Leaving Eastport, Maine, on June 6, 1833, he headed toward a cold, watery world of birds in Labrador.[64]

They found almost more than even Audubon had imagined. One of the most striking images he offers of the early days of the Labrador expedition came when his ship passed close by the "Bird Rocks," which were covered with thousands of Gannets, "a mass of birds of such a size as I never before cast my eyes on." The birds were so packed in next to each other that their striking whiteness made it appear "as if a heavy fall of snow was directly above us." When members of Audubon's crew shot into the mass of birds, most "remained seated on their nests quite unconcerned . . . but where the shot took effect, the birds scrambled and flew off in such multitudes, and in such confusion, that whilst some eight or ten were falling into the water either dead or wounded, others pushed off their eggs, and these fell into the sea by hundreds in all directions."[65] Here in Labrador, as in Florida or other coastal points in between, the avian abundance of the Atlantic environment almost always seemed an invitation to violence. A subsequent sighting of Arctic Terns, a bird Audubon had never seen before, caused him to become "agitated with a desire to possess it." Once again, he and his fellow shooters blasted away at the birds, and "one after one you might have seen the gentle birds come whirling down to the waters." So successful was the slaughter that Audubon would write of experi-encing momentary remorse—"Alas, poor things! How well do I remember the pain it gave me, to be thus obliged to pass and execute sentence upon them"—but he got over it in the pursuit of science and "excused myself with the plea of necessity, as I recharged my double gun."[66]

While on the Labrador journey, he also recorded a measure of regret when he performed a "severe experiment" on a pair of Black Guillemots. In another cruel-seeming practice—again, perhaps an ornithological equivalent of Poe's "The Cask of Amontillado"—he confined the two birds "in a fissure of rock for many days in succession," seeking to determine how long they could live without food. Rough seas made it impossible for Audubon to return to the rocks where the starving birds remained imprisoned, and "many a time I thought of the poor captives." When he finally got back to check on their condition, he found them, not surprisingly, "apparently in a state of distress," but still just strong enough to fly away from him as fast as they could. Audubon did not record the results of this experiment, and the birds' suffering apparently served no scientific purpose.[67]

Not all the suffering was inflicted by human hands, however, but also by birds themselves. Audubon wrote about a violence-filled world of death by avian predation, which he put in sometimes explicit comparison to the intrusion of men. The best illustration of the connections between the two came in the description of the one bird that came to dominate Audubon's account of his Labrador trip, the Great Black-backed Gull (see Plate 8). Gulls don't normally garner much narrative attention—they seem ubiquitous on the beach, as common as sparrows in the city streets—but this particular species set itself apart from the mass of shorebirds, even above them, by its aggressive and rapacious behavior. In the pages of his Labrador Journal, he portrayed the bird in bloody, almost grisly terms: "I saw a Black-backed Gull plunge on a Crab as big as my two fists, in about two feet of water, seize it and haul it ashore. . . . I could see the crab torn piece by piece till the shell and legs alone remained. . . . [The gulls] wash the blood off their bills by plunging them in water, and then violently shaking their heads."[68]

Later, in his chapter on the Great Black-backed Gull in *Ornithological Biography*, Audubon painted an impressive verbal picture of the bird's powerful presence: "High in the thin, keen air, far above the rugged crags of the desolate shores of Labrador, proudly sails the tyrant Gull, floating along on almost motionless wing like an eagle in his calm and majestic flight." The image of the gull flying "far above the rugged crags of the desolate shores . . . like an eagle" summons up an implicit comparison to Audubon's description of the Golden Eagle, who might have been allowed to "spread out his broad wings and sail away toward the rocks of his wild haunts." The blood on the gull's bill also recalls the blood on the eagle's talons and makes the "tyrant Gull" another agent of violence against other species: "Like all gluttons, he loves variety, and . . . without remorse, he breaks the shells, swallows their contents and begins leisurely to devour the helpless young."[69]

But beyond any comparison to the violence of avian behavior, the gull's egg-breaking proclivities also provide a striking connection to human predators who engaged in much the same sort of remorseless destruction. Audubon wrote with undisguised contempt about the "eggers" of Labrador, who plundered the nesting places of birds in a "rascally way" to take their eggs and feathers to sell in the city markets: "The eggers destroy all the eggs that are sat upon, to force the birds to lay again, and by robbing them regularly they lay until nature is exhausted and few young are raised."[70] In his description of these "ruthless and worthless vagabonds" in *Ornithological Biography*, he drew a harsh contrast between the eggers' environmental depredations and their eventual economic gain: "Look at them! . . . At every step each ruffian picks up an egg so beautiful that any man with a feeling heart would pause to consider the motive that would induce him to carry it off. But nothing of this sort occurs to the Egger. . . . The dollars alone chink in his sordid mind."[71]

But as much as Audubon disdained the eggers' destructive work, he and his own allies also collected eggs. In a matter-of-fact passage in his Labrador Journal, Audubon writes of landing on an island filled with guillemots just after eggers had begun gathering the birds' eggs. "We did the same and soon collected about a hundred. These men . . . had collected eight hundred dozen and expect to get two thousand dozen." But Audubon also notes one other threat to the eggs: "The Black-backed Gulls were here in hundreds and destroying the eggs of the Guillemots by thousands."[72]

Thus Audubon, the eggers, and the gulls all appear in the same paragraph, all engaged in the same predatory practice. There are differences of degree, not to mention motive, for the destruction—Audubon for science, the eggers for profit, and the gulls for basic sustenance—but Audubon could see the possible implications for the future: "This war of extermination cannot last many years more."[73] The birds could move on, and so could the eggers, but Labrador itself could not. For all the abundance he saw on its shores, Audubon came to the gloomy conclusion that "Nature herself seems perishing" there, "owing to man's cupidity." "When no more fish, no more game, no more birds exist," he lamented, "then she will be abandoned and deserted like a worn-out field."[74]

It may well be anachronistic and therefore unwise to call Audubon an environmentalist, at least in the modern sense, and it is historically unrealistic to force a nineteenth-century figure into a twenty-first-century framework. Still, we can certainly see Audubon, here and elsewhere, as a very close and perceptive observer of nature, including, at times, human nature. Sometimes, in fact, he employs human traits to explain avian behaviors, as is the case with the Great Black-backed Gull, often a "glutton" and a "coward," but above all a "tyrannical" bird. Perhaps especially in an oceanic environment, where the number of birds is so great and the violence so visible, the gull's aggressive invasions of other species' spaces might strike the reader as alarming enough. But the additional danger created by human intruders—the eggers, if not Audubon himself—seems likely to tip the fragile balance of nature toward ultimate destruction, toward "perishing." Nature is never static, and Audubon painted, more in words than in images, an unsettling environmental vision of the future. As Katherine Govier observes in *Creation*, her novel about Audubon's expedition to Labrador, "He came here a hunter and singer of birds and he will leave here a mourner of birds."[75]

But Audubon remained, above all, a collector of birds, first of their bodies, then their images. Toward the end of his voyage on the *Ripley*, he took account of his work thus far: "Twenty-three drawings have been executed, or commenced and nearly completed," and that would be the final number for the trip as a whole. On the way back to port in the United States, he gave himself a measure of credit for all he had done: "I am content, and hope the Creator will permit us to reach our country."[76] The yearning for "our country" aside, Audubon made the Labrador story part of his larger ornithological—and national—narrative. Labrador may not have been part of the United States of America, but birds could never be bothered by national boundaries; in this case, neither could Audubon. In an act of ornithological appropriation, he brought both Labrador and its birds into *The Birds of America*.

Perhaps equally important, he brought himself back into the fullness of his own self-image as the American Woodsman, an ornithologist who could suffer in the name of science and face nature as a man. After confronting sudden infirmity indoors, experiencing a "severe fit of Ilness" after killing a caged eagle in a Boston hotel room, Audubon relished the outdoor adventure of Labrador. Writing to a friend at the end of the trip, he recounted his rigorous pursuit of birds, "going down with Ropes from the top of the precipitous rocks" to find the specimens he needed. With no mention of danger or labor, he celebrated the self-refreshing effect of his exploits: "In a word we have had rare sport and plenty of it," he concluded, "and I am Audubon again!"[77]

Being Audubon again—and again and again, tramping through the forest, shooting on the seashore, hanging on a rope off the side of a rock face—had long been physically difficult and frequently dangerous work. As a man who made the American Woodsman his scientific signature, he could have it no other way. He could hardly enclose himself within the indoor comforts of the "closet philosopher." He had to be outdoors and on the move, exposing himself to the American environment and associating himself with the American people. As much as he relished pursuing and painting birds, Audubon took a much broader view of a much bigger picture, one that encompassed physical nature as well as human nature, incorporating both as important parts of his Great Work.

Chapter 7

~~

Putting People into the Picture

To render more pleasant the task which you have imposed upon yourself, of
following the author through the mazes of descriptive ornithology, permit me,
kind reader, to relieve the tedium which may be apt now and then to come upon
you, by presenting you with occasional descriptions of the scenery and manners of the land which has furnished the objects that engage your attention. The
natural features of that land are not less remarkable than the moral character of
her inhabitants.

—John James Audubon, "The Ohio," in *Ornithological Biography*

Audubon almost never included people in his paintings, but he seldom let them
get far from view in his writing.[1] He never intended *The Birds of America* to be
just a collection of bird pictures, and neither did he write *Ornithological Biography* to be just a book about bird behavior.

As early as 1830, when he had just begun working on *Ornithological Biography*, he noted the dual purpose of the book. "I am trying my best to render
the work equally highly scientific as popular," he wrote to a friend.[2] At the same
time, he explained to another that "I wish if possible to make a pleasing book
as well as an instructive one." For *Ornithological Biography* to be both popular
and pleasing, he sought to give it the appearance of being personal as well. "My
first volume will comprise an introduction and one hundred letters addressed
to the Reader referring to the 100 plates forming the first volume of my illustrations." Casting his writing as "letters" suggests the sort of direct relationship
Audubon hoped to develop between himself as author and his reader. "I will
even enter on local descriptions of the country," he continued. "Adventures
and anecdotes, speak of the trees & the flowers the reptiles or the fishes or
insects as far as I know."[3]

Those "descriptions of the country" did not apply to the landscape alone or to the other nonhuman life forms he noted. Wise as he was in the ways of nature, Audubon also knew enough about his readers to give them an occasional escape from "the mazes of descriptive ornithology" and let them also see some of the other people of America, playing their various parts in shaping "the moral character of her inhabitants." The chapters on birds often contained anecdotes and observations from ordinary people Audubon encountered, but beyond those, he eventually set aside sixty specific chapters on nonavian topics—or "Episodes," as they came to be called—in the first three volumes of *Ornithological Biography*, inserting one after every five chapters on birds. (By the time he got to the fourth and fifth volumes, he was in a hurry to complete his Great Work, and those last two volumes focused on ornithology alone.[4])

Audubon's descriptions of nature combined place and people, wonder and warning, presenting the reader with an environment in which natural forces could be beautiful and benign, but also strikingly violent. The human beings who lived in that environment could be much the same: innocent-seeming victims of nature's power, but also active perpetrators of violence themselves, both against nature and against each other. By looking beyond birds, then, Audubon did much more than merely "relieve the tedium" of ornithological inquiry. He provided an evocative portrayal of a wide range of human behaviors in the early American republic and, in the process, offered a surprisingly generous judgment of the American people. Taken on its own, *Ornithological Biography* could quite well stand as an important contribution to the literature of American culture and society in the new nation.

But like many other works in that genre, Audubon's work poses a challenge that every serious reader ought to engage with a combination of curiosity and caution: Audubon was no Tocqueville. He was not seeking to make a systematic examination of American society, to get somehow to the essence of America itself. Neither was he promising to offer a coherent narrative, with a strong sense of thematic development and chronological progression; just as his presentation of the plates in *The Birds of America* did not reflect a well-ordered taxonomic system, so, too, did his Episodes appear in a somewhat idiosyncratic fashion. Still, Audubon offers valuable and revealing material in these chapters, and it would be a mistake to see them as mere entertainment, much less as filler or fluff.

It would likewise be a mistake to accept them as fact. To be sure, one of Audubon's earliest and most thorough biographers, Francis Hobart Herrick, took a remarkably credulous and charitable approach to Audubon's authority: "Audubon was a keen observer of men and things as well as of birds and animal life," Herrick wrote, "and when writing down his experiences on the spot, as was his invariable custom after 1820, he was as truthful with his pen as with his pencil and brush."[5] Subsequent biographers have taken a somewhat more selective approach to Audubon's Episodes, taking some essentially at face value for descriptions of particular (and plausible-seeming) events, but ignoring others that might complicate the credibility of their portrait of the artist. Post hoc fact checking of these Episodes is virtually impossible, of course, and exact accuracy may not be the most important point anyway. Again, it makes good sense to take Audubon seriously without taking him literally, to read these Episodes for their larger effect in portraying the "scenery and manners" of America and, equally important, portraying Audubon's own sense of self as an American.

Teaching to See and Read

Anyone who creates a work of art—a painting, a book, or both—still hopes to know that someone sees it, but Audubon wanted his Great Work to be seen on his own terms. There was nothing passive or implicit in Audubon's approach to his reader. His exceedingly well-developed sense of self consistently came through in his paintings and writings, and he never let the viewer/reader ignore or forget the great man who created them. In the text to accompany the image of the Ruddy Duck, for instance, he takes the lead in appreciating the quality of his own work: "Look at this plate, Reader, and tell me whether you ever saw a greater difference between young and old, or between male and female, than is apparent here."[6] In the case of the Carolina Parakeet—one of his most famous paintings, showing a tree full of this now-extinct bird—he pays himself the compliment he knew must certainly come: "Doubtless, kind reader, you will say, while looking at the seven figures of Parakeets represented in the plate, that I spared not my labour. I never do, so anxious am I to promote your pleasure." Audubon was also anxious to promote the purchase of his work, and he urged the reader of *Ornithological Biography* to put his description of a bird together with the corresponding plate from *The Birds of America*, "which, if not already before you, I hope you will procure."[7]

Whoever did procure both works would find them to be a remarkable matched pair, companion pieces that worked in partnership to make a bibliographically innovative experience—and a challenging one as well. At thirty by forty inches and forty or more pounds, a Double Elephant Folio volume of *The Birds of America* could never rest easily on the lap, nor could the viewer even turn the pages without the risk of tearing them. To look at one of the plates in the fashion Audubon intended, one would presumably sit several feet away, have a corresponding volume of *Ornithological Biography* at hand, and shift one's gaze back and forth from book to book, from text to image—quite often with Audubon as an active, even didactic, guide. His written directions to go back and forth between book and plate represent what we would now call an interactive reading experience, an early version of a multimedia melding of word and image. At the time, it was also a way for Audubon to put—or keep—himself in the picture and assert his ornithological and artistic authority over the viewer/reader's experience.

Indeed, he occasionally made a very specific connection between image and text in *Ornithological Biography*, directing the viewer/reader to see a particular image in *The Birds of America* in the way Audubon, as artist, wanted it to be seen. In his description of the House Wren, he tells the reader to "look at the little creatures anxiously peeping out of hanging to the side of the hat, to meet their mother, which has just arrived with a spider, whilst the male is on the lookout, ready to interpose should any intruder come near."[8] If the viewer could not fully appreciate the family-centered avian activity taking place, Audubon supplied the words to make the point—and the picture—clear (see Plate 9).

He also drew attention to less admirable attributes of birds, as in his interpretation of a scene involving a pair of Great-footed Hawks (or Peregrine Falcon) "eating their *dejeune a la fourchette*." "One might think them real epicures," he wrote, "but they are in fact true gluttons. . . . Their appetites are equal to their reckless daring, and they well deserve the name of 'Pirates,' which I have above bestowed upon them."[9] It might not have been good science to impute human behaviors to birds, much less bestow unflattering labels on them, but Audubon used this disparaging description of avian species to engage human readers in a narrative that included more than scientific information.

He sometimes even invited his "kind reader" to become a vicarious companion on his own travels, joining Audubon in observing birds: "Permit me to place you on the Mississippi, on which you may float gently along," he wrote. "The Eagle is seen perched, in an erect attitude, on the highest summit of the tallest tree. . . . His glistening but stern eye looks over the vast expanse."[10]

Trying to be good to his word about his desire to "relieve the tedium" of "descriptive ornithology," Audubon sought to create a descriptive picture in words that would not just complement but accentuate the energy of his avian images. He used his pen to make sure that his reader could share his own fascination with the beauty and behavior of the birds and, in the process, see through Audubon's own eye as it looked over the vast expanse of the United States.

Blackbirds Baked in a Pie

Audubon knew that birds satisfied another sort of taste beyond the artistic and literary: People ate them. In the early nineteenth century, birds were a staple of the nation's cuisine, and the variety of avian specimens destined for the table included birds of all sizes and descriptions. Some of the larger, fleshier birds—turkeys, doves, ducks, geese, and chickens—would not strike the modern diner as at all surprising, but some others that appeared on American menus might. Cookbooks in the early republic contained no end of instruction on preparing birds of all sizes, right down to sparrows, larks, and buntings. A recipe for the Reed Bird (now more commonly known as the Bobolink) started out with "a lump of butter the size of a hickory nut put into the belly" and then went on to make a simple-seeming dish, with the bird rolled in egg yolk and bread crumbs, roasted and basted, and finally garnished with lemon slices.[11]

Pigeons, which were slightly bigger and certainly more common, "may be dressed in so many ways," noted an 1819 cookbook, whether roasted, broiled, stewed, or pickled. Pigeons in jelly, for instance, called for the cook to bring a bit of taxidermy to the table: "Leave the heads and the feet on," the recipe instructed. "The head should be kept as if alive, by tying the neck with some thread, and the legs bent as if the pigeon sat upon them." If done right, the resulting dish would provide "a very handsome appearance in the middle range of the second course."[12]

Audubon never suggested such elaborate preparation and presentation, but he did have a seemingly omnivorous appetite for avian cuisine. He had eaten many a Purple Finch, and could thus "consider their flesh equal to that of any other small bird, excepting the Rice Bunting." The Tyrant Fly-Catcher (or Eastern Kingbird), which gained so much hostility from farmers for eating so many bees, had flesh that Audubon found "delicate and savoury."[13] Audubon sometimes warned his readers that a bird's diet might well affect their own. While "the sportsmen of the Middle Districts" seemed happy to eat the Golden-winged Woodpecker (or Northern Flicker), "I look upon the flesh as very disagreeable, it having a strong flavour of ants."[14] Likewise, the food of the American Snipe consisted largely of worms and insects, "which tend to give its flesh that richness of flavor and juicy tenderness, for which it is deservedly renowned," and some people actually liked to eat their viscera, "worms and insects to boot, the intestines being considered the most savoury parts." Having himself "more than once found fine large and well-fed ground worms, and at times a leech" in the innards of a snipe, Audubon said he preferred his snipe "well-cleaned, as all game ought to be."[15]

Audubon offered quite a few such culinary comments about the ways different birds tasted, and he occasionally recorded contrary opinions from other Americans as well. He found Marsh (or Red-winged) Blackbirds "little better than the Starling of Europe, or the Crow Blackbird of the United States," but he allowed that they "are thought good by the country people, who make pot-pies of them."[16] Similarly, the Florida Cormorant was not much to his liking—"Their flesh is dark, generally tough, and has a rank fishy taste"—but he admitted that some "refined epicures" might "pronounce it excellent." "The Indians and Negroes," he continued, "salt them for food [and] I have seen them offered for sale in the New Orleans market, the poorer people there making gombo soup of them."[17] He had much the same negative reaction to the taste of the Great Northern Diver—"the flesh of the Loon is not very palatable, being tough, rank and dark colored"—but then "I have seen it much relished by many lovers of good-living, especially at Boston, where it was not unfrequently served almost raw at the table of the house where I boarded."[18] Audubon even made fine distinctions between closely related birds, expressing his preference for the Ruffed Grouse, which he deemed second only to the Wild Turkey in flavor, to the Pinnated Grouse (or Greater Prairie Chicken), which was much esteemed by "Eastern epicures" but which he found "dry and tasteless."[19]

Audubon gave other birds unequivocal praise for their taste, and in some instances he essentially set the table for the reader to experience his anticipation of excellent avian eating. In his description of the American Woodcock, for instance, he led the reader into a reverie of domestic delight:

How comfortable it is when fatigued and covered with mud, your clothes drenched with wet, and your stomach aching for food, you arrive at home with a bag of Woodcocks, and meet the kind smiles of those you love best. . . . [Y]ou will ere long see a dish of game . . . so white, so tender, and so beautifully surrounded by savoury juice; when a jug of sparkling Newark cider stands nigh; and you, without knife or fork, quarter a Woodcock, ah, Reader![20]

Audubon invited his reader, the oft-repeated "you," to share vicariously in this mouth-watering, cutlery-free food fantasy. So it went elsewhere in the pages of *Ornithological Biography*, which often began to read like an epicurean guide to a steady diet of the better-tasting birds, with Audubon happily serving as the experienced host.

Shooting Birds in Abundance

He also served as an expert hunter, a role that would surely find a receptive reading in Audubon's America. The United States was a nation of hunters who pursued birds for sustenance, commerce, and sport—three reasons that sometimes overlapped and, taken together, accounted for the deaths of untold numbers of birds. The image of a hunter bringing home a mere bag of woodcocks would be almost nothing, in fact. Audubon noted that the birds were commonly killed in "almost incredible numbers," and one might at times see "gunners returning from their sports with a load of Woodcocks, composed of several dozens; nay, adepts in the sport have been known to kill upwards of a hundred in the course of a day."[21] Audubon certainly considered himself one of the "adepts in the sport," and it would be difficult for anyone to disprove that.

The American Woodsman was a good shot who admired good shooting in others—even, on occasion, those who didn't use a gun. He took note of the Indians in the Mississippi region north of the Ohio River, who killed the Pinnated Grouse "with arrows whenever they chanced to alight on the ground or low bushes." Indians at the other end of the Mississippi, in Louisiana, impressed him with their skill in using blowguns to kill White-throated Sparrows and other small birds "at a distance of eight or ten paces." "With these blow-guns or pipes," he continued, "several species of birds are killed in large quantities, and the Indians sometimes procure even squirrels by means of them."[22]

For the most part, of course, Audubon wrote of his fellow Euro-American hunters who used guns to even greater advantage. Writing again about his time in Louisiana, he described being on a hunt for Golden Plovers with French gunners near New Orleans: "Every gun went off in succession, and with such effect that I Several times saw a flock of a hundred or more reduced to a miserable remnant of five or six individuals." Estimating some two hundred hunters in the field, and "supposing each to have shot twenty dozen," he calculated that "forty-eight thousand Golden Plovers would have fallen that day."[23] Audubon even invited his reader along on a fictive hunting expedition during the fall migration season, "having constantly within your view millions of birds on their way to the south . . . and gazing in astonishment at the multitudes of feathered travelers." Audubon alerts his "good-natured reader" when "all of a sudden a larger birds attracts your eye" as it "sweeps along in the stillness of the autumnal evening with a rapidity seldom equaled, creating confusion, terror, and dismay along the whole shores." The larger bird, a Pigeon Hawk (or Merlin), seeks out a single smaller bird in the mass, and Audubon calls upon his reader to join the human hunters in a much larger kill: "Now is your time. . . . Pull your trigger, and let fly, for it is impossible, should you be ever so inexpert, not to bring down several birds at a shot."[24]

The abundance of birds in the American environment posed both a sporting opportunity and an unsettling question: Although it was a boon to hunters, even those "ever so inexpert," it posed a concern about the eventual effect of their shooting on the avian populations. Audubon took note of both. Like other observers of birds in early America, he marveled at the massive flights of Passenger Pigeons that darkened the skies for days at a time, so numerous that it seemed impossible to count them. People tried, though. Alexander Wilson once did a rough calculation of a flight he witnessed in Kentucky:

> If we suppose this column to have been one mile in breadth (and I believe it to have been much more), and that it moved at the rate of one mile in a minute, four hours, the time it continued passing, would make its whole length two hundred and forty miles. Again, supposing that each square yard of this moving body comprehended three pigeons, the square yards in the whole space, multiplied by three, would give two thousand two hundred and thirty millions, two hundred and seventy-two thousand pigeons—an almost inconceivable multitude, and yet probably far below the actual amount.[25]

Always seeking to equal or exceed Wilson in all things ornithological, Audubon made his own mathematical measurement:

> Let us take a column of one mile in breadth, which is far below the average size, and suppose it passing over us without interruption for three hours, at the rate mentioned above of one mile in the minute. This will give us a parallelogram of 180 miles by 1, covering 180 square miles. Allowing two pigeons to the square yard, we have one billion, one hundred and fifteen millions, one hundred and thirty-six thousand pigeons in one flock.[26]

A billion-plus, perhaps two billion or more—who could really tell? And with such massive numbers of birds, whatever they might be, who could really worry?

What mattered more to most people on the ground was the opportunity to kill pigeons by the thousands, and accounts of the eager anticipation of the pigeon shoot became almost a literary staple of antebellum nature writing. The most notable came in a famous passage from James Fenimore Cooper's *The Pioneers* (1823), when "the heavens were alive with pigeons" and when every inhabitant of the town of Templeton turned out to shoot them—and turned every weapon skyward, from sticks to bows and arrows to muskets to the town's old cannon. Natty Bumppo famously fretted about the aggressive excess of the townspeople—"it's wicked to be shooting into flocks in this wastey manner"—but his worries about the all-village barrage were very much a minority report. "'Sport!' cried the Sheriff . . . 'every old woman in the village may have a pot-pie for the asking.'" Alexander Wilson likewise described the massive slaughter of pigeons with nets and guns, echoing "the thundering of musketry . . . perpetual on all sides from morning to night." For days thereafter, "pigeons become the order of the day at dinner, breakfast and supper, until the very name becomes sickening." Audubon, too, observed that the incessant blasting away at the birds created such a "scene of uproar and confusion" that it was impossible to hear anything, and useless to try to speak: "Even the reports of the guns were seldom heard, and I was made aware of the firing only by seeing the shooters reloading." The shooting stopped when the surviving pigeons passed on, and then the eating began. For his own part, Audubon considered the Passenger Pigeon to be only "tolerable eating," but other people eagerly gorged themselves on the bounty that fell from the sky: "The population fed on no other flesh than that of Pigeons, and talked of nothing but Pigeons."[27]

In talking about the Passenger Pigeon himself, Audubon did pause to look to the future and consider the eventual implications of the slaughter, that "such dreadful havock would soon put an end to the species." He reassured his reader, though, that Passenger Pigeons multiplied so quickly, doubling or even quadrupling their population each year, so that extinction seemed hardly likely to happen: "I have satisfied myself, by long observation, that nothing but the gradual diminution of our forests can accomplish this decrease." Audubon published those words in 1831 and therefore never lived to know that the species would indeed become extinct, in 1914, when the last Passenger Pigeon died in the Cincinnati Zoo.[28]

Proper Hunters and Pot Hunters

Audubon's various descriptions of hundreds of hunters shooting thousands of birds might well have touched a sensitive social nerve among some of his readers, particularly those gentlemen "sportsmen" who had strong notions about who should be doing the shooting, and for what purpose. For all its celebration of the Common Man and its ethos of increasing equality, the early American republic remained a society in which class still mattered, and bird shooting served as one of many measures of social identity. Moreover, for all their ostensible desire to distinguish themselves from their aristocratic British brethren, American sportsmen could still look sometimes longingly at the genteel standards established in England, where pigeon shooting had become a more refined-seeming competition among well-bred and well-equipped members of the upper classes.

American gentlemen who read the local sports page could follow the British shooting news easily—and with some envy, no doubt. In June 1833, the *New York Sporting Magazine* reported on a contest between a "Mr. Edge of the Manchester club, and Mr. Deaton of the Sheffield club, commenced at the Hyde Park Cricket Ground, Sheffield, . . . well attended by all the sportsmen in the neighborhood." Although Mr. Deaton had home-field advantage, the betting among the two thousand spectators soon favored Mr. Edge by two-to-one, and he hung on to win, beating Deaton by a seven-bird margin after two days of shooting.[29] But leaving such sporting exploits to men on the far side of the Atlantic could hardly be considered sufficiently exciting, and American shooters had begun to stage their own competitions, which borrowed from the British model and likewise made their way into print in sportsmen's publications. The *American Turf Register and Sporting Magazine*, a Baltimore-based horse-racing sheet that first appeared in 1829, generally devoted a few pages to gentlemanly hunting. One of its early issues had the report "Pigeon Matches Near Baltimore," in which two teams of five shooters each—including three U.S. Army officers, the British consul, and a member of Maryland's most prominent family, Charles Carroll Jr.—fired away at pigeons released from a trap. At the end of the shooting, there remained "scarcely a bird that was not struck," and since "it was the first time that many of the gentlemen had ever shot at a bird from a trap, it may be considered good shooting." So pleased were the participants, in fact, that "the party adjourned to meet again next summer, to catch trout and shoot deer and wild turkeys in Alleghany!"[30] Dead birds made for male bonding, giving the well-bred men in Baltimore a sense of camaraderie that seemed destined to endure, at least until the following year.

On the other hand, too-obvious aping of the English seemed an unsatisfactory practice for gentlemen on this side of the Atlantic, and some American shooters took pains to make a case for the superiority of their own version of the sport. A report on a pigeon-shooting contest in New York took note not only of the skill of the day's best gunner, a Mr. Henry M. Boughton, but of the man's English-made gun, "a double barrel, Westley Richards—percussion—length of barrel 2 feet 6 inches—calibre 5/8 in.—and weight 7 pounds 4 ounces. The shot No. 6, (Youle's patent) charge 1 1/2 ounces—the powder, Pigou and Wilke's, Dartford, and charge 1 1/2 drachms." Such obsessive attention to the weapon mattered, explained the writer, because "in reference to those made use of in England for pigeon matches, there is nothing in the annals of . . . shooting that can be compared with this."[31] Put simply, the message seemed clear: Americans could outshoot the British with their own guns.

They also had better dogs and, in a way, better birds. A correspondent to the *American Turf Register* wrote glowingly of the shooting of George Mason of Virginia, who once "killed forty-nine partridges without missing a single shot." Mason's aim seemed even more impressive when one took into account the degree of difficulty he faced, at least in comparison to British shooters: "In reading the accounts of the English sportsmen," the correspondent wrote, "it should be borne in mind that the English partridge is larger than ours, more abundant, as the game laws afford them protection, is not so shy, and do not take such rapid flights—of course are easier killed." As much as the writer praised George Mason, however, he gushed with almost equal admiration for

Mason's American dog. "Mr. M. hunted two pointers," he noted, "one an imported English dog, and his dog Pluto." The English dog was good—"had a high character, was perfectly trained"—but "Pluto proved greatly his superior." After giving Pluto credit for his incomparable "fleetness or staunchness, in ranging or finding single birds," the author of Pluto's profile went into considerable detail about the dog's long and distinguished lineage: "Pluto was out of a setter bitch, called Phillis . . . and got by Ponto, who was descended from the first stock of pointers introduced into Virginia."[32] If there were a canine chapter of the First Families of Virginia, Pluto would no doubt qualify.

The comparison went well beyond dogs and birds, however. In the minds of many American sportsmen, the United States needed its own way of distinguishing itself with a socially appropriate approach to hunting. In 1827, a writer styling himself "A Gentleman of Philadelphia County" supplied just such an American alternative, the *American Shooter's Manual*. The book began with its own declaration of independence, arguing that hunters in the United States deserved "to have a treatise on the subject, better adapted to our own country, and at a more reasonable charge, than the British publications which are occasionally imported." It contained no end of good advice to the gentlemen-hunters of America, ranging from the practice of "*habitual coolness*" during the hunt ("not to be agitated when the dog is at a *stand*, nor flurried when the birds are flushed"), to the safe handling of a gun ("On no account, nor under any circumstance, point a gun at another, unless you intend to take his life"), to the antidote to the inexplicable "bad shooting" day that could discourage nineteenth-century hunters ("I would recommend the shooter to stop shooting and to take his seat by the side of a bank, or at the foot of a tree, and then, by completely abstracting his mind from the object of his pursuit, in a short time, get the better of his evil genius").[33]

On a good day, though, hunting could be a source of salvation—even, as one writer put it, of the divinely predestined kind. "A man must be a born shot as he must be a born poet," claimed a sportsman using the pseudonym J. Cypress Jr. "Reading and writing are inflicted by schoolmasters, but a crack shot is the work of God." Another writer stressed the more basic benefits of hunting, particularly its ability to "drive off *ennui, malaise,* or any other moping malady with a French name, which fashionable flesh is heir to." Those suffering from such maladies seemed to come most frequently from men whose minds had become "overtasked by business, . . . the thousands of persons engaged in the arduous pursuit of professions, which require intense abstraction, and who

would invariably break down if deprived of their usual relaxations in the shooting seasons."[34] Even with an allowance for some measure of humorous excess,
the point seemed clear: Hunting could do wonders for men in the urban professional classes, whose lives revolved around the countinghouse or office, whose
physical and mental health could be best restored by going to the field and
shooting a few ducks and partridges, presumably in the company of their
equally pressured peers.

In that regard, these celebrations of the benefits awaiting the American
sportsman represented only one side of the page. The equally important purpose of such essays was to draw a clear—and class—distinction between the
proper hunter and everyone else, between those who would exhibit "a taste for
genteel and sportsman-like shooting" and those who hunted for allegedly lesser
ends—such as meat for the market or even food for the table. The problem, so
the "Gentleman of Philadelphia County" argued, was that, unlike England,
the United States remained "a country destitute of game laws, and almost
without any restrictions, in regard to its destruction." In such a destructive
democracy, "every man, whose leisure, or circumstances will permit, may
become a shooter," and while so many did, "few indeed, will be found entitled
to the appellation of sportsmen; they are very generally game-killers, and
nothing more." The "Gentleman of Philadelphia County" turned his considerable ire on those who engaged in "that abominable poaching, game-destroying,
habit of ground-shooting, trapping, and snaring, which prevails throughout
our country, in the neighbourhood of all cities and large towns." His increasing
concern was that game birds were becoming less abundant in the United States,
and he offered several reasons for the decline: the growth and geographic
expansion of the population, severe winters, and animal predators such as "the
hawk, fox, mink, weasel, and other vermin." No sooner had he finished this
list of possible causes than he turned his animosity toward what he took to be
an especially troubling human threat: the "sneaking pot hunting shooter, who
at one fell sweep, will destroy a whole covey."[35] Coming immediately after the
term "other vermin," the placement of the "pot hunting shooter" made the
estimation by the "Gentleman" clear enough.

The hostility was not all his alone. Other writers complained about the
prevalence of "pot hunting"—that is, the practice of procuring birds for food,
to be cooked in the pot for a meal—criticizing not so much the motive as the
method. Pot hunters, so the argument went, took little heed of season, breeding habits, and habitat; they just killed for their own benefit, sometimes even

just for fun. One ornithological observer on Long Island noted the autumnal custom of shooting flocks of robins "as a matter of amusement," which was "only becoming to juvenile sportsmen." Another writer decried a more serious issue in the same area, the "absolute extinction of that noble bird the heath-hen, or pinnated grouse, on Long Island, where within the memory of our elder sportsmen they might be taken in abundance at the proper season, but where not a solitary bird has been seen for years." The fault, he argued, lay largely at the feet of the "ruthless pot-hunter." Yet another excoriated those who failed to understand, much less respect, the breeding and fledging patterns of birds and carried on an "exterminating warfare" out of season: Such a hunter "deserves the name of worse than poacher."[36]

One writer offered what initially appeared to be an inclusive-seeming appeal to the cross-class benefits of hunting. "In the city," wrote E. J. Lewis in his *Hints to Sportsmen*, "people of different ranks stand scowling and apart," but "in the fields, they are fellows . . . and if all mankind would study Nature, all mankind would be brothers." Lewis took note of various callings among those scowling urbanites—the "wan-faced student of science," the "learned counselor," the "skilful physician," the "anxious, upright, and enterprising merchant," and, not to be left out, "even the industrious artisan"—and urged them to come together in the hunt: "To the fields, then!" It might have been a self-serving social illusion to imagine that the artisan would be invited to join the lawyer, doctor, or merchant in a foray into the field, but in Lewis's eyes they could be fellows. On the other hand, like others who wrote primarily to engage the "gentleman sportsman," Lewis specified one sort of hunter to whom he would defiantly deny any right to be in the field: "Of all the disagreeable characters that a well-bred sportsman is likely to be thrown in contact with, that of a *pot hunter* is the most disgusting, the most selfish, the most unmanly, the most heartless, a being who alone can pride himself in a ruthless desire to destroy." The pot hunter, Lewis concluded, should essentially be forced to bear a hunter's version of the mark of Cain, "as a warning to all young sportsmen to shun his company, and detest his vices."[37]

To their ecological credit, the "Gentleman of Philadelphia County," E. J. Lewis, and other writers who attacked pot hunting, trapping, netting, and other such forms of bird killing did so with a self-proclaimed concern for conservation. Despite the abundance and variety of game of all sorts, "whether of fin, fur, or feather," the United States seemed to be fast becoming a land "in which the gentle craft of venerie is so often degraded into mere pot-hunting" that "as a natural consequence, the game that swarmed of yore in all the fields and forests, in all the lakes, streams, bays, and creeks of its vast territory, are in such peril of becoming speedily extinct."[38] The environmental threat seemed evident, even imminent. Equally evident, though, seemed the social origins of that threat. Proper hunting had its proper place in the social order, and the genteel distinction between those who practiced the "gentle craft of venerie" and those who engaged in "mere pot-hunting" defined the difference between the upper and lower levels. So it went in various books and periodicals aimed at gentlemen sportsmen of the early American republic. No matter how many birds teams of gentlemen might shoot in a day's sport, the finger of blame for the feared decline of bird populations was increasingly pointed at hunters who had neither the resources nor the leisure of the gentlemen-hunters. Poorer people who hunted for the sake of sustenance became readily branded as environmental vandals.

The American Woodsman's Broader Embrace

To his democratic credit, Audubon did not add his voice to the class-based associations that came to surround shooting birds. To be sure, he sometimes relied on the writings of urban sportsmen to supplement his own descriptions of appropriate hunting practices, and some of his favorite fellow shooters—John Bachman and Edward Harris, above all—had a decent claim to the skill and gentility that such authors expected of the well-bred hunter.[39] Audubon also wrote, as we have seen, with his own special disdain for the avaricious activities of "eggers" and other such predatory people. But he could still quite comfortably rub shoulders with ordinary people who procured birds in ordinary ways, even those that might not reflect the sort of sporting etiquette that the "gentle craft of venerie" seemed to require. The American Woodsman extended his embrace to Americans of all classes—and almost all hunting practices.

Audubon wrote without apparent criticism of the common practice of capturing birds in "snares, common dead-falls, traps, and pens," not to mention "the principal havock . . . effected by nets." Again, in his chapter on the Passenger Pigeon, Audubon mentioned in a matter-of-fact manner, that he "knew a man in Pennsylvania, who caught and killed upwards of 500 dozens in a clap-net on one day, sweeping sometimes twenty dozens or more in a single haul." Partridge netters, he noted, often took as many birds in a day, but they would typically "give liberty to a pair out of each flock, that the breed may be continued." Catching American Snipe in snares could be dismissed as child's play, as "a good number are thus obtained by the farmers' children." The male Painted Finch (or Painted Bunting), sold by the dozens as ornamental birds in the New Orleans market, could easily be captured by being lured into a trap with a stuffed bird, and Audubon took a few lines to describe the method "that it may afford some amusement."[40] He even wrote rather dispassionately about the practice of luring birds to their death as another source of amusement, this one apparently common among woodcutters in Maine "with what they call 'transporting the carrion bird.'"

> This is done by cutting a pole eight or ten feet in length, and balancing it on the sill of their hut, the end outside the entrance being baited with a piece of flesh of any kind. Immediately on seeing the tempting morsel, the Jays alight on it, and while they are busily engaged in devouring it, a wood-cutter gives a smart blow to the end of the pole within the hut, which seldom fails to drive the birds high in the air, and not unfre-quently kills them.[41]

Just as Audubon could make light of the Kentucky sharpshooters' entertain-ment of "barking squirrels," so could he look past the gratuitous cruelty of the woodsmen's game in Maine.

Audubon never took an absolutist stance against the many destructive human practices he found in his search for birds. He could be critical on some occasions, but more often he remained indulgently neutral, seldom subjecting ordinary people to socially superior judgments, much less class-based con-demnation, even when their behavior toward birds sometimes bordered on meaningless violence. His self-conscious identification with his fellow Americans led him to see them as part of the larger picture of nature in the new nation. To appreciate the place of ordinary people in that picture, in fact, Audubon sometimes took birds altogether out of it.

"The Scenery and Manners of the Land"

When Audubon promised to take his "kind reader" beyond "the mazes of descriptive ornithology," he got right to it, taking off on a travelogue in the early pages of *Ornithological Biography*. The very first of his Episodes, "The Ohio," offered a rhapsodic view of the autumnal landscape of the "queen of rivers": "Every tree was hung with long and flowing festoons of different varieties of vines, many loaded with clustered fruits of varied brilliancy, their rich bronzed carmine mingling beautifully with the yellow foliage, which now predominated over the yet green leaves, reflecting more lively tints from the clear stream than ever landscape painter portrayed or poet imagined." Using his painter's eye for color to guide his pen, Audubon created a scene in which the bounty of the natural world seemed a welcoming place for human habitation, where "here and there the lonely cabin of a squatter struck the eye, giving note of commencing civilization."[42] So it went throughout *Ornithological Biography*, with repeated pleasing descriptions of an inviting landscape throughout Audubon's travels up and down the United States, from a maple sugar camp in Pennsylvania ("Now, good reader, . . . should you find yourself by the limpid streamlets that roll down the declivities of the Pocono Mountains to join the Lehigh, and there meet with a sugar-camp, take my advice and tarry for a while") to the Florida Keys ("The surface of the waters shone in its tremulous smoothness, and the deep blue of the clear heavens was pure as the world that lies beyond them") and even out to sea in the Gulf of Mexico ("Vast numbers of beautiful dolphins glided by the side of the vessel, glancing like burnished gold through the day, gleaming like meteors by night"). Sometimes the beauty of the scenery gave way to a sense of the sublime, leaving Audubon uncharacteristically inarticulate in describing the immensity of nature: "Go to Niagara, reader, for all the pictures you may see, all the descriptions you may read of these mighty Falls can only produce in your mind the faint glimmer of a glowworm compared with the over-powering glory of the meridian sun."[43]

However vivid the descriptions, such scenes might seem unexceptional, even a bit saccharine, were it not for Audubon's locating them in a larger ensemble of more arresting Episodes. Instead of merely adhering to standard conventions of prettified, placid-seeming pictures of nature, Audubon more often introduced the specter of violence, sometimes lurking just below the surface and sometimes right in the center of the story. The pages of *Ornithological Biography* would occasionally burst forth with sudden shocks to the landscape—"A Flood," which presented "a splendid, and at the same time an appalling spectacle"; "The Earthquake," which created "this awful commotion in nature"; "The Hurricane," which "howled along in the track of the desolating tempest"; "The Force of the Waters," which rolled rapidly through a gorge, "the emblem of wreck and ruin, destruction and chaotic strife"; and the "Breaking Up of the Ice," which opened the Ohio River to flow suddenly into the Mississippi, so that "the two streams seemed to rush against each other with violence," throwing up huge fragments of ice that then crashed back down, "as the wounded whale, when in the agonies of death, springs up with furious force, and again plunges into the foaming waters."[44] Even when the landscape itself seemed to be placidly at rest, the suggestion of death sometimes intruded into the scene, as in Audubon's reverie about the beauties of a spring evening in Kentucky: "How calm is the air! The nocturnal insects and quadrupeds are abroad; the bear is moving through the dark cane-break, the land crows are flying towards their roosts, their aquatic brethren towards the interior of the forests, the squirrel is barking his adieu, and the Barred Owl glides silently and swiftly from his retreat, to seize upon the gay and noisy animal." The prowling owl is but one of many predatory birds, of course, and Audubon's depictions of avian life, both written and drawn, portrayed dozens of violent encounters between birds and mammals, birds and reptiles, birds and fish, and even birds and other birds.[45]

The violence inherent in animal life also found its reflection in the violence facing human beings in the American environment. The vast expanses of uncharted swamps and forests could themselves be a menace, leaving people lost and exhausted. Audubon told the tale of a "live-oaker" in the St. John's River region of Florida who became confused among the faint, overgrown trails, "wandered like a forgotten ghost that had passed into the land of spirits," spent his days isolated in the woods, and his nights surrounded by a terror that "rendered him almost frantic." Finally, after forty days—thus reaching the biblical equivalent of suffering in the wilderness—the live-oaker emerged near a river, heard some nearby boaters, and raised "his feeble voice on high;—it was a loud shrill scream of joy and fear." With the "Lost One" found, Audubon took note to tell his reader that "this is no tale of fiction"—although like all of Audubon's Episodes, it could well be, or at least a bit embellished—but it served to demonstrate the dangers nature could pose to an ever-vulnerable individual.[46]

Danger seemed to be everywhere in Audubon's America, and it came in many different forms. Bears, cougars, and animals of all sorts, even the occasional polecat, could pose threats to people in the woods, and the American landscape also proved to be a place with its share of potentially menacing human inhabitants—perhaps pirates, wreckers, eggers, and various others who emerged from "the refuse of every other country" whose "evil propensities find more free scope" on the ungoverned fringes of the American frontier.[47]

Audubon chronicled his own unhappy encounter with such rough folk in "The Prairie," in which he sought a traveler's rest one night in the cabin of a gruff-voiced woman and her two besotted sons. She offered him food and a place to sleep—"Such a thing as a bed was not to be seen, but many large untanned bear and buffalo hides lay piled in a corner"—but she also took a fancy to his gold watch, "saying how happy the possession of such a watch should make her." Her interest in the watch made Audubon nervous, and as he bedded down for the night, next to an Indian who also happened to be staying in the cabin, he made himself ready for whatever might happen next: "Never until that moment had my senses been awakened to the danger which I now suspected about me." Sure enough, the woman and her sons came to kill Audubon, and he was ready to shoot them, too—but then, just in the nick of time, in classic melodramatic fashion, the cabin door burst open, "and there entered two stout travelers each with a long rifle on his shoulder." Audubon and the others quickly tied up the woman and her sons, marched them into the woods to suffer some undisclosed fate, and burned down their cabin.[48]

The moral to the story, such as it was, had less to do with the perils of the prairie or the suddenness of frontier vigilante justice, but more with a form of American exceptionalism. Audubon gave the Episode a happy ending by noting that "during upwards of twenty-five years, when my wanderings extended to all parts of our country, this was the only time at which my life was in danger from my fellow creatures." In the case at hand, in fact, those "fellow creatures" who sought to do him harm were probably not fellow citizens anyway. "I can only account for this occurrence by supposing that the inhabitants of the cabin were not Americans." For all the violence and danger he might find in nature—and he found it everywhere—Audubon took care to speculate that when it came to human nature, violence must come from someone other than his fellow Americans—from "the refuse of every other country," perhaps, but not the good people of his adopted country. He could even consign his personal brush with danger to the past, putting it in the distant shadows of national development: "Will you believe, good-natured reader, that not many miles from the place where this adventure happened . . . large roads are now laid out, cultivation has converted the woods into fertile fields, taverns have been erected, and much of what we Americans call comfort is to be met with. So fast does improvement proceed in our abundant and free country."[49] From one bad night in a crude cabin, he could thus extract a positive picture of progress and give his "good-natured reader" a promising reassurance about the upbeat benefits of being American. That defined, more broadly, Audubon's approach to American history.

Audubon's American History

In what must be the longest single sentence in all of *Ornithological Biography*— with 263 words and eight semicolons—Audubon offered a reflection on all that had taken place during his time in Kentucky and, by extension, in the United States. Cumbersome though the passage may be, it encompasses a catalog of social, economic, and environmental change in the new nation, and, as Audubon's most succinct survey of American history, it deserves to be rendered in its entirety:

> When I think of these times, and call back to my mind the grandeur and beauty of those almost uninhabited shores; when I picture to myself the dense and lofty summits of the forest, that everywhere spread along the hills, and overhung the margins of the stream, unmolested by the axe of the settler; when I know how dearly purchased the safe navigation of that river has been by the blood of many worthy Virginians; when I see that no longer any Aborigines are to be found there, and that the vast herds of elks, deer and buffaloes which once pastured on these hills and in these valleys, making for themselves great roads to the several salt-springs, have ceased to exist; when I reflect that all this grand portion of our Union, instead of being in a state of nature, is now more or less covered with villages, farms, and towns, where the din of hammers and machinery is constantly heard; that the woods are fast disappearing under the axe by day, and the fire by night; that hundreds of steam-boats are gliding to and fro, over the whole length of the majestic river, forcing commerce to take root and to prosper at every spot; when I see the surplus population of Europe coming to assist in the destruction of the forest, and transplanting civilization into its darkest recesses;—when I remember that these extraordinary changes have all taken place in the short period of twenty years, I pause, wonder, and, although I know all to be fact, can scarcely believe its reality.

In this passage, Audubon seems initially wistful in his nostalgia for a lost "state of nature," where "Aborigines" and buffalo roamed in the "unmolested" beauty of "the dense and lofty summits of the forest," only to be removed as a result of violent human conflict stained by "the blood of many worthy Virginians," the intrusion of new technologies that brought the disruptive "din of hammers and machinery," and the ongoing "destruction of the forest" in the wake of increasing Euro-American settlement. But then, with scarcely a pause between paragraphs, he pulls back from taking a position on the matter: "Whether these changes are for the better or for the worse, I shall not pretend to say."[50]

Audubon had quite a tendency not to say anything about better or worse in several of his Episodes. He repeatedly wrote about topics that any enlightened observer might consider among the most critical issues of the era: economic development and environmental degradation, to be sure, but also racial conflict and slavery. Rather than render any real judgment on any of them, however, Audubon consistently declined to take a stance, occasionally even evading altogether the seemingly obvious implications of the very issues he had addressed. Nowhere was that more evident than in Audubon's discussions of the "other" Americans: Native Americans and African Americans. In those few times when he brought them into his Episodes, he gave them a brief appearance on the page, but then quickly erased them, thus also removing the need to answer any unsettling questions that might linger among mainstream Americans.

In one of his earliest Episodes, "Kentucky Sports," Audubon celebrated his supposed friend Daniel Boone, saying he "probably discovered" the region, and the sharpshooting settlers who followed him. In doing so, though, Audubon offered a brief, contradictory, and ultimately evasive history of what he acknowledged to be the "conquest of Kentucky," which, he noted with considerable understatement, "was not performed without many difficulties." Before the arrival of white "intruders," the native inhabitants "looked upon that portion of the western wilds as their own, and abandoned the district only when forced to do so, moving with disconsolate hearts farther into the recesses of the unexplored forests . . . dismayed by the mental superiority and indomitable courage of the white man." A few paragraphs later, Audubon resolved and essentially dodged the issue by adopting a position of quasi-pacifism: "I shall not describe the many massacres which took place among the different parties of White and Red men," he wrote, "because I have never been very fond of battles, and indeed have always wished that the world were more peaceably inclined than it is." Leaving the point by simply saying that "in one way or other, Kentucky was wrested from the original owners of the soil," Audubon then chose to change the subject and "turn our attention to the sports still enjoyed in that now happy portion of the United States"—sports that involved the skillful use of a gun, as had the "many massacres" that led to the pacification of this "happy portion" of American territory.[51] He pacified the process even further in the second volume of *Ornithological Biography*, when he noted that the Pinnated Grouse "abandoned the state of Kentucky, and removed (like the Indians) every season farther to the westward, to escape from the murderous white man."[52] The parenthetical association of Indians with birds made both part of nature, and Indian removal seemed almost to be a necessary step of migration. Thus Boone's Kentucky became Audubon's Kentucky.

When Audubon looked beyond Kentucky, he put the whole Indian question safely in the past. In writing of his 1832 trip to the St. John's River in Florida, he described how his party "spied a Seminole Indian approaching us in his canoe," apparently hoping to sell them some fish he had caught. Audubon used the occasion to take a literary leap into an elegy for this "poor dejected son of the woods." "Alas! Thou fallen one, descendant of an ancient line of freeborn hunters, would that I could restore to thee thy birthright, thy natural independence, the generous feelings that were once fostered in thy brave bosom. But the irrevocable deed is done." The Seminole soon becomes a "vanishing" Indian in the most literal sense, selling some trout and turkeys to the white men, then "without a smile or bow," taking off in his canoe "with the speed of an arrow from his own bow." At that point in the narrative, Audubon shifts immediately to a lively description of hunting alligators. The "irrevocable deed" of Indian dispossession, not to mention the violence that occasioned it, never appears in Audubon's Episodes again.[53]

Writing "The Runaway," Evading Slavery

Audubon could be equally evasive when it came to the greatest source of systemic violence in antebellum America: slavery. His sole detailed discussion of the peculiar institution occurred in the second volume of *Ornithological Biography*, in an Episode he called "The Runaway," about an escaped slave he claimed to have encountered in Louisiana.[54] Lucy Audubon was the first to note the success of her husband's storytelling efforts in this instance. Writing from Edinburgh in October 1834, when she and Audubon were furiously working on completion of the text, she assured her older son, Victor, that "the *runaway* reads well and is thought by some here to be most interesting."[55] Interesting it certainly was, but not just for the reasons Lucy apparently had in mind. However well it might read, Audubon's one extended narrative foray regarding slavery needs to be read for its disturbing but ultimately revealing insights into his larger sense of history and, within that, his sense of self, particularly his relationship with the question of race in American society, which had been a critical issue in his own identity since the day he was born.

Long before Audubon put "The Runaway" in print, he occasionally wrote about personal encounters with African Americans, almost always taking care to identify himself as being different from them—usually implicitly, but sometimes insistently—as a white man. While he was in New Orleans in 1820–1821, as we have seen, he played down the sexual attractions of the "Citron hue" of the mulatto women he saw there, writing in his journal about his preference for white women. He may have written that to placate his wife, Lucy, who he knew would later be reading his journal, but he may also been mindful of a price that could not be measured in dollars: To be secure about his own whiteness, he had to remind himself that he found women of color "disgusting" and that he only had eyes for women with "rosy Yankee or English cheeks" or, more bluntly, "*White Ladies* and Good Looking ones."[56]

In the case of black men, Audubon could separate one out from the mass and see admirable qualities, but mostly to note an exception. In the same journal of his Mississippi trip, for instance, he described visiting a *"Monsieur St Armand's* Sugar Plantation,"* where the owner had about seventy slaves engaged in the grueling work of cutting cane while he and Audubon went out shooting. Audubon looked at the slaves only long enough to say that the "Miserable Wretches at Work begged a Winter Falcon We had killed, saying *it Was a great treat for them*." (What Audubon did not say was whether or not he gave them the dead bird.) Instead, he turned his eye to their overseer, "a Good Looking Black Man," who had been in his position for eight years and had gained "so much of his Master's Confidence, as to have the Entire Care of the Plantation." The overseer "spoke roughly to his under servants but had a good indulgent Eye, and No doubt does what he Can to Accomodate, Master and All." The greater accommodation, of course, was Audubon's to the institution of slavery: the "Good Looking Black Man" in the overseer's position, with his "good indulgent Eye," allowed Audubon to see a reasonable human mediation between the master and the "Miserable Wretches" in the cane fields, and thus to see a good or certainly acceptable system that put a black man in charge of other black people and allowed the master, St. Armand, to go out bird hunting with his "richly ornemented Double Barrelled Gun"—and with Audubon, his recently arrived hunting partner.[57]

Hunting could define, in fact, Audubon's racial crucible. In writing a British friend about an expedition in Florida in December 1831, Audubon noted that he set out in a boat with "six hands, and '*three white men.*'" The quotation marks and emphasis are Audubon's, and they serve to make the "six hands" unmistakably distinct from the "three white men," who soon became the heroes of the story. Audubon and his party became stuck "fast in the mud about 300 yards from a marshy shore," where they had no chance of building a fire to ward off the nighttime cold, so they suffered through the night until the next morning, when the dawn found "all hands half dead, and masters nearly exhausted as the hands." Finally, though, they rowed their way through the marsh, and eventually landed on the shore, whereupon "on reaching the margin of the marsh, two of the negroes fell down in the marsh, as senseless as torpidity ever rendered an alligator, or a snake; and had we, *the white men*, not been there, they certainly would have died."[58] Once again, Audubon seemed especially insistent about underscoring, both on the page and in his reader's mind, the identity of "*the white men*," and the "we" in apposition emphasizes Audubon's own inclusion among them. The whole story of the overnight ordeal set the "masters" further apart from the "servants" or "hands," by twice describing the superior physical and mental strength of the white men: The masters were "nearly" as exhausted as the hands, but their presence—and perhaps presence of mind—saved the black men from death. Indeed, the fact that Audubon not only described these black men as "servants," but also associated them with "an alligator, or a snake" makes them still more distinct from their "masters"—and less human. Even without the emphasis on "*the white men*," Audubon made clear his assumption of the fundamental inferiority of the "servants" or, less euphemistically, slaves.

Audubon could accept slavery because he himself had been a slaveowner at times, even though he seldom said much about that in print. In his unpublished autobiographical account, "Myself," he made a few references to the slaves he and Lucy once owned back in their Kentucky days, most notably telling his sons about how he disposed of one of them to a Spanish-born adventurer he called General Toledo, who was "then on his way as a revolutionist to

South America." When Toledo tried to recruit Audubon to join him on the expedition, he declined, but he apparently didn't see any problem with sending one of his slaves off to be a "revolutionist"—albeit presumably an unfree one: Toledo "purchased a young negro from me, presented me with a splendid Spanish dagger and my wife with a ring, and went off overland to Natchez, with a view of there gathering recruits."[59] He had to sell his remaining slaves at the end of his Henderson years, when he faced financial ruin in the Panic of 1819. He and Lucy acquired more slaves in the 1820s, though, and when he came back to the United States to get her to join him in England, in 1830, the couple sold a woman and her two sons to friends before departing.[60]

Thus divested of his own human property, he still seemed singularly unimpressed with the growing abolitionist movement on both sides of the Atlantic. In 1833, he wrote Victor that "in England little can now be expected, The English are agog on Emancipation reform, &c &c." A year later, he wrote Lucy about reading a newspaper account "of some disturbances in the British West Indies which did not surprise me, but rendered Me Surer than ever, that in giving Freedom to the Slaves of Those Islands, the British Government had acted imprudently and too precipitously."[61] If it occurred to Audubon that the situation in the British West Indies hit rather close to home—quite literally, close to his birth home in Saint-Domingue—he never made the connection in writing. To the extent that he wrote anything about emancipation and liberation, he did so only with skepticism and even disdain. It is with rather indulgent understatement, then, that one of Audubon's recent biographers has observed that "Lucy and John Audubon took no stand against the institution of slavery."[62]

That personal background provides the context for Audubon's story of "The Runaway," in which he did take a stand against slavery—in his own mind, at least, and on his own terms. Indeed, as the story unfolds, the relationship between Audubon and the runaway tells us as much about Audubon as about the slave, and it certainly underscores Audubon's sense of identity as being separate from, and certainly superior to, the unfortunate fugitive—a man who remains nameless throughout the story.

Audubon encountered the runaway quite by surprise, he recounts, while returning from a day of collecting bird specimens, in this case six ibises, in a menacing environment, the "Louisiana swamps pregnant with baneful effluvia." As he waded across a bayou with his dog, Plato, he suddenly noticed that the animal began "exhibiting marks of terror, his eyes seeming ready to burst from their sockets, and his mouth grinning with the expression of hatred, while his feelings found vent in a stifled growl." Thinking that the dog's agitation came from sensing another animal, perhaps a wolf or bear, Audubon had no sooner raised his gun than he heard a "stentorial voice" command him to "stand still, or die!" He then found himself standing face-to-face with a "tall firmly-built Negro" (perhaps modeled on the "Good Looking Black Man" he had earlier encountered as the overseer on a Louisiana sugar plantation) who had a gun aimed at Audubon's breast. Audubon had a gun at the ready, too, though, and in the first instant of this armed confrontation, he immediately thought of shooting: "Had I pressed a trigger, his life would have instantly terminated." But Audubon quickly observed that the man's gun was "a wretched rusty piece, from which fire could not readily be produced," and therefore no threat, so he "did not judge it necessary to proceed at once to extremities." Instead, Audubon lowered his gun, calmed his dog, and asked the man what he wanted.[63]

After grabbing his reader's attention with the sudden specter of an armed, "firmly-built" black man—a fearsome image sure to play to the anxious fantasies of antebellum whites—Audubon quickly restored his sense of superior racial and social authority to the scene: "My forbearance, and the stranger's long habit of submission, produced the most powerful effect on his mind." In an instant, the armed runaway became a simpering servant again, addressing Audubon as "Master" and begging his mercy: "For God's sake, do not kill me, Master!" Given the late hour of the afternoon, he invited Audubon to come to his camp in the swamp, where he promised him protection for the night and assistance the next morning, even at considerable risk to himself: "I will carry your birds, if you choose, to the great road." Audubon seemed impressed with the man's offer, not to mention his sudden reversion to type, both of which he accepted with a combination of relief and noblesse oblige: "Generosity exists everywhere. . . . I offered to shake hands with the runaway. 'Master,' said he, 'I beg you thanks,' and with this he gave me a squeeze, that alike impressed me with the goodness of his heart, and his great physical strength. From that moment we proceeded through the woods together." By that moment, in fact, Audubon's narrative rendered the runaway completely pacified, a strong but deferential-seeming slave who would twice beg his "Master" for mercy and thanks. Indeed, it even transformed his racial identity at one point, with Audubon complimenting his new companion for being "a perfect Indian in the knowledge of the woods, for he kept a direct course as precisely as any 'Red-skin' I ever traveled with." Audubon's dog still had his doubts, though, and he sniffed the strange being whose sudden appearance had set "his mouth grinning with the expression of hatred, while his feelings found vent in a stifled growl." But as if to assure the reader that his own human generosity stood superior to the dog's animal instincts, Audubon again calmed the dog as they walked toward the runaway's camp.[64]

When they got there, Audubon met the rest of the runaway's family, a wife and three children. The man clearly loved his family, Audubon noted, and "with an expression of gentleness and delight, when his beautiful set of ivory teeth seemed to smile through the dusk of evening," he talked of them with admiration and pride: "'Master,' said he, 'my wife, though black, is as beautiful to me as the President's wife is to him; she is my queen, and I look upon our young ones as so many princes.'" To Audubon, the children seemed a bit less princely, and they soon "retired into a corner like so many discomfited raccoons." On the whole, Audubon's description of the runaway family—the man, whose ivory-toothed smile could have come right out of a minstrel show; the woman, whose beauty, "though black," even her husband felt the need to qualify; and the children, whose status Audubon changed from little princes to little "raccoons" in the space of a few lines—cast them as common, almost comical, racial stereotypes. Yet still Audubon reminded his reader to consider them potentially dangerous: "Only think of my situation, reader! Here I was, ten miles at least from home, and four or five from the nearest plantation, in the camp of runaway slaves, and quite at their mercy."[65]

Audubon might have considered the nearest plantation a safe haven, but for the runaway slaves, the plantation was the place they had eagerly escaped— or so, at least, the reader might expect. But that difference of perspective on the plantation never really came into play in defining the ultimate outcome of the whole story—which was, after all, Audubon's to tell.

The slaves' story, in Audubon's telling, was of going from one plantation to another, but under the worst of circumstances. On the first plantation, where the runaway "had ever been treated with the greatest kindness," the master had fallen onto financial hard times and "was obliged to expose his slaves at a public sale." The runaway and his family were all valuable slaves—the man brought "an immoderate price," his wife fetched eight hundred dollars, and the children, "on account of their breed, brought high prices"—but they went to different owners on distant plantations. Pining away over the loss of his loved ones, the runaway "feigned illness" and waited for a chance to find his family members. Then, on a suitably dramatic dark and stormy night, "when the elements raged with all the fury of a hurricane, the poor Negro made his escape." Making his way to the other plantations, he eventually "succeeded in stealing" his wife and children, "until at last the whole objects of his love were under his care." Together they fled into the canebrake, where, with the furtive assistance of slaves on the original plantation, they managed to live in secrecy and avoid detection—until Audubon happened upon them.[66]

After telling Audubon their story around the firelight of their camp, the fugitive man and woman tearfully appealed to their unexpected visitor: "Good master, for God's sake, do something for us and our little children." Audubon admitted to being affected by their story, for "who could have heard such a tale without emotion?" Agreeing to sleep on the situation and suggest a plan in the morning, Audubon then bade them good night and went to bed, while the couple "both sat up that night to watch my repose." True to his word, Audubon awoke the following day with a sense of optimism that, indeed, the reader might be just on the point of sharing.[67]

But then the story took a turn back toward the South's status quo: "I scarcely doubted of obtaining their full pardon . . . and promised to accompany them to the plantation of their first master." Obtaining their pardon, that is, meant taking them back to be slaves again, not helping them escape for good. In the end, off they all went to the old plantation, "the owner of which, with whom I was well acquainted, received me with the generous kindness of a Louisiana planter," undoubtedly a true gentleman in Audubon's eyes: "He afterwards repurchased them from their owners, and treated them with his former kindness; so that they were rendered as happy as slaves generally are in that country." Exactly how happy that was, Audubon didn't say. He did, however, conclude the story on a supposedly upbeat note: "Since this event happened, it has, I have been informed, become illegal to separate slave families without their consent."[68] And there the story ended, with the family reunited but reenslaved, the original owner again in possession of his human property, the other owners compensated for the loss of theirs, and Audubon apparently departed from the scene, no longer directly involved but apparently distantly aware ("I have been informed") of some modest reform of the slave system— a reform that, in fact, never happened.[69]

"The Runaway" gave readers a gripping melodrama, and different publications on both sides of the ocean almost immediately appropriated it for their own pages. Perhaps not surprisingly, William Lloyd Garrison's fiery abolitionist newspaper, *The Liberator*, ran "The Runaway"—but only up to a point.

Garrison's version of the narrative stopped three paragraphs short, just as the runaway slave had told of making his escape, rescuing his family members from other plantations and reuniting them in his hideout, and somehow avoiding being recaptured in the canebrake, which was "daily ransacked by armed planters." *The Liberator* left off the last part about Audubon's returning them to the plantation of their original master, where "the Runaway and his family were looked upon as his own." For Garrison and his readers, it was the slaves' resistance that defined the heart of the tale. Any notion of the first master's comparatively kind indulgence, not to mention any alleged amelioration of Louisiana's slave laws about breaking up families, had no place in an abolitionist newspaper.[70] Other newspapers ran Audubon's episode in its entirety, all the way through to the return of the slaves to their former master to be "as happy as slaves generally are in that country," or at least added a reassuring sentence or two to make the story end as Audubon had originally written it. The Newark (NJ) *Daily Advertiser*, for instance, ran the same truncated version of "The Runaway" as *The Liberator*, but it added an additional note at the end, bringing the story to Audubon's upbeat-seeming conclusion: "We may conclude this interesting anecdote by telling the reader that the Runaways were, by the intervention of our author, placed in a state of comparative comfort, under the protection of a Louisiana planter."[71]

Audubon never objected to the ways other publications used "The Runaway," and he never had to defend the story as being absolutely true in the first place. But again, literal, even rather elastic, truth seemed hardly the point. What no doubt made the story of the runaway interesting to an antebellum audience was the range of reactions a reader might find after following Audubon through his canebrake encounter with this "tall, firmly-built Negro." The initial fear of facing an armed black man gave way to the immediate relief that came with the slave's simpering pacification in the presence of his "good master." The apparently sympathetic hearing of the runaway's separation from his loved ones led only to a pragmatic-seeming decision to return the family to their former master. The implicit appreciation of the fundamental cruelty of slavery became buried under a tepid, off-handed report of reform. In general, from confronting the implications of the runaway slave's situation to conforming to the "manners" of the slaveholding South, "The Runaway" invited its nineteenth-century readers to look askance at the worst aspects of slavery but, in its original version at least, also ultimately allowed them to accept the peculiar institution's continued existence. For

Audubon and, indeed, for most white people in the new nation, what could be more American than that?[72]

And what could serve Audubon's own Americanness better than that? Whatever he knew about his own origins in Saint-Domingue, and whatever he might have feared could emerge from his association with such a racially complicated birthplace, Audubon had consistently, even insistently, established his identity on the white side of the racial divide in antebellum America. *Ornithological Biography* may not have explicitly been the ornithologist's autobiography, but it did provide enough of a first-person account to establish his identity as the American Woodsman, a man of science who could claim his rightful place as both an exceptional observer of birds and a man of the people who could celebrate his place among ordinary—and white—Americans. As much as Audubon used his written work to show readers of his own era how to see birds, he also used it to show them how to see him as well.

Chapter 8

~~~

Exploring the Ornithology of Ordinary People

I am persuaded that you love nature—that you admire and study her. Every individual, possessed of a sound heart, listens with delight to the love-notes of the woodland warblers. He never casts a glance upon their lovely forms without proposing to himself questions respecting them.

—John James Audubon, "Introductory Address," in *Ornithological Biography*

"Every individual, possessed of a sound heart"—Audubon had to hope that such people must still exist and that they might have something to tell him. He also knew that the study of nature needed all the eyes and ears it could muster. Paying attention to the natural world and proposing questions about what one sees and hears form the basic tasks of scientific inquiry, and every individual could indeed do that. Audubon's challenge, then, was to involve all those eyes and ears in ways that could be useful to his own scientific inquiries and, given the conflicted context of science at the time, credible within the larger scientific community.

Audubon was by no means the first naturalist in early America to understand that. Peter Kalm, the noted Swedish naturalist who came to the American colonies in 1748 and traveled extensively through the mid-Atlantic region, conducted his observations of the American environment with considerable assistance from the local inhabitants. Kalm had studied in his own country under the eminent naturalist Carolus Linnaeus, and he thus had an impeccable pedigree in eighteenth-century science. In America, though, he sometimes grumbled that the people he encountered there took "little account of Natural History, that science being here (as in other parts of the world) looked upon as a mere trifle, and the pastime of fools." But even if some Americans did seem to take little account of natural history as a field of study, they still took account of nature itself. Kalm's narrative of his travels contains numerous references to the information he gained from common people—"Some old people here told me . . ."; "A very creditable lady and her children told me . . ."— and he not only took care to note their observations, but also generally accepted them as worthy enough to incorporate into the more formal discourse of natural history.[1]

Alexander Wilson similarly noted that he had been "honoured with communications of facts, from various quarters of the United States," which he deemed to be absolutely necessary to the naturalist's task. With so many species spread over such an extensive territory, he wrote, "the observations of one man . . . are altogether insufficient to embrace the whole; and unless assisted by the experience and observations of others, a thousand interesting facts and minutiae of character, would unavoidably escape him."[2] Wilson's emphasis on the challenges of collecting information in a country with so much space, so many species, and so few sources of formal scientific support underscored the particular importance of engaging allies in observation.

Audubon's sometime friend, William Swainson, grasped the issue of accessibility equally well, offering a modest but meaningful observation about the role ordinary people could play in the investigation of nature: "Natural history has this peculiar advantage—that it can be prosecuted, in one shape or other, by almost every body. . . . It is as much within the reach of the cottager as of the professor; . . . it embraces questions which can be solved by the former, just as well, and frequently much better, than by the latter."[3]

There were, in fact, precious few professors at the time. The boundaries of science were reasonably elastic and porous, before scientific research became divided into specific disciplines, before ornithology became a distinct academic discipline, before scientists came to reside in university- and government-sponsored research labs. (The term "scientist," in fact, did not come into the English-language lexicon until the 1830s.[4]) Particularly in the period when the questions of natural history could still be pursued "by almost every body," the traditions of commonsense science among ordinary people—what some anthropologists now call "folkbiology"—remained a significant part of scientific discourse and defined, as one scholar has put it, "the negotiation between popular and formal knowledge."[5]

Never was that negotiation more apparent and more significant than in Audubon's era, and no one made it more of his scientific signature than Audubon himself. In the decades following American independence, as Audubon joined other American naturalists in seeking to define both natural history and even nature itself in distinctively American terms, the reliance upon, even alliance with, the American people seemed central to the naturalists' enterprise. For all the vitriol Audubon and his ornithological antagonists spewed onto each other in their enervating internecine spats, these "gentlemen of science" knew they could not fully define the study of nature as their exclusive domain. The wiser ones, Audubon certainly included, learned to pay attention to the wisdom they might derive from their fellow Americans. Any serious student of science had to embrace the notion that ordinary people could demonstrate a deep, surprisingly detailed understanding of the natural world that sometimes rivaled—or certainly supplemented—their own. As a man of science who styled himself the American Woodsman in an unending pursuit of birds in the American republic, Audubon could hardly afford to dismiss anyone as a possibly useful source of observation and information.

"I Have Heard It Asserted"

On those few occasions when he failed to pay attention to ornithological information from ordinary people, he regretted it. While visiting his friend John Bachman in Charleston, South Carolina, Audubon admitted he almost ignored the observations of a young boy who claimed to have seen a pair of flycatchers "differing from all others with which he was acquainted." Audubon and Bachman "listened, but paid little regard to the information," declining even to visit the spot where the boy claimed to have seen the new birds. The boy persisted, however, and a week later Audubon and Bachman agreed to follow him to the nest, only to find that it had been destroyed by some other boys. "The birds were not to be seen," Audubon wrote, "but a Common King Bird happening to fly over us, we jeered our young observer, and returned home." Audubon soon left Charleston, but sometime later he received a letter from Bachman telling him that he had in fact seen the new species, the Pipiry Flycatcher (or Gray Kingbird), near the spot where the boy had taken them. Appropriately embarrassed at having mocked the observant boy, Audubon admonished himself in print by making the moral of the story "the propriety of never suffering an opportunity of acquiring knowledge to pass, and of never imagining for a moment that another may not know something that has escaped your attention."[6]

On the other hand, Audubon still had to uphold his own scientific authority as a man of science in a republican society. As much as he came to appreciate the information ordinary people might give him about birds, he still had occasion to question, if not completely dismiss, some of what they told him. Although his own descriptions of birds could sometimes contain lively stories that a modern ornithologist might dismiss as unscientific fluff, Audubon still took great care to provide the kind of close observation about physical characteristics, habits, and habitat that could stand up well enough to scientific scrutiny. He might be quite willing to consider ornithological reports from almost anyone, but he often accepted them only with skepticism and even subjected them to the test of experiment.

He wrote, for instance, about having "spent much time in trying to ascertain in what manner the Chuck-will's-widow removes her eggs or young." This bird, which does not make a nest as such but scratches out a space in dead leaves on the ground, moves her eggs or young birds if she senses a threat. Audubon had never witnessed this behavior for himself, but other people apparently had. Some farmers had told him that the bird carried her offspring under her wing. But the farmers had apparently offered their information "without troubling themselves much about the matter," and he gave it little consideration. Audubon turned, then, to their slaves: "The Negroes," he noted, "some of whom pay a good deal of attention to the habits of birds and quadrupeds, assured me that these birds push the eggs or young with their bill along the ground." Given the two sources, then—black slaves, whose observations came from paying "a good deal of attention" to birds, and white farmers, who offered theirs "without troubling themselves much"—Audubon concluded that "the account of the Negroes appearing to me more likely to be true than that of the farmers, I made up my mind to institute a strict investigation of the matter."[7]

Thus he revealed the importance of scientific (and racial) authority inherent in the situation: Even though he accorded the slaves a measure of respect for their habits of observation, and even though he acknowledged that their account seemed "more likely" than that of the white men, he could not (or would not) accept their information without testing the observation for himself. Throughout his descriptions of bird behavior in *Ornithological Biography*, in fact, he repeatedly noted observations that he had received from other people but had not verified himself—"I have heard it asserted ... but have never met with an instance of the display of this alleged faculty"; "I have never witnessed any thing of the kind, and therefore cannot vouch for the truth of the assertion"; "I was told . . . but, as I have not myself made the experiment, I cannot speak of this as fact"—and if he could not convince himself, he would not try to convince his reader.[8]

In that sense, the apparent acceptance of the ordinary American as a fellow investigator could conceal a somewhat problematic relationship. It certainly did not mean complete intellectual equality, at least from Audubon's perspective. In writing about ornithology, he quietly but clearly assumed ultimate authority over what he wrote, particularly when it came to classifying and naming species and describing bird behavior. He had to, of course. In the end, he knew it was his reputation he had to develop—and, equally important, defend. Throughout *Ornithological Biography*, Audubon might generously give ordinary people a place on the page, but he never let anyone forget that that it was, after all, his page.

The Ornithology of Ordinary People

If Audubon had had to name the most popular bird among the American people, he might well have picked the Purple Martin (see Plate 10).

> Its notes are among the first that are heard in the morning. . . . The industrious farmer rises from his bed as he hears them . . . [and] renews his peaceful labours with an elated heart. The still more independent Indian is also fond of the Martin's company. He frequently hangs up a calabash on some twig near his camp, and in this cradle the bird keeps watch, and sallies forth to drive off the vulture. . . . The humbled slave of the Southern States takes more pains to accommodate this favourite bird. . . . It is, alas! to him a mere memento of the freedom which he once enjoyed; and . . . he cannot help thinking how happy he should be, were he permitted to gambol and enjoy himself day after day, with as much liberty as that bird.[9]

Purple Martins could be valuable allies for everyone, not least for very pragmatic reasons: Martins feed on flying insects, especially swarms of mosquitoes and gnats and the like, and they generally help keep human habitations free of other flying pests that can be even more bothersome, including vultures, hawks, and crows. That sort of protective service could be especially useful in a society in which most people took their living from the land, whether the "industrious farmer," the "independent Indian," or the "humbled slave."

But Audubon did more than simply describe the behavior and benefits of this particular bird. He also described the reactions of human beings to the bird, offering the reader examples of the ways human beings found both useful information and hopeful imagination in the birds around and above them. Putting ordinary people into the picture made an important, if inadvertent, point: No matter what their race or condition, people did pay attention to birds, and some watched birds closely and carefully enough to recognize their songs, habits, and usefulness. To a considerable degree, it made sense to take note of these creatures that flew over the earth, marked the changes of the seasons, and visited, even protected, human habitations. That attention to birds also manifested itself in a variety of aesthetic and emotional responses, ranging from the farmer's "elated heart" to some of the deepest of human desires, perhaps chief among them the slave's longing for the freedom to "gambol and enjoy himself day after day."

Audubon likewise looked forward to seeing the Purple Martin, and no matter where he lived in the United States, he typically took note of the bird's arrival—the first few days of February in New Orleans, as early as the fifteenth of March at the Falls of the Ohio, around the tenth of April in Philadelphia, then Boston about the twenty-fifth of that month, whence they "continue their migration much farther north, as the spring continues to open." By keeping an eye out for this always-welcome bird, he was doing what many Americans had been doing for decades.[10]

He was also taking his place in an ornithological conversation that had likewise been going on for decades, pitting "good" birds versus "bad" ones, with American farmers playing a role as both interested spectators and aggressive combatants. Some of the leading participants in the debate were, indeed, quite extraordinary farmers, older and more prominent Americans than Audubon himself. In March of 1789, for instance, James Madison took a respite from pondering the implications of human nature that seemed so critical to the new federal Constitution he had worked so hard to help produce and promote. Instead, he turned his attention to other forms of nature that surrounded him on his Virginia plantation, Montpelier. On the twentieth day of that month, he wrote in his meteorological journal that the day dawned chilly and clear, and that "black birds first appeared." A week later, on a day when the sky was clear enough at dawn for him to see the aurora borealis, he noted that

"Martins first appeared." And so it went in March for several years in a row, with Madison acknowledging the advent of "Crow Black birds," followed a few days later by the appearance of Purple Martins, and occasionally a sighting of "Wild Geese Passing."[11] Madison, the celebrated framer, was hardly a common farmer—after all, he had enslaved people at Montpelier to do the agricultural labor—but he understood, perhaps as well as his slaves, the relationship between crows and martins, knowing that as soon as one arrived in his fields, he would soon look forward to seeing the other. The coming of crows always occasioned considerable worry, because these plentiful predators could ravage a field long before the plants had a chance to mature and be harvested. On the other hand, the near-concurrent arrival of martins served as something of an ecological equalizer, providing a measure of relief by helping keep the crows at bay. Clearly, Madison knew that some birds could be the farmer's friend, and he likewise knew that making a living from the land also required taking occasional note of the skies.

Madison and other American farmers also knew that birds likewise took note of the farmer's labors, particularly in transforming forest into farmland. "Singing birds do not frequent the deep woods," wrote Jeremy Belknap in his *History of New Hampshire* (1792), and in the forest only "the squalling of a jay" might reach—and perhaps offend—the listener's ear.[12] His New England neighbor, Samuel Williams of Vermont, agreed, noting that the "musical birds do not deign to dwell in such places" because they might be "disgusted with so gloomy a scene, or dislike the food in the uncultivated lands." The only way to enjoy the beauty of bird songs, which he called "the only natural music," would be to transform nature itself: "No sooner has man discharged his duty, cut down the trees, and opened the fields to the enlivening influence of the air and the sun, than the birds of harmony repair to the spot, and give it new charms by the animating accents of their music."[13] Williams's notion that birds might be so "disgusted" with the scenery or cuisine in the woods that they would not "deign" to stay there depicted them almost as if they were disappointed tourists, and he would not be the last naturalist to commit the scientific sin of describing animal life in human terms. Neither would he be the last to describe the human impact on animal life, noting how his fellow Americans discharged their "duty" to cut down trees in the service of agricultural and national improvement.

Certainly no other human activity in the pre-industrial era had as great an impact on the American environment in general and on bird populations in particular than clearing the forests for farming. Indeed, in the eyes of many observers of the American landscape, none was more beneficial, both to the nation and to nature. "In the advancement of civilization," wrote Gilbert Imlay in 1793, "agriculture seems to have been in every country the primary object of mankind—Arts and sciences have followed, and, ultimately, they have been relevant to each other."[14] Birds followed, too. If, as Imlay argued, arts and sciences followed agriculture, the science of ornithology had to be among the first in line.

What these proponents of clearing forests for the benefit of both birds and farmers neglected to say, however, was that cultivation brought more than songbirds to the fields, and farmers had more to do than merely listen. Crows may have been the most obvious bane of every farmer's existence—"noxious birds," Peter Kalm called them, "for they chiefly live upon corn"—but they alone did not account for all the agricultural depredations that farmers faced. Kalm included other blackbirds and grackles in the category that he called, as was the custom in his Swedish homeland, "maize thieves." These birds came and increased in number as a logical consequence of the farmer's life and labors: "As [human] population increases, the cultivation of maize increases, and of course the food . . . is more plentiful." On the other hand, Kalm also pointed out that people sometimes planted grain that had been dipped in "a decoct of the root of the *veratrum album*, or white hellbore," so that "when the maize-thief eats a grain or two, which are so prepared, his head is disordered, and he falls down: this frightens his companions, and they dare not venture into the place again."[15] On the whole, Kalm did not question the wisdom or effectiveness of these folk approaches to bird control but simply included them as part of his dispatches from the field in the long-contested war over corn.

Still, when farmers persisted in trying to kill the various birds that caused them trouble—not just crows, grackles, and other blackbirds, but catbirds, woodpeckers, and flycatchers—they were harming their own crops, many naturalists insisted. Like the so-called maize thieves, these other birds did the farmer a service by consuming worms and grubs and other insects that would otherwise damage fresh-planted crops in the spring. Killing them indiscriminately, Alexander Wilson warned, could "desolate the country with the miseries of famine!" People needed to understand the benefits of these birds within the basic principles of ecology, which he cast in quasi-religious terms: "Is not this another striking proof that the Deity has created nothing in vain; and that it is the duty of man, the lord of creation, to avail himself of their usefulness, and guard against their bad effects as securely as possible, without indulging in the barbarous and even impious wish for utter extermination."[16] In a sense, once the farmer had carried out his "duty," as Samuel Williams had called it, to clear the land, Wilson posited another sort of duty: the need to find a reasonably balanced approach that allowed the farmer to raise crops and the birds to eat worms. All birds were useful, Wilson argued, even the Turkey Vulture, which, for all its "filthy habits," he still described as "gregarious, peaceable and harmless"—surely the most charitable characterization of that bird that anyone had yet made.[17]

The "Narrow-minded Farmer" in Nature

Recognizing the useful as well as the harmful side of bird behavior became a consistent theme among early American naturalists, and Audubon took it up repeatedly in the pages of *Ornithological Biography*. He acknowledged that ordinary farmers sometimes seemed to understand the point—but not well enough. In writing about the Red-Winged Starling (now called the Red-winged Blackbird), Audubon observed that "farmers do not molest them in spring," because they understood how valuable the birds could be in consuming grubs and caterpillars and the like. He even described a scene that seemed almost emblematic of a peaceable kingdom, in which the birds apparently sensed their safety in the presence of a grateful farmer: "They then follow the ploughman, in company with the Crow Blackbird, and as if aware of the benefit which they are conferring, do not seem to regard him with apprehension."[18]

But a few pages later, Audubon reverted to the more familiar theme of decrying the unsophisticated self-interest of the farmer. In a passage about the Tyrant Fly-Catcher (now called the Eastern Kingbird), he argued that even though this admirable bird deserves to "remain unmolested," the farmer typically finds "a single fault sufficient to obliterate the remembrance of a thousand good qualities ... [and] persecutes the *King Bird* without mercy." It was the farmer, then, that thus became the tyrant in Audubon's account, not the bird, which in this reversal of roles became one of his "subjects," persecuted without mercy for the farmer's self-aggrandizing greed:

This mortal hatred is occasioned by a propensity which the Tyrant Fly-catcher now and then shews to eat a honey-bee, which the narrow-minded farmer looks upon as exclusively his own property, although he is presently to destroy thousands of its race, for the selfish purpose of seizing upon the fruits of their labours, which he does with as little remorse as if nature's bounties were destined for man alone.[19]

By taking account of the supposed thousand-to-one rate of destruction of bees, and then the appropriation of the honey they had worked to produce, Audubon not only put the farmer in the role of a voracious thief, but also accused him of environmental arrogance, "as if nature's bounties were destined for man alone." In a subsequent discussion of the American Crow, Audubon likewise took note of "the bounty of Nature in providing abundantly for the subsistence of all her creatures," suggesting that his reader should "admire all her wonderful works, and respect her wise intentions, even when her laws are beyond our limited comprehension." Even in his image of the Purple Grackle, where he showed what he described as the "nefarious tendencies" to devour the "tender, juicy, unripe corn on which he stands," Audubon held the bird blameless in the larger scheme of things: "This is the tithe our Blackbirds take from our planters and farmers; but it was so appointed, and such is the will of the beneficent Creator."[20]

Although this notion of Nature's "wise intentions" was not a position Audubon would develop fully or promote consistently, it did posit a broader, perhaps loftier view of the environment that distinguished himself as naturalist—and perhaps his reader—from the "narrow-minded farmer," whose possessive notion of nature's bounty as "his own property" created such an apparent penchant for avian devastation:

> Wherever within the Union the laws encourage the destruction of this species, it is shot in great numbers for the sake of the premium offered for each crow's head. You will perhaps be surprised, reader, when I tell you that in one single State, in the course of a season, 40,000 were shot, besides the multitudes of young birds killed in their nests. Must I add to this slaughter other thousands destroyed by the base artifice of laying poisoned grain along the fields to tempt these poor birds? Yes, I will tell you of all this too.[21]

Taking a "tithe" of corn according to "the will of the beneficent Creator" put the bird's behavior in quasi-religious terms; the farmer, by comparison, had only the law of the state on his side.

Still, Audubon was a close-enough observer of human nature as well as bird behavior to know the risks of going too far out on a limb: Perhaps not wanting to alienate such a large portion of the American population, he refrained from attacking American farmers too harshly for their protective practices in the corn wars. To farmers with a continuing crow problem, Audubon offered a simple and conciliatory-seeming solution: "I would again address our farmers, and tell them that if they persist in killing Crows, the best season for doing so is when their corn begins to ripen."[22] He could even bring humor into the picture, as he did in the story he told "from the mouth of an honest farmer" who had trouble with Wild Turkeys. Too many turkeys eating too much just-emerged corn "induced him to swear vengeance against the species," Audubon wrote, and the farmer came up with an ingenious-seeming solution. He dug a long trench and filled it with corn, then rigged up "a famous duck gun of his, [and] placed it so as that he could pull the trigger by means of a string, when quite concealed from the birds." One day, when the farmer saw the trench "quite black with the Turkeys," he whistled loudly, causing all the birds to raise their heads just above the level of the trench, whereupon he pulled the trigger of his heavily loaded gun and blasted the surprised birds, sending them "scampering off in all directions, in utter discomfiture and dismay." He found nine dead birds in the trench, and the ones who escaped seemed to get the message: "The rest did not consider it expedient to visit his corn again that season."[23]

Native Americans and Natural Historians

When Audubon placed the image of the "independent Indian" right after that of the "industrious farmer" in his passage on the Purple Martin, he introduced yet another set of eyes he knew could be a useful source of ornithological observation and information. Indeed, when he went on an exploratory trip to Labrador in 1833, he hoped that "I should meet with numberless Indians who would afford me much information . . . and who, like those of the far west, would assist me in procuring the objects of my search." Unfortunately for Audubon, the native population had already been reduced along the Labrador coast, and he could only lament "alas! how disappointed was I when . . . I scarcely met with a single native Indian, and was assured that there were none in the interior."[24] While implicitly invoking the image of the "vanishing" Indian, Audubon still explicitly acknowledged the role native people played in the collection of information that was so crucial to the work of natural history.

They had been doing so for at least two centuries, in fact, and the cross-cultural communication between Native Americans and Europeans had long been a recurring feature of natural history discourse. Thomas Morton, whose *New England Canaan* (1637) was one of the earliest English works to try to list the birds in the American environment, wrote about enlisting the help of local Indians in counting turkeys: "I have asked them what number they found in the woods, who have answered Neent Metawna, which is a thousand that day."[25] Mark Catesby, the early eighteenth-century artist-naturalist, likewise relied on native people to help with specimen collection, as did John Lawson, Jean Bossu, and other Europeans who investigated nature in North America. Bossu, a French traveler in the lower South in the 1760s, wrote that the Indians near the Tombekbe (Tombigbee) River in Alabama were "like ferrets in the woods": They helped him discover the nest of a *"royal eagle"* (or Golden Eagle), which they found full of "a great quantity of game of all kinds . . . such as fawns, rabbets, wild turkies, grous, partridges, and wood pigeons . . . indeed it was to be looked upon as a *Manna* sent by Providence, which favored us with these desarts." No one knew the birds better than the Indian inhabitants of a particular region, and early European observers could hardly have collected specimens without them.[26]

But counting and killing birds proved to be only part of the services native people provided to naturalists in frontier regions of North America. They also became crucial to identifying birds and describing their habits. In any form of field identification, the first sort of information a naturalist might ask of a native would be the most basic questions: What is that? What do you call it? What does it do? Native inhabitants frequently gave newly arrived naturalists that sort of ornithological information quite freely, which the naturalists in turn did not disparage or dispute. When Audubon's allies John Kirk Townsend and Thomas Nuttall reported on their exploration of the Columbia River region in 1834, for instance, they noted a wide variety of information about bird behavior they had gained from the local Indians. Native informants told them of the best method to obtain waterfowl ("The Indians have adopted a mode of killing them which is very successful"), the predatory habits of the Arkansas Flycatcher ("The Indians of the Columbia accuse him of a propensity to destroy the young, and eat the eggs of other birds"), and the migratory patterns of the Canada Jay ("The Indians . . . say, that they are rarely seen, and that they do not breed hereabouts").[27]

Sometimes, of course, information could easily get lost in translation or even purposeful obfuscation. When Alexander Wilson traveled among the Chickasaw and Choctaw peoples in the lower Mississippi region with a pet Carolina Parrot (now called a Carolina Parakeet, although now also extinct), the Indians apparently took delight in the bird, "laughing and seeming wonderfully amused with the novelty of my companion." The Chickasaws knew the bird and "called it in their language '*Kelinky*;' but when they heard me call it Poll, they soon repeated the name."[28] Here two names came together in one encounter, the apparently species-specific Chickasaw name "Kelinky" and Wilson's more personal (and highly unoriginal) name "Poll." Wilson recorded the Chickasaws' name, they repeated his, but whether or not "Poll" later became part of native parlance, Wilson had no way of knowing. He likewise had no way of knowing if the Chickasaws could distinguish between the calls of a Broadwinged Hawk and a Whip-poor-will. When he asked them, "They always assumed a grave and thoughtful aspect; but it appeared to me that they made no distinction between the two species."[29] It is entirely possible, of course, that the Indians knew the differences between the two birds quite well and preferred, for their own reasons, to withhold the information from their inquisitive visitor.

Yet whatever contributions native peoples made to ornithological discourse, there was one aspect of their input that came increasingly to be discarded: Indian naming practices. Native Americans had their own names for birds, of course, terms that generally reflected some knowledge of the bird's songs, physical characteristics, or habits. Today, in fact, many ornithologists and other students of nature have come to understand that "knowledge of vernacular names allows biologists to plumb the detailed knowledge that traditional peoples possess about local species." Indeed, they also understand that such naming practices often contain their own systems of classification based on "the wisdom that our forebears accumulated over millions of years of hunting/ gathering existence." For most of human history, all people had been on an essentially equal footing, able to hear and see and name things according to their observations of the immediate environment, encountering the visible and audible without the assistance of telescopes, microphones, and other sophisticated technologies.[30]

A few naturalists in early America seemed comfortable enough with that sort of scientific equality, willing to accept the names Indians gave birds more or less without question. Jean Bossu, traveling through the country of the

Peoria Indians in the 1760s, reported that "the savages who were with me" showed him some small birds that "feed on the strawberries in this meadow, which is red all over with them in the season." Perhaps understandably, the natives called the birds "strawberry bills," and Bossu did not suggest another equally expressive name as an alternative.[31] By the same token, Jonathan Carver, the eighteenth-century British explorer, discussed the differences in the names given to a nocturnal bird by the native inhabitants, who called the bird "Muckawiss," and the European colonists, to whom the bird was better known as "Whipperwill." Since the bird's nocturnal habits meant that it was more often heard than seen, the common tendency to identify it by its call became especially important in this case, and Carver himself seemed to be reasonably open to multiple auditory possibilities. Both peoples, he noted, named the bird for the sound it made—or for the sound they thought they heard it make—and the same nonhuman sound might indeed be rendered differently in human speech, so that one name (Muckawiss) might be just as descriptive, and therefore as "accurate," as another (Whipperwill). "The words, it is true, are not alike," Carver admitted, "but in this manner they strike the imagination of each; and the circumstance is a proof that the same sounds, if they are not rendered certain by being reduced to orthography, might convey different ideas to different people."[32] Even Audubon's enemy George Ord could sometimes accept the reasons behind native names. The Tell-tale Godwit (or Snipe) in Hudson's Bay, he reported, was "called there, by the natives, from its noise, *Sa-sa-shew*," and he went on to admit that "sa-sa-shew," if pronounced quickly, gave a good approximation of the snipe's whistle; that connection between sound and name, he argued, should be "a proof of the advantage of recording the vulgar name of animals, when these names are expressive of any peculiarity of voice or habit."[33]

But taking note of the "vulgar" name did not mean taking it seriously enough to make it a commonly accepted term. Even the common names given birds by Euro-Americans often met with derision and ultimate rejection. Wilson might look upon some common names with a bit of humor, as in the case of the Night Heron, which was sometimes called the "Qua bird," because it uttered "in a hoarse and hollow tone the sound *Qua*, which by some has been compared to that produced by the retchings of a person attempting to vomit." Similarly, he wrote with wry amusement about the White-eyed Flycatcher (or White-eyed Vireo), which builds its nest out of wood, weed stalks, and scraps of newspaper, "an article almost always found about its nest, so that some of my friends have given it the name of the *Politician*." But Wilson simply could not abide with equal indulgence the case of the Night Hawk, which also carried what he decried as "the ridiculous name *Goatsucker* . . . from a foolish notion that it sucked the teats of goats, because probably it inhabited the solitary heights where they fed." The name had originated in Europe but had crossed the Atlantic, and it annoyed Wilson so much that he refused to use it in his *American Ornithology*. "There is something worse than absurd," he complained, "in continuing to brand a whole family of birds with a knavish name, after they are universally known to be innocent of the charge. It is not only unjust, but tends to encourage the belief in an idle fable that is totally destitute of all foundation."[34]

Defending the family honor of birds was not really the main task, though. The greater challenge for Wilson, Audubon, and other nineteenth-century students of natural history was to assert, if not defend, their own honor and identity as the ultimate national authorities for identifying and naming birds and other elements of the natural world. Rather than adopt "vulgar" or vernacular names for birds, American naturalists almost always considered it necessary to reject the native name and give the bird a new one. Naming meant more than just reproducing a birdcall or describing a distinctive feature of appearance or behavior: It also defined a means of classifying a part of nature and, in a sense, of taking intellectual possession of it, all in the name of science—and even nation. Even though Peter Kalm noted that, to his Swedish ear, the Whip-poor-will "does not call *Whipperiwill*, nor *Whip-poor-will*, but rather *Whipperiwhip*," the sound did not ultimately determine the common name the bird carried in the American colonies: "The English change the call of this bird into Whip-poor-will, that it may have some kind of signification."[35]

That "signification" was significant. As Jonathan Carver had observed, the deciding factor in ornithological naming practice was the notion that the name could indeed be "rendered certain by being reduced to orthography"—to writing—a source of authority that Euro-Americans possessed and therefore privileged over almost all others. For their work to seem accurate and authoritative, they insisted on assigning birds common names in English, then further appropriating them linguistically with the additional authority of Latin binomials. Thus Native American names like "Kalinky," "Muckawiss," and "Sa-sa-shew" never made it into the formal lexicon of ornithology. Neither did "Fout-sah," the name Townsend and Nuttall heard Chinook Indians call a small warbler when the two naturalists made their expedition to the Columbia

River region. Instead, Townsend named the warbler for his friend Audubon, giving it the binomial nomenclature *Sylvia Audubonii* to distinguish it from a very similar eastern species, *Sylvia coronata*, thus assigning it the authority of a Latin term in the Linnaean classification of nature.[36]

The Foundations of "Idle Fable" in Avian Observation

By the beginning of the nineteenth century, the United States had become what one student of natural history has called a "Democracy of Facts"—that is, a culture that "eschewed theorizing about nature" and instead relied on close observation based on "clear-headed, commonsense use of one's faculties as well as a sound mind."[37] But to get at those facts, American naturalists often had to confront the anarchy of fable, the pervasive and exceedingly unsystematic scattering of superstition throughout American society, the widespread residue of religious stories, folktales, and other sorts of popular narratives that still had credence in the culture—enough, at least, to attract the attention of American naturalists. Within this fabulous mass of misinformation lay a considerable body of belief about birds, who took on a variety of symbolic roles, from ominous signs of doom to impressive sources of power. Before dismissing popular ornithological lore completely out of hand, gentlemen of science sometimes stopped to think twice about some of the more outlandish-seeming stories that circulated throughout society, even taking time on occasion to put them to the test of scientific experiment. They had no corner on ornithology, and they had to pay attention to what their fellow Americans thought they knew about birds—even if that turned out to be ridiculously wrong.

One of the best ornithologist stories about trying to address such bird beliefs came from Alexander Wilson, when he was staying with some Chickasaws in the lower Mississippi. Early one evening, Wilson paused on his horse near the Indians' camp to listen to a Northern Mockingbird "pouring out a torrent of melody." It was the first mockingbird he had seen or heard in the area, and he was so happy to hear it that he dismounted, tethered his horse, and stood almost transfixed by the beauty of the bird's song: "I think I never heard so excellent a performer." A gunshot suddenly interrupted his reverie, however, and he "looked up and saw the poor mocking-bird fluttering to the ground," and then turned and realized that "one of the savages had . . . barbarously shot him." Outraged at the silencing of the songster, Wilson immediately stalked over to the Indian and started berating him, telling him that "he was bad, very bad!" and that "the Great Spirit was offended with such cruelty, and that he would lose many a deer for doing so." At that point, the Indian's father-in-law intervened to defend his son-in-law from Wilson's tongue-lashing— which was being translated and delivered, as it happened, by an African American interpreter on the scene. The elder Chickasaw explained to Wilson that "when these birds come singing and making a noise all day near the house, *somebody will surely die.*" Wilson had heard that belief before, he noted, "exactly what an old superstitious German, near Hampton, in Virginia once told me."[38]

In less than a page, Wilson's story about the death of the mockingbird presents the reader with a remarkable cross-cultural tableau: the Scottish-born Wilson yelling, apparently in English, at a seemingly uncomprehending Indian that the Indian's god will punish him for killing a bird; a black man, perhaps a companion of Wilson's or a member of the Chickasaw group, trying to translate the threat; an older Indian understanding enough of the harangue to tell Wilson, again apparently through the black interpreter, that the young Indian man had good reason to shoot the bird in order to avert a certain death in his own household (a belief seemingly seconded, if not corroborated, by an unseen German back in Virginia). The incident pits Euro-American against Native American, with an African American as the means of intercultural communication. It also pits one spiritual belief system against another, with Wilson trying to frighten the Indian by appropriating the Indian's own god, a supposedly vengeful Great Spirit, only to be countered by another Indian arguing another version of supernatural certainty. It even pits one Euro-American against another, with the "old superstitious German" taking the side, albeit in absentia, of the Chickasaw father-in-law. In writing the story, Wilson no doubt intended to underscore the brutish behavior and superstitious beliefs of people who had no regard for the beauty of a mockingbird's song and, in doing so, to assert his own superior sensibility and rationality. In reading the story, however, we can see simply different beliefs and sensibilities coexisting, even competing, in a common ground—and in the natives' home ground, at that. But above all, the apparent alliance between the elder Chickasaw in the Mississippi region and the old German in the Tidewater region suggests that the bird beliefs at issue in Wilson's story might well extend far beyond the immediate context in both place and time.

The supernatural power of mockingbirds was just one such belief, one not confined to Mississippi Indians alone. Dr. Benjamin Smith Barton, of the

University of Pennsylvania, faced a similar issue when he set out to put to rest some widespread—and still spreading—misinformation about bird migration in his home state of Pennsylvania. In a pamphlet entitled *Fragments of the Natural History of Pennsylvania* (1799)—which has been called "the first ornithological monograph written by an American," but which Barton himself called, with some degree of self-deprecation, "fragmentary rubbish"—Barton identified several issues he felt could be "thrown upon the public with some degree of confidence:—with confidence, merely because it regards a country, the natural history of which has hitherto been so little attended to."[39]

He noted, for instance, that the "*Psittacus carolinensis*, Carolina Parrot, has been occasionally observed . . . within twenty miles of the town of Carlisle," a sighting that contradicted the extent of the migratory range given by both the much-esteemed William Bartram and the much-maligned M. de Buffon. "The arrival of these birds in the depth of winter was, indeed, a very remarkable circumstance." But what seemed a "remarkable circumstance" to Barton as a man of science could be remarkable in quite a different way to some of his fellow Pennsylvanians. "The more ignorant Dutch settlers were exceedingly alarmed," he wrote. "They imagined, in dreadful consternation, that it portended nothing less calamitous than the destruction of the world."[40] Therein lay the quandary: In a country in which natural history "has hitherto been so little attended to," someone interested in accurate information, as Barton surely was, could not always trust the hypotheses of self-styled naturalists, much less the doom-filled fantasies of "ignorant Dutch settlers" or perhaps any other superstitious people.

Barton's study of avian migration also took him into another realm of popular bird belief, the then-current controversy about the wintertime whereabouts of swallows. Ever since the time of Aristotle, some people had argued that swallows did not migrate away from cold regions but submerged themselves underwater or in mud and went into a state of torpidity throughout the winter, after which they would emerge again, dirty but ready to fly, in the spring. This swallow submersion theory had been debated for centuries in Europe, and by the end of the eighteenth century it had crossed the Atlantic and become well established in North America: "The question concerning the Swallows is not yet settled," Barton observed, "and in this country the notion which I deem an erroneous one with respect to these birds is gaining ground."[41] So extensive was the belief in the ability of swallows to spend the winter in the mud, in fact, that Barton could not dismiss it out of hand. He went on for over four pages discussing, with complete seriousness and remarkable intellectual generosity, the merits of the swallow submersion theory, duly noting other forms of fauna that went into a torpid state for the winter (lizards, frogs, snakes, and other cold-blooded animals) or hibernated in a semitorpid state (bears and some squirrels), all the while still considering the possibility that some birds might do the same. But in the end, the notion that swallows submerged themselves in mud seemed too much, and Barton tilted toward migration over submersion.

Then, in classic academic fashion, he called for further research. "How much it is to be wished, that some intelligent naturalists would furnish us with a list of the migratory birds of Mexico, Brasil, the West-India-Islands, &c. noting down, with care, the times of their disappearance from those countries, and the periods of their return to them. This would throw great light upon the difficult question which I am examining." The ultimate answer would have to come from someone else, though, because Barton had no intention of pursuing the issue himself: "I want leisure." What he needed—what the nation needed, in fact—was "some happy genius, qualified by that union of talents, of leisure, and enthusiastic ardour, which is necessary to form the character of a genuine naturalist."[42]

Wilson and Audubon may not have qualified in the "happy genius" category, but Barton could have taken comfort in the knowledge that they stood right by him on the swallow torpidity issue, albeit with a much less measured tone. "Away with such absurdities!" Wilson wrote. "They are unworthy of a serious refutation."[43] Barton died too soon, in 1815, to meet Audubon, but he would have been gratified to know that Audubon's own observation "puts a compleat *Dash* over *all* the Nonsense Wrote about their Torpidity during Cold Weather."[44]

Despite their rejection of the swallow story, Wilson and Audubon could not wholly ignore other forms of apparent nonsense and absurdity without subjecting them to scientific inquiry. Once again, the Carolina Parakeet served as the species under investigation, this time an inquiry into what Wilson described as the "general opinion . . . that the brains and intestines of the Carolina Parakeet are a sure and fatal poison to cats." Declaring himself "determined to put this to the test of experiment," he collected the brains and innards of more than a dozen Carolina Parakeets to prepare for the experiment, but his initial attempt suffered from an unfortunate absence of a feline subject: "Mr. Puss was not to be found." Instead, he found a "respectable lady near the town of Natchez, on whose word I can rely," who had apparently conducted such an experiment on her own and assured Wilson that "the cat had actually died on the succeeding day." Rather than accept her conclusions, however, Wilson remained intent on conducting the experiment himself. This time he took a cat and two kittens, shut them up in a room "with the head, neck, and whole intestines of the Parakeet," to pass the night. The next morning he found almost all the parakeet parts eaten, but the cat "exhibited no symptom of sickness; and at this moment, three days after the experiment has been made, she and her kittens are in their usual health."[45]

Audubon, always aware of Wilson's work, tried a similar experiment, but this time from the canine side, using his dog, Dash, who had just given birth to a litter of puppies. "We Boiled 10 Parokeet to night for Dash, Who has had 10 Welps—purposely to try the effect of the Poisonous effect of their hearts on Animals." Audubon never revealed the results of this findings—he was on his Mississippi flatboat trip at the time, not the best environment for writing up research reports—nor did he discuss the ethical or emotional implications of using his own dog as an experimental subject. From Audubon's silence on the outcome of this experiment, one might hopefully assume that Dash survived her master's science long enough to raise her pups and live out a dog's life in peace.[46]

Audubon also turned his scientific eye to other investigations of folk belief about birds. One of the more striking descriptions of the experimental approach appeared in the pages of the *American Turf Register and Sporting Magazine*—hardly a learned scientific journal, to be sure, but a popular sporting periodical that carried, in addition to its many reports on racehorses, occasional pieces about other forms of animal life. In a communication entitled "A Remarkable Fact in Natural History," a correspondent signing his name "Sigma" described an experiment conducted by a Major John Pillars of Missouri to test a curious but apparently common belief: that the down from the underside of a vulture's or buzzard's wing could cure blindness. Major Pillars had apparently first heard of the restorative powers of buzzard's down from a hundred-year-old slave named Joseph, and to see for himself, Pillars and some associates took a buzzard and "with a small shoemaker's awl, *ripped open its eyes, so that no part of the ball of either remained.*" According to the report by "Sigma," the blinded bird put its head under its wing for a few minutes, and when the head reappeared, it contained "two good sound eyes—free from blemish, and possessing in every degree the power of vision!" The author of the piece added that he had mentioned this experiment to "several persons,—practical, uneducated men," who confirmed that buzzard down could cure blindness not only in buzzards, but also in horses and even human beings. Rather than let the last word lie with these "practical, uneducated men," though, "Sigma" concluded with a seemingly authoritative deposition from Major Pillars himself, who swore to the accuracy of his experiment, noting that he had conducted it "fifty times, and once, at a log rolling, ten times in one day."[47]

On one level, this whole story borders on being a tall tale, an extended joke put before (or played on) the readers of a sporting periodical, making a parody of the experimental method: Surely such a typically alcohol-soaked event as a log rolling attended by "practical, uneducated men" did not lend itself to being a suitable laboratory for serious scientific inquiry. But on another level, this sort of story made the journalistic rounds enough to be taken seriously, or at least seriously enough for further scrutiny. In writing about some of his own experiments on vultures, Audubon noted "a story, widely circulated in the United States, through the newspapers" about buzzards and the cure for blindness. Some of the men with him, "medical gentlemen who were present," performed experiments to test the buzzard question, and they came away convinced of the "absurdity" of the idea.[48] To call the story an "absurdity" certainly made sense, but Audubon could do so with proper confidence only after the "medical gentlemen" had completed their scientific investigations.

Just as Audubon relied on the "medical gentlemen" to bury the buzzard story, so, too, did he turn to an equally serious student of natural history, his good friend John Bachman, to help dispel another equally dubious supposition. For decades stories had circulated on both sides of the Atlantic that Soras not only could live for extended periods of time underwater but also, indeed, could not be drowned (see Plate 11). Bachman put the Sora question to a test, and Audubon quoted in full his friend's account of the experiment:

> I once, in company with some naturalists of Philadelphia, tried two experiments with two Soras that had been slightly wounded in the wing, to ascertain how long they could live under the water. They were placed in a covered basket, which was sunk in the river. One remained fifteen, the other eight minutes, under water; and on being taken out, they were both found dead. We placed them in the sun for several days, but, I need hardly say, they did not revive.[49]

Here again, despite the apparent absurdity of the scene—of drowning birds and then leaving them out to dry and hoping for them to come back to life—the credibility Audubon accorded his friend Bachman gave credence to the scientific intent, and even validity, of the experiment.

On the whole, all these experiments, brutal and stupid though some of them may seem now, still formed a serious part of the work of natural history in the early nineteenth century. No matter what source the bird stories came from—a farmer, a slave, a "respectable lady," even a correspondent to a horse-racing sheet—the bird lore that lurked in the corners of popular culture had just enough currency to attract the naturalist's attention.

It certainly got Audubon's attention. In a nation still awash in misinformation—the residue of long-standing superstition and old folktales, spurious observations, and outright pseudo-scientific hokum—the task of investigating any aspect of nature, birds or whatever it might be, remained a necessary and compelling endeavor. Even if the ultimate result was to reject the stories as absurd, Audubon could still give them a hearing and put them to the test of science. It helped, of course, that the Sora test had been conducted by Audubon's friend Bachman with the assistance of "some naturalists of Philadelphia," the center of American science at the time, just as Audubon's own buzzard-based research had the imprimatur of some "medical gentlemen." Again, Audubon might have been willing to take some aspect of popular bird lore seriously enough to consider it worthy of experiment, but he remained more likely to accept the experimental findings of people he considered his scientific and social peers—especially the white male gentleman-naturalists who not only had an apparent passion for making their own formal contributions to natural history, but also carried the social status adequate to endow their ornithological observations or experimental findings with scientific authority.

Audubon Reaches Out to the Reader

Still, whatever his skepticism, whatever his insistence on experiment, whatever his preference for naturalists with some measure of social standing, Audubon always had questions about birds, and no other naturalist sought answers as widely and directly as he did. He repeatedly solicited assistance from an unseen ally, his "kind reader," his "gentle reader"—a term that recurs in some form or other over three hundred times in the five volumes of *Ornithological Biography*. He never seemed to tire of reaching out to the reader in an almost personal fashion—"Can you, reader, solve the question?"—and the queries appear throughout Audubon's writings:

> Why do not the Turkey Buzzard, the Fork-tailed Hawk, and many others possessing remarkable ease and power of flight, visit the same places? . . . Why do not the Pigeons found in the south ever visit the State of Maine, when one species, the *Columba migratoria*, is permitted to ramble over the whole extent of our vast country? And why does the small Pewee go so far north, accompanied by the Tyrant Flycatcher; while the Titirit, larger and stronger than either, remains in the Floridas and Carolinas, and the Great Crested Flycatcher . . . seldom travels farther east than Connecticut? Reader, can you assist me?[50]

In passages like this, how should the reader, either then or now, read Audubon? Today, his occasional admission of ornithological ignorance and his direct question to the reader may seem little more than an act of literary artifice, a democratic-seeming gesture at first glance, but really a request addressed only to people prosperous enough to purchase his work. Perhaps, in fact, his apparent appeal to the reader served as a disingenuous device for subtly asserting his own authority by raising questions he knew most readers could not answer. In reading Audubon, it always makes sense to take careful account of his self-conscious construction of his own place on the page: Whenever he seems to be talking to the reader, he is most often talking about himself. But still, when we consider Audubon in his own scientific context, at a time when many parts of the continent yet remained relatively unexamined by naturalists and when many questions of natural history yet remained unanswered, it might also make sense to take Audubon's pleas for assistance seriously.

Some of his readers certainly did. A couple of precocious college students—William H. Edwards at Williams College and Spencer Fullerton Baird at Dickinson College—took the initiative to write to Audubon with answers to questions he had posed in *Ornithological Biography*. Both of them would go on to become prominent men of science in their own right, but at the time they were both just teenagers, who had encountered Audubon only in print and who took seriously the notion that Audubon's appeal to his "kind reader" might apply to them.[51] Audubon himself took the initiative to develop a network of other correspondents who could send him information and bird specimens, and sometimes he reached out to them individually. In 1836, before the final volume of *The Birds of America* had been finished, Audubon wrote to a South Carolina subscriber, R. O. Anderson, asking for whatever last-minute information might be available: "My very great anxiety to enhance the Science of our Ornithology, and that of completing my Work to the utmost of my exertions and capabilities embolden me to beg of you any Specimens of Bird which you do not Identify in the Three Volumes in your possession." Should Anderson know of any birds in his area that did not yet appear in the Great Work, Audubon asked him to send him specimens soaked in common whiskey.[52] Even on the verge of victory, Audubon could happily value help with a few last-minute specimens.

But beyond birds, Audubon seemed also to value quite happily the identity his relationship with the "kind reader" could give him in his adopted but beloved country. In a new nation that increasingly promoted the Common Man as a political and cultural icon, Audubon openly embraced and repeatedly celebrated being a man of science who would reach out to the ordinary American. He defined himself in print, after all, as one of them, the gun-toting ornithologist who dedicated his life to collecting and depicting all the birds of America, who knew the American wilderness as much as anyone possibly could, and who made his self-styled image as the American Woodsman emblematic of a manly—and decidedly American—approach to science. In creating that very masculine, very American identity, Audubon enhanced both the standing of natural history and his own standing in American society, defining the cultural connection in a self-conscious way no one else had ever done, or even dared, before.

Chapter 9

≈

Forging a Legacy, Finding a Disciple

Though my strength is not what it was twenty years ago, I am yet equal to much, and my eyesight far keener than that of many a younger man, though that too tells me I am no longer a youth.

—John James Audubon, "Missouri River Journals"

In early 1840, about two years after the final plates of *The Birds of America* appeared and a year after the final volume of *Ornithological Biography* had come into print, a writer for the *North American Review* speculated on the feelings Audubon must have had as his Great Work reached completion. On the one hand, the author observed, Audubon "had raised an imperishable monument to commemorate his own renown," while in the process overcoming a host of "anxieties and fears," the doubts of his "overprudent friends," and the "malicious hopes" of his enemies, "for even the gentle lover of nature has enemies." In the end, Audubon "had secured a treasure of rich and glowing recollections, to warm his own heart in his declining years." On the other hand, though, the enterprise was over, and "whatever works he might afterwards engage in, the great work of his intellectual life was finished." There could be other good work to come, of course: Audubon was still "of such an age that we may hope for many advances in his favorite science yet to be made by his energetic and inquiring mind; meantime he has called forth many others by his spirit and example, who will follow out his researches with that enthusiastic interest which the study is sure to inspire."[1]

Audubon would have easily agreed on all counts: the conflicted sense of achievement and completion, the dogged commitment to continuing his labors, and the hopeful notion of inspiring someone else to do something more, perhaps even what he himself could no longer do. Taking on nature had taken its toll, and Audubon knew it. Throughout the 1830s he had still had the stamina for new adventures—the Florida Keys, Labrador, then Texas—but by the time he was bringing his Great Work to a close, he also suspected that he would soon be closing that chapter of his own life. In July 1837, he wrote his good friend John Bachman about his mixed feelings about coming to completion: "What a strange realization of a Dream this finishing of a Work that has cost me so many Years of Enjoyment, of Labour and of Vexations, and yet a few more months will I trust see it ended aye ended and Myself a Naturalist no *longer!*"[2] Indeed, in the later volumes of *Ornithological Biography*, Audubon occasionally wrote the reader into his bird descriptions as a fictive companion to accompany him on expeditions that he now took mostly in memory: "Although several years have elapsed since I visited the sterile country of Labrador, I yet enjoy the remembrance of my rambles there; nay, Reader, many times have I wished that you and I were in it once more, especially in the winter season. . . . But alas! I shall never spend a winter in Labrador."[3] As he was beginning to feel himself in the winter season of his own life, Audubon implicitly passed the ornithological torch to his readers, reminding them that despite all the work that he and others had done, there were still birds that remained to be discovered: "Therefore, Reader, I would strongly advise you to make up your mind, shoulder your gun, muster all your spirits, and start in search of the interesting unknown." Regretfully, he added, "I cannot more go in pursuit,—not for want of will, but of the vigour and elasticity necessary for so arduous an enterprise."[4]

And yet he did go again in pursuit, of both bird specimens and human subscribers. Even as he saw *The Birds of America* and *Ornithological Biography* come into print as such impressive companion volumes, Audubon could still not allow himself to quit adding to his intellectual—and financial—legacy. It would be far too easy to take it easy, and therefore far too out of character. Having completed his Great Work, Audubon still needed to add the finishing touches to another great work—Audubon himself.

Still Searching for Subscriptions

In late 1837, a London newspaper took sympathetic note of the difficulties Audubon had faced in the past decade of making and marketing *The Birds of America*, particularly given the subscription process Audubon had chosen to follow: "When only a few Numbers of this Work had been published," the writer explained, "many noblemen and gentlemen, as well as a considerable number of Natural History and other societies, libraries, &c, were desirous of possessing it." But with the long time projected for its completion, many feared "the probability of its ever being finished, therefore, so remote, [that] they determined to await its completion before they subscribed." What the report failed to mention was the immediate impact of the Panic of 1837, the economic crisis that stretched to both sides of the Atlantic and caused some of Audubon's subscribers to let their subscriptions lapse. But while subscribers became cautious, even skeptical, about the economic future, Audubon pressed on: "Mr. Audubon, therefore, feels desirous . . . to announce that seventy-eight Numbers have now appeared, and that with seven more it will be completed . . . on the 1st of April or May next."[5] Audubon made good on his commitment—he delivered the last drawings in mid-April 1838, and two months later, Havell's shop completed the last plate—but even though the Great Work may have been finished, the work of selling it had decidedly not.

And the prospects for doing that work in the larger context of the transatlantic collapse had become decidedly discouraging. In July 1837, Audubon wrote from New York to Havell that "the times here are hard passed reckoning—There is in fact no business of any importance carried on, and therefore little or no money to be seen (much less to be had) and no new subscribers can be expected *for some time.*"[6] By then, Audubon had already devoted over a decade to all aspects of the production and distribution process: overseeing the workers engraving and coloring the plates; making sure the completed plates actually got to their purchasers; keeping an eye on lazy agents and booksellers; making trips across Great Britain, France, and the United States seeking new subscribers; and trying to make sure earlier subscribers actually made good on their commitments. He even succumbed to doing something he had always disliked—putting his work on display in public. In 1839, he exhibited several hundred paintings in New York, getting good reviews but losing $55 for his efforts. He drew the line, however, at doing a subsequent show in Philadelphia: "If I had an extraordinary fat hog to show, and should place him in a large room on an elevated pedestal, with a comfortable bed of straw, I could draw thousands from far and near; but paintings, however beautiful or well done, will not attract enough people to cover the expense."[7]

The Birds of America was only part of the task, of course. Since 1830, Audubon had also worked steadily on *Ornithological Biography*, occasionally coordinating his labors in night-and-day shifts with William MacGillivray ("I rise at four or earlier, he at ten," he wrote in 1838, "but I go to bed at eleven, he at two"). Their work on that book came to an end in May 1839, when the fifth volume came into print. That done, Audubon must have sighed with some satisfaction, and in the last pages of that last volume, he proudly provided a list of all the individual and institutional subscribers to date—82 in the United States, 79 in Great Britain and Europe—for a total of 161 in all.[8]

A year later, a Massachusetts newspaper did its own summing-up of the subscription list, using a much higher number, which tilted, so the paper concluded, "somewhat in favor of America." Of a total of 281 subscribers, well over half resided outside the United States, with Great Britain and Ireland accounting for 177. Still, the 91 subscribers in the United States represented a greater achievement, so the argument went, based on the proportional size and comparative wealth of the respective populations. London, for instance, had a population of two million inhabitants, among whom could be counted many wealthy members of the British nobility, and the great metropolis had accounted for 31 subscriptions. On the other hand, Boston, "with eighty thousand, has 15, or half as many as London." Similar arithmetic put Salem, with three subscribers among a population of 15,000, proportionally ahead of Manchester, whose 29 subscriptions had come from a much larger population of 300,000. This exercise in subscriber scorekeeping thus gave the United States "more than her share according to the population of the two countries" and perhaps also provided an answer to earlier complaints about any national indifference to Audubon's work.[9] America had embraced both the American Woodsman and his birds, and the nation could proudly hold its head higher in the transatlantic standings of artistic commitment and taste.

The newspaper's higher count of subscriptions to *The Birds of America* might have seemed an encouraging number, but that didn't mean Audubon would ever consider it enough. He had to start seeking subscribers for another business venture as well, a long-planned new edition of *The Birds of America,* a smaller, more affordable version that could supplement sales of the elegant but expensive Double Elephant Folio volumes. Even at a much-reduced price, he predicted as early as 1833, the smaller edition would provide "the greatest profits we are to derive from our Publications" and thus be a big boost to the Audubon family income.[10] It would indeed be a family enterprise, just as the work on *The Birds of America* had been. Audubon's younger and more artistically gifted son, John Woodhouse, would use a camera lucida—a prism-based optical device that facilitated tracing of an image onto paper—to reduce the original bird pictures from large size to small, and Victor would again serve as occasional overseer of production and solicitor of subscriptions. Lucy had already done essentially everything she could to support her husband's work, but she would continue to look on with intense interest.

By 1839, work on the new version got under way, and it was indeed new in several significant ways. In addition to its different dimensions (octavo-sized, or an eighth of a standard folio sheet, a little over 10 x 6 inches), the small work offered a more orderly organizational scheme (birds arranged by genus and species, according to Audubon's recently published *Synopsis*) and text to accompany the images (abridged descriptions from *Ornithological Biography*, with the human-centered Episodes omitted).[11] Audubon also employed a new printer, the Philadelphia-based John T. Bowen instead of London's Havell, and a new technology, lithography instead of engraving. Lithography allowed a picture to be drawn directly onto a prepared stone surface and then, after a chemical process to "fix" the image, hundreds of accurate reproductions could be printed onto paper. Finishing the image would still be the work of colorists—typically young women, and frequently the focus of complaints about lack of consistency—but the smaller size of the pages allowed for a faster pace of production. Beginning with Audubon's initial order for just a few hundred copies of the lithographs in late 1839, Bowen soon met the growing demand as Audubon again took to the road to sell subscriptions.[12]

Audubon could never dawdle in his dealings with subscribers, some still prospective, some already problematic. A prime example of the latter was Audubon's enervating experience with Massachusetts Senator Daniel Webster, a famous figure who had become tardy, even a bit evasive, in making his payments. Audubon first met Webster in September 1836, while on a subscription-hunting trip to Boston, and the two seemed to get along famously at first. Audubon wrote enthusiastically to Lucy that the politically well-placed senator thought it "likely a copyright of our great work might be secured," and with that bit of business aside, he and Webster "took tea, talked of ornithology and ornithologists; he promised to send me specimens of birds, and finished by subscribing to my work." Webster also gave Audubon an all-purpose letter of introduction, expressing his "great respect for Mr. Audubon's scientific pursuits." Audubon was ecstatic, concluding that "I feel proud, Lucy, to have that great man's name on our list, and pray God to grant him a long life and a happy one."[13]

Webster may indeed have been enough of a "great man" to deliver on some of the promised bird specimens, but not so great on fulfilling his subscription. Audubon had to keep pestering Webster to pay up. In the late summer of 1840, while on yet another tour of New England in pursuit of subscribers for his works, Audubon met a woman who said she had visited Webster's home, "where she has seen *my* great work several times." She referred to the volumes, understandably, as Webster's, but Audubon took exception in the pages of his journal: "*His* Copy??? When paid for!" A few months later, in November, Audubon sought out Webster in Boston, first trying his hotel ("not in left a card") and then at his office, where he found the senator. "He was greatly surprised that I have not received a Dollar yet on a/c of what he owes us for the Copy of The Large Work to which he subscribed years ago," Audubon wrote, but then Webster assured him "that he would attend to that business at once, and indeed settle it to my satisfaction by Wednesday next." Audubon had already seen enough to be wary of Webster's promises, and he reacted with his own Francophone skepticism: "Nous verrons!" Several days later he called on Webster again and finally "got $100 Dollars from him on a/c of the Large Work, and a memorandum authorizing me to draw upon him at Three different dates for the balance he now owes us." Webster also subscribed to "the little work" and promised Audubon to send letters of introduction to friends in other New England towns within a few days. Again, Audubon had his doubts: "Nous verrons!"[14]

Webster did deliver on his promise to provide letters of introduction, and they helped open doors when Audubon went westward across Massachusetts. He wound up in Worcester one Saturday morning in December 1840, where he went out "in good spirits" to see the sights (a "very handsome place," he said, "and in the summer must be quite beautiful") and, as always, to solicit sales for *The Birds of America*, one size or the other. He and his local agent, Clarendon Harris, "the Bookseller," visited the town's lunatic asylum ("found it kept in the very best of order"), and then they went to the American Antiquarian Society ("saw its curious old Books, paintings, etc., etc."). There they encountered the librarian, Mr. Samuel Haven, who claimed to be too busy to see Audubon at the time but instead told him that another local notable wanted to subscribe to his work—Elihu Burritt, "the famous learned Black Smith," who would become one of the town's most renowned reformers. Audubon and Harris quickly went to see Burritt, got a subscription for the Royal Octavo edition of *The Birds of America*, then went on to seek out Isaac Davis, a prominent local lawyer. Audubon stayed at the Davis mansion long enough to enjoy a sociable drink ("2 glasses of good wine"), admire Mrs. Davis ("an interesting Lady"), and try to entice Davis to buy the big version of *The Birds* rather than subscribe to the smaller one ("I hope it may be the first"). After the second glass of wine with the Davises, Audubon and Harris hurried back to the antiquarian library to see Mr. Haven—but too late: "He was gone, we not having been punctual with him, and thereby have lost a subscriber."[15] (The American Antiquarian Society likewise laments its own opportunity lost, which "causes pangs of regret to this very day."[16])

Audubon might have started the day "in good spirits," but the work he did in Worcester wore on him. There he was in a midsized Massachusetts town, knowing himself to be a well-connected, well-reviewed artist, a member of several scientific societies on both sides of the Atlantic, yet still having to get introductions and go door-to-door as a salesman, gamely hoping he could find just another handful of subscribers who would promise to put down a thousand dollars for his big book—or even commit to a hundred for the smaller one. Even when he moved on, the work of selling the Great Work to reluctant locals lowered his spirits considerably:

> How curious it is to me to see how eagerly the non employed loungers about Taverns peep in to the *book of names*, and again stare at the poor but honest, and I would almost say *modest* "American Woodsman." There are indeed times when I wish I could leave the Earth and fly away from the staring gaze of these Idlers, but as I cannot fly, I must sit still & silent and give them the fairest of chances to gratify their appetites.[17]

For the next two years, in fact, throughout 1841 and 1842, Audubon spent much of his time on the road, going from New England down to Philadelphia, Baltimore, and Richmond; up to Montreal and Quebec City in Canada; and several times to Washington, DC. In the nation's capital he once again called on Daniel Webster, "who gave me a check for $100. on a/c! (The times are hard here)."[18] They were hard for Audubon, too, and would only get harder.

Building a New Home, Then Leaving It

By 1841, despite all the trouble he had faced in soliciting and tracking down subscribers to *The Birds of America*, Audubon had made enough money to be

Figure 11. Minnie's Land. From *Homes of American Authors* (New York, 1853).

able to buy thirty acres of land and build a house in the distant upper reaches of Manhattan's west side, adjacent to the Hudson River (roughly the area now bounded by 155th and 158th Streets, between Riverside Drive and Broadway). After so many years of renting living accommodations on both sides of the Atlantic, Audubon could hope that Minnie's Land (the name he gave his estate, based on Lucy's nickname) would be his dream home.

A visitor described going there in 1842, walking down "a rustic road which led directly down towards the river," to find a "secluded country house . . . simple and unpretending in its architecture, and beautifully embowered amid elms and oaks." No doubt embellishing the scene more than a bit, the writer then took note of the animal life on the lawn: "Several graceful fawns, and a noble elk, were stalking in the shade of the trees, apparently unconscious of the presence of a few dogs, and not caring for the numerous turkeys, geese, and other domestic animals that gabbled and screamed around them." It seemed only fitting that such a sylvan setting would be the home of Audubon, whose "white locks fell in clusters upon his shoulders, but were the only signs of age, for his form was erect, and his step as light as that of a deer."[19]

Audubon would not be there for long, however. "Indeed," the visitor continued, "he was even at that advanced period of his life on the eve of an excursion to the Rocky Mountains, in search of some specimens of wild animals of which he had heard, and the following year he passed the summer on the upper Missouri and Yellow Stone rivers."[20] Whatever appearance of erect form and deerlike step the writer might have found in visiting Audubon at Minnie's Land in 1842, the following year would change that. Audubon's trip to the West would be a severe test of his own health and stamina—one he would barely pass, in fact, and one that might well have hastened his aging and led to his death.

The Last Expedition

While the Royal Octavo edition of *The Birds of America* moved through Bowen's shop, Audubon had a parallel project in the works, a collection of images and text on North American mammals, specifically viviparous quadrupeds (that is, four-footed animals that give birth by delivering live offspring, not eggs). This new endeavor would likewise employ lithography to create folio-sized images, which Audubon initially said he would draw onto the stones himself. In fact, the work would not be completely by himself: He would rely on his son John for help with the images, and he had enlisted his fellow naturalist Bachman to work with him as coauthor of the text. At the beginning of January 1840, he wrote Bachman about his high hopes for the collaboration. "Such a publication will be fraught with difficulties innumerable," he admitted, "but *I trust* not insurmountable." He immediately laid out an exuberant expression of their partnership. "We Join our Names together, and you push your able and broad Shoulders to the Wheel, [and] I promise to you that I will give the very best figures of quadrupeds that ever have been thought of or expected." With such good work by both men, the book should be finished within three years. "Only think," Audubon wrote with anticipatory triumph, "of the quadrupeds of America being presented to the World of Science, by Audubon and Bachman." In celebrating their scientific partnership, though, Audubon could not help taking yet another gratuitous shot at the other partnership that had weighed on him for so long. Would the publication of the quadruped work move "our good and most learned Friend George Ord" to gratitude? If so, "how many copies of that magnificent Work Will that most generous Friend of poor Alexr Wilson will subscribe to?"[21] Audubon hardly had to say it, but both he and Bachman knew that the chance of getting George Ord to subscribe lay somewhere between a pipe dream and a joke.

Audubon and Bachman also knew that the work would have to reach well beyond the eastern states to the West, and an expedition of that extent would take considerable planning and, above all, fundraising. Audubon had drawn a few sample images of mammals, so he packed them up with his bird books and took them along on his subscription trips between 1840 and 1842. He had the good fortune to get financial support from his friend and fellow naturalist Edward Harris, who would not only share the expenses, but also accompany him in his adventures out West. Audubon also hoped to get funding from the United States government, and soon after moving into Minnie's Land in April 1842, he went down to Washington in July to see what money he could muster. None, it turned out.

He did, however, succeed in soliciting a packet of letters from prominent people in the upper reaches of DC society, all of them attesting to the importance of his work and the quality of his character. Daniel Webster, his elusive subscriber and the newly named secretary of state, set the tone: "Know Ye," began Webster, "that the bearer hereof, John James Audubon, a distinguished naturalist and native citizen of the United States, has made known to me his intention of travelling on the continent with the view principally of aiding the cause of science by extending his researches and explorations in natural history" and should thus be accorded all the safety and support "which becomes the hospitality of civilized and friendly nations." A British diplomat, Lord Alexander Baring Ashburton, chimed in with his endorsement ("a very distinguished naturalist"), as did General Winfield Scott ("our distinguished countryman"), Secretary of War John C. Spencer ("an eminent naturalist"), and, in the political pièce de résistance, President John Tyler ("a naturalist of eminent acquirements and estimation, a man of character and honor and worthy of all personal respect and regard").[22]

Once he got going, Audubon never really needed all these good words. He and his companions—his friend Harris, taxidermist John G. Bell, artist Isaac Sprague, and general assistant Lewis Squires—set out for the West in mid-March 1843. They traveled by various means of generally uncomfortable conveyance—train, stagecoach, and steamboat—from Philadelphia to St.

Louis, and along the way Audubon encountered a few old friends who knew him quite well enough without a letter of introduction. Leaving St. Louis on April 25, Audubon and his expedition partners took a second steamboat, the *Omega*, up the Missouri River, and three weeks later, when they arrived at a U.S. Army encampment on the prairie above Ft. Croghan, Audubon showed his credentials to the officer in charge, only to be told he needn't have bothered. Captain John Henry K. Burgwin, a young West Point graduate, "smiled, and politely assured me that I was too well known throughout our country to need any letters." Two months later, farther upriver at Fort Union, in what is now North Dakota, Audubon ran into a man from Charleston, South Carolina, "who knew the Bachmans quite well, and who had read the whole of the 'Biographies of Birds.'"[23] Audubon seemed pleased to find a few other people on the prairie who already knew him from his reputation, but he also discovered that most others—ordinary soldiers, traders, engagés, or Indians—probably didn't care who he was or what he had done, distinguished naturalist or not. They had their own work to do.

So did Audubon. In addition to searching the West for quadruped specimens, he kept up a very active writing life, producing a voluminous journal that covered the whole expedition in commendably thorough, sometimes dramatic, detail. Early on in the trip, he ended a long day by sighing, "And now that I am pretty well fatigued with writing letters and this journal, I will go to rest, though I have matter enough in my poor head to write a book."[24] Write a book he might, and Audubon's journal served not just to keep a record of his research, but to provide material for a new work that could serve as an encore to *Ornithological Biography*, with human-centered "episodes" and all. He told the story of the trip from his own perspective, of course, but on several occasions he turned the journal over to other men, who added to Audubon's pages their own tales of life and death in the West—the ravages of smallpox among the Indian population, a battle with Indians, a dangerous encounter with a bear, and "an account of Squires' Buffalo hunt . . . which he has kindly written in my journal and which I hope some day to publish."[25] Back in Philadelphia, Audubon's fellow ornithologist John Kirk Townsend had just published his own account of the extended western expedition he had taken with Thomas Nuttall, and Audubon no doubt considered himself too good a writer to let Townsend have the field to himself: He clearly had the material, and he knew how to get it to market.[26]

Even though the trip had one main purpose, acquiring and drawing quadruped specimens, Audubon could never resist focusing on his first love, birds, which he happily found in abundance from the beginning. His journal entry for May 4, for instance, records over fifty species, including various sorts of thrushes, warblers, and woodpeckers, a one-day list that a modern-day birder would consider enviable indeed. About a week later, while presenting his credentials to Captain Burgwin, Audubon literally jumped when he "heard the note of a bird new to me . . . and saw the first Yellow-headed Troupial alive that ever came across my own migrations." He "suddenly started, shot at the bird, and killed it," creating such an abrupt interruption that Burgwin "thought me probably crazy, as I thought Rafinesque when he was at Henderson." Audubon could add a note of self-deprecating humor to the scene by comparing himself to Rafinesque, the eccentric M. de T. of the story from almost a quarter-century earlier; unlike Rafinesque, though, he actually got a new species.[27]

Audubon's sense of discovery took on a very different cast when he came into contact with native people along his way to the West. Whatever positive-seeming words he had written some years earlier about the "independent Indian" or the "naked Indian, who has, for unknown ages, dwelt in the gorgeous but melancholy wilderness" soon vanished.[28] Notions of independence and innocence gave way to language of disparagement and disgust. In the first days of the expedition, steaming up the river from St. Louis, Audubon noted "a good number of Indians in the woods and on the banks, gazing at us as we passed; these are, however, partly civilized, and are miserable enough." A few days later, he described some Indians on a sandbar as "destitute and hungry as if they had not eaten for a week, and no doubt would have given much for a bottle of whiskey." Soon enough, Audubon would denigrate the "Indian rascals" as a "thieving and dirty set, covered with vermin." And so it went for the rest of the trip, with repeated expressions of disdain for the native inhabitants of the West.[29]

Indians might be, at best, subjects for scientific inquiry. Following several descriptions of searching for birds' nests, for instance, Audubon wrote an altogether matter-of-fact account of taking an Indian's coffin out of a tree, in order to remove the skull inside—presumably in the name of science. Skull collecting, or craniology, had become yet another form of scientific collecting among nineteenth-century naturalists, built on the belief that cranial capacity could be an indicator of human intelligence—or the lack thereof—and therefore yet another means of reinforcing notions of racial superiority and inferiority. The ethnographic research of the day focused largely on Native Americans and African Americans in the United States, and desecrating the burial sites of nonwhites became almost commonplace.[30] In taking the coffin from the tree, Audubon engaged in some desecration of his own: "The coffin was lowered, or rather tumbled, down, and the cover was soon hammered off.... The head had still the hair on, but was twisted off in a moment, under jaw and all.... We left all on the ground but the head." Some of Audubon's companions later tried to return the coffin and the remains to the tree, but "the whole affair fell to the ground, and there it lies; but I intend to-morrow to have it covered with earth." Audubon went on to add a few sentences about the dead Indian, a chief named White Cow, who had died, apparently by his own hand, three years earlier. "He was a good friend to the whites," Audubon noted, with absolutely no sense of irony about what sort of friends to Indians he and the other white men with him had been in turn.[31]

Audubon's generally dim view of native peoples brought him into a quasi-scientific, albeit decidedly one-sided, rivalry with another artist-naturalist of the era, George Catlin. Catlin had first come to the region just over a decade earlier, in 1832, and, like Audubon, made a point of painting as much as he could—in Catlin's case, sketches and portraits of the native peoples of the Great Plains. In the course of five trips, between 1832 and 1836, he produced hundreds of images of Indians and, again like Audubon, put his work on display in exhibitions for appreciative (and paying) urban audiences on both sides of the Atlantic.[32] Catlin became for Indians what Audubon became for birds, but Audubon could never appreciate the comparison. Instead, his comments on Catlin reflected Audubon's decades-long treatment of Alexander Wilson, both of whom served as tempting targets for frequent, often carping, criticism, typically wrapped in a concern for scientific accuracy, but just as often released as spiteful jealousy. Audubon began taking shots at Catlin early on in the trip, first with a sarcastic "Ah! Mr. Catlin, I am now sorry to see and read your accounts of the Indians *you* saw—how very different they must have been from any that I have seen!" A little later, when he again questioned Catlin's depiction of Indians, he offered a half-hearted exception: "But different travelers have different eyes!" He repeatedly saw enough differences, though, to become altogether uncharitable, calling Catlin's work "altogether a humbug" and casting Catlin himself aside: "Poor devil! I pity him from the bottom of my soul."[33] What Audubon failed to note in any of his criticisms, however, was that Catlin came to the region a few years before the smallpox epidemic of 1837–1838 had spread devastation among the native population. Different travelers might not only have different eyes, but also find very different circumstances among the people they would see.

Audubon certainly saw the effects of smallpox among the Indians he encountered, and he also heard about the widespread devastation of the recent epidemic. On a rainy Tuesday in June, "to fill the time on this dreary day," he asked a longer-term resident of the region, François Chardon, to give him an account of the smallpox scourge, and the report fills several pages of Audubon's journal. Yet whatever he saw and heard, he remained unmoved by the human suffering that had reigned throughout the region, and he offered no larger lesson from the demographic catastrophe that had swept the native population. In writing about the survivors, Audubon typically used "dirty," "disgusting," and other dismissive terms to describe the native people he saw. He said little better about the dead. Coming upon some mounds of earth that covered the remains of "poor Mandans who died of the small-pox," he noted that "perched on the top, lies, pretty generally, the rotting skull of a Buffalo," which struck him as "a kind of relation to these most absurdly superstitious and ignorant beings."[34]

By contrast, he reacted much more sympathetically to the plight of the buffalo (or more accurately, bison) population. Like most antebellum travel narratives of expeditions to the West, Audubon's journal abounds with accounts of the hunt. As a newcomer to the region, he had the good sense to heed the warning "that the hunting of Buffaloes, for persons unaccustomed to it, was very risky indeed; and that no one should attempt it unless well initiated, even though he may be a first-rate rider." Audubon, who no doubt considered himself a first-rate rider, could hardly resist the thrill of the chase, not to mention the ritual of the kill. When he shot his first bison bull, he refrained from joining some of his comrades in eating the warm brains directly from the animal's skull ("The very sight of this turned my stomach"), but he did accept the trophy when it was offered him ("I cut off the brush of the tail and placed it in my hat-band"). At the same time, his sense of himself as a man of science caused him to take the measurements of another dead bison: "To cut up so large a bull, and one now with so dreadful an odor, was no joke; but with the will follows the success, and in about one hour the poor beast had been measured and weighed." The bull turned out to be a big one—131 inches long from nose to tail, 67 inches high at the shoulder, and almost 1,800 pounds in weight—but not one worth taking: "The flesh was all tainted, and was therefore left for the beasts of prey."[35]

The real beasts of prey were Audubon and his fellow human beings, and he drew a puzzled contrast between the massacre of the animals ("the immense numbers that are murdered almost daily on these boundless wastes called prairies") and the apparent persistence of the massive herds ("we see so many that we hardly notice them more than the cattle in the pastures about our homes"), only to come to the disconcerting conclusion that "this cannot last; . . . before many years the Buffalo, like the Great Auk, will have disappeared; surely this should not be permitted." But, again, as Audubon so often did when considering some sort of human-created catastrophe, he abruptly changed the subject: "our boat is going on, and I wish I had a couple of Bighorns."[36]

On the other hand, Audubon's contemplation of the eventual demise of the bison at least led him to reflect on his own decline. The connection seemed especially clear, in fact, when he had to give up going on a hunt, "for alas! I am now too near seventy to run and load whilst going at full gallop."[37] He was actually only fifty-eight, not so near seventy in actual time, but perhaps close enough to the biblical "threescore and ten" to start thinking about the inevitability of the aging process. Throughout his trip in the West, Audubon's journal contains recurring references to his loss of energy and endurance, almost always looking back on an earlier version of himself and taking regretful note of what he could no longer do. Once, after trying to cross a muddy stream, he admitted, "I was nearly exhausted; which proves to me pretty clearly that I am no longer as young, or as active, as I was thirty years ago." Two days later he bumped the age comparison upward a bit, noting that "walking eight or ten miles through the tangled and thorny underbrush, fatigues me considerably, though twenty years ago I should have thought nothing of it." Less than a week later, he split the difference: "How I wish I were twenty-five years younger!" When it came to personal endurance, he could still work in a way that might seem exceptional in almost anyone else in his age cohort, but Audubon measured himself only against his highest standard—himself. On a day when he arose to begin drawing at five in the morning and kept at it for ten hours straight, not stopping until three in the afternoon, he nonetheless felt frustration when he quit, "regretting I could no longer draw twelve or fourteen hours without a pause or thought of weariness." But by the end of the summer, when he knew his western expedition would soon have to yield to the oncoming cold weather and come to an end, he had a burst of renewed energy and didn't want to go back East just yet: "My regrets that I promised . . . I would return this fall are beyond description. I am, as years go, an old man, but I do not feel old." Still, he could not avoid reckoning the reality of pushing sixty, and even though he might claim that "I am yet equal to much, and my eyesight far keener than that of many a younger man," he still had to face the fact that "that too tells me I am no longer a youth."[38]

Three days after writing that, on August 16, 1843, Audubon boarded a barge at Ft. Union and left the Upper Missouri for what turned out to be a cold, uncomfortable trip back home. He and his compatriots hunted all the way—shooting ducks, deer, bears, elks, and bisons—and as always, Audubon kept an eye on the skies ("These birds are now travelling south"). So was he, at least as far south as St. Louis, which he reached after two months on the river, on October 19. Three days after that, he settled into the more comfortable quarters of a Mississippi River steamboat bound for Cincinnati, and just over two weeks later he would close his Missouri River account by writing "reached home at 3 P.M. November 6th, 1843, and thank God, found all my family quite well."[39]

"An Old Head . . . on Young Shoulders"

In addition to his family, Audubon also found a brief, almost breathless letter welcoming him back to the "civilized part of the country" and eagerly seeking inside information about "what new birds & beasts you have discovered."[40] The letter came from Spencer Fullerton Baird, a young man who had first contacted Audubon three years earlier, in 1840, and had quickly become a frequent correspondent and decided devotee, someone who had already earned the respect of a direct response. In answering Baird's immediate inquiry, Audubon offered an overview of all the new and rare western species he and his colleagues had encountered—"15 *New* Species of Birds, and a Certain Number of Quadrupeds"—even as he downplayed the dangers of the trip: "Why, only think that *I* saw not one Rattlesnake and heard not a Word of bilious fever, or of anything more troublesome than Muschietoes and of those by no means many!" Audubon closed, however, on the note of fatigue that had run through his journal earlier: "I feel quite assured that much remains to be done, and all I regret is that I am not what I was 25 Years ago, Strong and Active, for willing I am as much as ever."[41] He no doubt envied the student's youth—Baird was only twenty at the time—and knowing that "much remains to be done," he had to assume that Baird would be one of the ones to do it. In time, in fact, Baird would do as much as anyone else to extend Audubon's scientific legacy well beyond the aging naturalist's lifetime.

Theirs became a mutually beneficial May-December relationship based on a shared fascination with ornithology. In his first 1840 letter to Audubon, the then-seventeen-year-old Baird deferentially described himself as "but a boy, and very inexperienced." Having "taken (after much hesitation) the liberty of writing to you," young Baird raised a question about a particular flycatcher, a *Tyrannula*, that he had not found in any published work, including Audubon's, and he said he had three specimens of the bird—two stuffed, one kept in a bottle of spirits—the latter of which he offered to send if he could. The boy went on to describe the bird in considerable ornithological detail—"Bill large, depressed, decidedly convex in its lateral outlines," and so on—and he noted that he had other specimens in his collection as well. Fearing that he might have "trespassed too much on your patience," Baird closed the letter by offering Audubon any assistance he could, promising that "although others may tender it more ably, yet none can more cheerfully." It was a bold move for a young boy to bother a celebrated man of science, but it turned out to be a very good move for both.[42]

Apparently impressed by the boy's work, Audubon made good on the invitation he had offered to the unseen "Reader" of *Ornithological Biography*, responding to Baird with enthusiasm and encouragement: "Although you speak of yourself as being a youth," Audubon wrote a few days later, "your style and the descriptions you have sent me prove to me that an old head may from time to time be found on young shoulders!"[43] Baird returned a compliment of his own, writing Audubon that he had no doubt that *Ornithological Biography* would "do more to spread a love of Natural history, than any work ever published. For my part I read the descriptions of birds and the episodes . . . with the same motive of pleasure as I used to read a favorite novel."[44] Thus began an exchange of letters, then bird specimens, and eventually personal visits between Audubon and Baird, a cordial connection that lasted for the final decade of Audubon's life.[45] Baird became one of Audubon's most faithful correspondents in that time, first signing his letters "Your Obedient Servant" but then taking on a more appropriate role as "your affectionate Pupil."[46]

Baird had just finished being a pupil of a different sort. Five days after he first wrote Audubon, he received his undergraduate degree from Dickinson College in Carlisle, Pennsylvania, which he had entered at thirteen, following

Figure 12. Spencer Fullerton Baird, by unknown artist, ca. 1850. Smithsonian Institution Archives, Image #2002-12179.

two older brothers. Immediately upon graduation, he spent his vacation time, as one biographer put it, "in reading, study, excursions on foot and in the saddle, with young lady friends of the town and other young people, in the natural and appropriate amusements of the age."[47] But the young lady friends would almost always take a back seat to Baird's first love, natural history. As he wrote Audubon a couple of years later, "I know that you will give me credit for saying that I would rather get a new rare Quadruped or Bird than kiss all the pretty girls in creation."[48] Coming from a teenaged male, that seemed quite a commitment to a budding career in science.

Indeed, according to family legend, Baird had been a student of nature from his toddler days, accompanying his father, Samuel Baird, on walks in the woods or following him around the garden and helping with the weeding at his childhood home in Reading, Pennsylvania. Once, when the elder Baird, a successful lawyer, had to be away on business in Boston, the eight-year-old Spencer dutifully wrote him to report that the "grapes flowers and the Garden are in a Good state."[49] Samuel Baird died in 1833, when Spencer was only ten, and then the family moved to Carlisle, where young Baird grew up with his mother, grandmother, three brothers, three sisters, and a large extended family in the town. In addition to a supportive kinship network, Carlisle offered Spencer Baird easy access to an outdoor education in nature. Like Audubon's original Pennsylvania home near Norristown, Carlisle lay on a major migratory route for birds, and Spencer grew up with a good eye for birds, especially when looking down the barrel of a gun. By the time he finished college, he and his older brother William already had an extensive collection of hundreds of preserved birds, including the flycatcher he brought to Audubon's attention.[50] No one could have been better prepared, by personal inclination as much as formal education, to become Audubon's acolyte.

The family plan, however, called for Spencer Baird to become a doctor. Within six months of graduation from Dickinson, he dutifully—although not enthusiastically—began to read medicine with a local physician, and a year after that, in the fall of 1841, he went to New York City to study with several doctors and attend lectures at Bellevue Hospital. Medicine never moved him, though. During his time in New York, he also spent time going to meetings of the New York Lyceum of Natural History and meeting some of the men of science in the city, always talking more of ornithology than medicine. By the end of 1842, his medical studies came to an end, and his brother William, who had been supporting him financially, accepted the fact that Spencer had "a distaste for a medical life . . . [and] perhaps it would be best not to pursue the study, for no one ever succeeds in a profession of which he is not fond." Knowing Spencer's fondness for birds (which he himself shared), William added a word of fraternal advice: "As it is necessary for you to do something, you ought to make up your mind. No means of livelihood, however, is to be obtained in America from ornithology."[51]

By that time, however, Spencer Baird had come under the scientific sway of the one man in America who had long made a livelihood of ornithology: Audubon. Soon after arriving in New York, in fact, Baird made a point of going to see Audubon and, as he put it in his journal, "found him very unlike my preconceived idea of him." (Whatever that may have been, he didn't say.) Audubon gave Baird lessons in art and ornithology and, even more generously, gave him some rare bird skins and a set of his five-volume *Ornithological Biography*, which Baird declared "a pretty clever present."[52]

Baird's first few visits with Audubon seemed a highlight of his life, and his return to Carlisle felt like an exceedingly disappointing denouement. "After a trial of two weeks," he wrote Audubon in early 1842, "I begin to find that I am getting over the shock caused by the sudden transition from the bustle of Broadway to the lifelessness of Carlisle.... Philadelphia was dull, but Carlisle was death itself." All Baird's friends wanted to know all about Audubon—"how he looked? What was his age, whether the idea they had formed of him from his writings was correct"—and Baird might have taken some solace in being, at least briefly, the center of attention among the other Audubon devotees in town. But he found immediate relief in going out shooting with his gun, which he named "Long Tom," after Audubon's own favorite piece, and told Audubon how he used it to bag a Goldeneye Duck. (He considered the bird "not at all eatable now," so he gave it to a friend to try.) On the topic of "eatable" possibilities, Baird wrote in a postscript that he had had a fine steak of broiled wild cat, which "tasted like a tender piece of fresh pork," and he closed by saying, "I intend to taste all the Quadrupeds inhabiting this part of the country"—a culinary benefit, perhaps, ancillary to scientific research.[53]

Audubon wrote back reassuringly, reminding Baird of the delights of small-town life and the "kindest of welcomes" the young man must undoubtedly have received from family and friends. "Surely all this was fact, and being so, would you not after all prefer *Little Carlisle* than *Great New York* with all its humbug, rascality, and immorality?" Audubon even engaged in a bit of nostalgia for Norristown, his own small-town home in the region, writing that "still my heart and mind oftentimes dwell in the pleasure that I felt there, and it always reminds me that within a few miles of that village, my Mother did live." (Of course, neither Audubon's birth mother in Saint-Domingue nor his adoptive mother in France ever lived anywhere near Norristown. Even with Spencer Baird, his precocious new pupil, he once again indulged himself in a

seemingly gratuitous falsehood about his parentage and, indirectly, his geographic origins.) In the end, Audubon sent encouragement for Baird's ongoing
efforts at drawing birds and mammals, advising him to "work slowly, and
constantly, that is, whenever you can!" If Baird did that, then "your attention
to *Drawing* will soon enable you to 'copperplate.' Go ahead!"[54]

Audubon offered an even better way to go: He invited Baird to come with
him on the 1843 expedition to the West. Baird wanted desperately to go along,
and he wrote Audubon that "nothing would delight me more than to go, if I
can afford it." That last qualifier loomed large for Baird, though, because the
economic troubles of the times had severely depleted the family fortune, such
as it was, and the cost of getting outfitted for such an expedition seemed daunting. Still, Baird saw impressive possibilities in the adventure, an opportunity
for "increasing the general sum of knowledge in every department of science,
Physical as well as Natural."[55] Audubon wanted Baird to go on the trip, too,
and he shared much of the young man's enthusiasm for this expeditionary
effort to "open the Eyes of naturalists to *Riches untold*, and *facts hitherto untold*"
by exploring "the portions of the country through which it is my intention to
pass, never having been trodden by white Man previously."[56] Audubon initially
offered to pay half of Baird's expenses once they left St. Louis and also to let
him keep half of the bird skins they might take, along with a quarter of the
mammalian specimens, "reserving to myself all that is new or exceedingly
new." (There had to be, of course, some privileges of mentorship.) He later
sweetened his first offer by saying that since he understood "that rascally *cash*
is the cause which prevents you from going along," he would pay all of Baird's
expenses, "excepting your clothing and your gun or guns, as you may have
them." Above all, he said, "my principal wish is to employ you as my Secretary,
in friendship, and for the sake of that Science to which I have been devoted
many years, and yourself several."[57]

Such a financially and intellectually generous gesture from a man as famous
as Audubon must have seemed remarkably appealing to a young man who had
just turned twenty, but Baird had just turned twenty, after all. His mother and
other relatives pressured him into giving up the trip, partly for the sake of the
family finances but also for his health, which had recently taken a turn for the
worse, with palpitations of the heart. Audubon had to accept the situation, and
he wrote Baird that he had finally "concluded to take a Young Gentleman in
your stead who is a Neighbor of ours, but who alas is no Naturalist, though a

tough, active, and very willing person." In parting, Audubon urged Baird that should he find specimens of "Quadrupeds that you consider new or very rare, please to save for us, taking notes of their exact measurements, localities, and dates of capture."[58] Even if he stayed home, Baird could be a useful scientific assistant on Audubon's new project.

Baird's Own Projects

Although Baird could not go west with Audubon, he had quite a bit to do on his own in the East, first in Carlisle and then in Washington. In Carlisle, he continued to cultivate his intellectual and cultural development, studying French, German, and calculus; joining the local musical society and the church choir; teaching a small group of young girls; and always going into the field to find more birds for the collection he and his brother William had begun. Spencer and William also coauthored a scientific paper on two new species of flycatchers they had found, and after a couple of publication delays, the paper appeared in the *Proceedings* of the Academy of Natural Sciences in Philadelphia in September 1843. "Your paper reads well," his fellow naturalist friend John Cassin complimented him, "and I will see that copies of this number are sent to most of the Naturalists, in this country at least."[59] But writing with William was hardly Spencer's preferred fraternal endeavor. When William was about to set out on a trip to Cape May, New Jersey, in the summer of 1843, Spencer wanted to go with him. Once again, though, the cost of the trip, around $16, seemed just out of his immediate reach. "How I wish that I was going along," he wrote William, with a combination of disappointment and vicarious enjoyment. "Would'ent we walk into the birds. Shooting and stuffing. I would think nothing of sitting up till twelve every night Stuffing what we had been all day in shooting. If I had the money I would start off to morrow."[60]

A few months later, in November, he did start off on a trip, this time to Washington, traveling by stage and, not surprisingly, keeping very careful track of his transportation costs: Carlisle to York, $2.00; York to Baltimore, $2.00; Baltimore to Washington, $2.50. In the nation's capital, he lodged with his uncle and made contact with some of the most prominent men of science, perhaps most usefully James Dwight Dana. Dana had been a member of the United States Exploring Expedition, led by Navy Lieutenant Charles Wilkes, which had gone to the Arctic region and had collected thousands of specimens of all sorts, from seeds to plants to birds. In the time Audubon had been on his own expedition to the Yellowstone region, Baird at least had the opportunity to get his hands on the Wilkes Expedition treasures and, in the process, to impress Dana with his scientific and curatorial skills. It proved to be a connection that would help shape the long-term future of Baird's career.

In the meantime, the still-young Baird went back to Carlisle and began to settle into adult life. In an 1845 letter to Audubon about, among other things, prairie hares in Texas, he added, almost as an aside, that "I forgot to say that I have been elected professor of Natural History in Dickinson College." Like most new professors, he groused a bit about the terms of the job—"The situation is entirely nominal, however, nothing to do & no salary whatever"—but it was a start. A little over a year later, he wrote Audubon of "settling down steadily in my Professional chair" at Dickinson, where he was "teaching Animal Physiology, Natural Theology & Mathematics." He also noted another "pretty important" event, his marriage to Mary Helen Churchill, the daughter of Col. Sylvester Churchill, Inspector General of the U.S. Army, who had been based in Carlisle. Mary Churchill had certainly learned what to expect of her ornithologically obsessed suitor. Baird gave her a picture of Audubon two years before their marriage, and soon after their wedding he gave her "a lesson or two in taxidermy." Taxidermy for two might not seem like everyone's idea of honeymoon romance, but Baird's new wife apparently accepted her husband's passion with good grace. Her daughter later wrote that "my mother never had any especial taste for Natural History, although always very much interested in anything which my father was doing." Baird claimed a bit more enthusiasm on the part of his new bride. "She suits me exactly," he wrote Audubon, "being as fond of birds & snakes & fishes etc. as myself."[61]

The job at Dickinson suited him better, too. He had a salary of $400, which he still deemed small, but which "I hope will be larger hereafter." (It was, rising to $650 in 1847 and $1,000 in 1848.)[62] Baird became a professorial fixture at Dickinson. "I like the business of teaching very much," he wrote William, "& believe I am a favorite with the Students." Indeed he was. He took them on weekend walks to experience the sort of outdoor field research he loved so much, and they in turn took to calling him the "Prof." He also took on curatorial duties at Dickinson, getting institutional support for building cabinets for all the specimens he collected. In a report to the college on his museum work, he noted that he had been "engaged for some time in arranging his collection of North American birds with the intention of placing it in the museum," and he proudly asserted that it "now composes the largest and most extensive collection of the Kind in the world," with some 450 species and about 2,500 specimens, many of them "unique, being the only one procured, and consequently not to be found in any other collection."[63] Clearly, Dickinson College had quite an asset in Baird, and he had a promising position in the college. For the twenty-three year-old "Prof," the future was beginning to look very good.

Within a few months, it began to look even better. In early 1847, Baird contacted Audubon again, saying that his friend James Dana was urging him to apply for a curatorship at the new Smithsonian Institution in Washington, DC, a job that would pay $1,500 a year plus rent. Happily surprised by the opportunity, Baird asked Audubon to "make out a flaming recommendation for the place" and, if he could, to exert "any personal influence you may have on the board of Regents." In turn, Baird promised to do what he could for Audubon if he got the job: "When there I would hope to be materially useful to you in your labors."[64]

Getting the job, however, turned out to be a three-year ordeal for Baird, with numerous delays based on funding problems, the slow pace of constructing the Smithsonian building, and the cautious institutional inertia of the Smithsonian's Regents. The Smithsonian's first secretary, Joseph Henry, told Baird it might be another five years before they would fill the position. In the meantime, Baird continued teaching at Dickinson, corresponding with some of the most eminent men of science in America—Audubon, to be sure, but also Louis Agassiz, Arnold Henry Guyot, and George Perkins Marsh—and always expanding his collections of natural history specimens. He had used his collections, in fact, as a sweetener in his application to the Smithsonian, promising that "should I go to Washington, my collections would of course accompany me." At the time, this young man in his mid-twenties had the personal wherewithal to provide a solid foundation for the nation's natural history collection, with "specimens of North American Birds, Quadrupeds, Reptiles and Fishes, Complete Skeletons, Crania, numerous Vertebrata, and Forest trees." His three thousand bird skins alone would be worthy of note, and he admitted to surpassing the master: "My ornithological collection is probably the richest in N. American species of any in the world, containing with very few exceptions all those figured and described by Audubon, with many others unknown to him."[65]

Baird had become too good to be kept waiting too long. In July 1850, he received a letter from C. C. Jewett, a member of the Smithsonian's Board of Regents, telling him that, at the urging of Joseph Henry, he had been unanimously chosen "assistant secretary and Keeper of the Cabinet." Henry wrote a few days later to underscore his commitment to Baird, assuring him that "the office will afford you an opportunity of prosecuting your favorite study to the best advantage, while it will enable you to render important service to the cause of knowledge in our country."[66] Baird kept up his end of the bargain, too. On October 2, 1850, he shipped two boxcar loads of specimens to Washington, where they immediately formed an important part of the Smithsonian's natural history collection—and where they still reside today.

If Audubon or anyone else had ever needed proof of the possibilities that could be created by inviting the "Dear Reader" into the work of natural history, Spencer Baird provided a prominent human embodiment of what that reciprocal relationship could produce. Baird was quite a rare bird, of course: a young man of ordinary-seeming origins who turned out to be quite extraordinary in his study of natural history. Baird took seriously Audubon's invitation to respond to the various inquiries in *Ornithological Biography* and, in turn, gained the recognition and respect from the most celebrated naturalist of antebellum America. Given Audubon's sometimes uneasy relationships with other fellow members of the scientific community, he must have felt himself fortunate to have found such an engaging and agreeable student as Baird.

Best of all, Baird learned his lessons well. Building on Audubon's early encouragement, not to mention his own industrious efforts as a collector, he himself eventually emerged as one of the most significant scientists of his own era. This student so inspired by Audubon when he was "but a boy, and very inexperienced" went on to spend the rest of his adult life at the Smithsonian, making a thirty-seven-year career there, starting as curator and assistant secretary in 1850, and then, after Joseph Henry died, in 1878, as secretary until his own death, in 1887. Throughout his time at the Smithsonian, Spencer Baird became a generous mentor to the rising generation of naturalists, paying forward the sort of support he himself had experienced under Audubon's useful tutelage.[67]

Last Days at Minnie's Land

Audubon never lived to witness the full trajectory of Baird's scientific career. By the time Baird began taking up his position at the Smithsonian, Audubon had begun drifting downward. Wearied by travel back and forth several times across the Atlantic to produce his Great Work and seek out subscribers, then weakened by his last years of exertion and exposure in the wilds in pursuit of specimens out West, Audubon faded quickly in the late 1840s, suffering another stroke and losing much of his mental focus and some of his eyesight. Rumors of his decline were sometimes a bit exaggerated, especially in the press. One journal claimed that the "Nestor of our naturalists" had been jolted from his quiet enjoyment of his home at Minnie's Land by the advent of the railroad, "when the surveyors and engineers, with their charter-privileges, invaded his retreat, built a road through his garden, destroyed forever his repose, and—the melancholy truth is known—made of his mind a ruin."[68] Another writer, the poet and journalist Park Benjamin, took a brief newspaper note—"John James Audubon, the great American naturalist, has completely lost his sight"—as his cue to write about the painful, if obvious, irony of blindness in a man who opened so many eyes to nature:

> Blind—blind! Alas! he is bereft of light
> Who gave such pleasure to the sense of sight.
> His eyes, that, like the sun, had power to vest
> All forms with color, are with darkness prest:
> Sealed with a gloom chaotic like the deep;
> Shut in by shadows like the realm of sleep.

Audubon was not, in fact, fully blind; he just disliked wearing his glasses.[69]

Friends who visited him, however, did see signs of deterioration. Baird saw him in the summer of 1847 and "found him much changed." Bachman came to Minnie's Land the following spring and wrote ruefully, "Alas, my poor friend, Audubon, the outlines of his countenance and his form are there, but his noble mind is all in ruins." Audubon himself apparently felt rueful at the prospect of not being able to paint, but otherwise he lived in comparative comfort, supported by caring family members.[70] He died, quietly and in their company, on January 27, 1851, just three months short of his sixty-sixth birthday.

Figure 13. *John James Audubon,* portrait by John Woodhouse Audubon (1843), Art Survey
no. 1403. Courtesy of the American Museum of Natural History Library.

Figure 14. John James Audubon, ca. 1847–1848, by Matthew Brady. Cincinnati Art Museum, Ohio, USA Centennial gift of Mr. and Mrs. Frank Shaffer Jr./Bridgeman Images.

Most of the obituaries would soon say he was ten years older, and in some ways he was. He had lived about as full a life as anyone could, crossing oceans and continents, pursuing art and science, building a family business and battling unfriendly critics, and perhaps most notably, bringing a stunning view of nature to the nation. The American Woodsman had created a Great Work, indeed, and in the process he had also created himself as a brash and brilliant embodiment of his era.

Chapter 10

≈

Bringing Audubon Back to Life

> Now, Reader, farewell! may you be successful in all your undertakings! may
> you be happy abroad and at home! and may the study of the admirable produc-
> tions of Nature ever prove as agreeable to you as it has been to me.
> —John James Audubon, "Introduction," in *Ornithological Biography*

The obituaries spread quickly across the Atlantic, many of them copied from one
newspaper to the next, further circulating the misinformation about Audubon's
age ("seventy-six years") and origins ("Louisiana, his birth place"), but all
essentially coming to a common point: The late naturalist had been a "celebrated
man," with a "fondness for natural history, which he pursued through life with
unabated enthusiasm" and whose remarkable achievement in *The Birds of
America* "placed him in the first rank as an artist."[1] He was a "simple-minded,
yet truly great man," said a British obituary, while an American one went quite
a bit beyond that, maybe even overboard, in calling him "one of the pioneers
of Western civilization."[2] Most concurred that he had "few equals in fame
and merit among the naturalists," or perhaps "no equals." At least one gave
the final verdict that Audubon had always wanted: "He far excelled even
Wilson himself."[3]

Beyond merely taking note of Audubon's death and saying a few kind words
about his life, some death notices went into much lengthier detail about his
physical condition and moral character. "In person he was tall, and remarkably
well made," said one writer who claimed to have known him. "His whole head
was remarkably striking. The forehead high, arched and unclouding . . . chin
prominent, and mouth characterized by energy and determination."[4] The
American Phrenological Journal went one step further, offering an especially
thorough postmortem personality evaluation (based, apparently, on an accom-

panying image rather than a hands-on examination). Using something of the same directive approach Audubon employed in showing the reader some of the details in his bird images, the *Phrenological Journal* likewise led its own readers through a close examination of the contours of Audubon's head: "See that bold projection at the root of the nose, between the eyebrows—the location of INDIVIDUALITY . . . and we get the great secret of his remarkable genius as a naturalist; the close observation, the ready perception, the critical knowledge of forms, colors, and arrangement of all the minute and varied phenomena of Nature's works." He had many other positive traits, to be sure—Firmness, Locality, Eventuality, Comparison, Constructiveness, and Hope among them—and his "moral organs were large, particularly BENEVOLENCE and VENERATION." On the other hand, "CAUSALITY does not appear large, and . . . he was much more of an observer than a philosopher."[5] Audubon, who had earlier applauded the positive results of a phrenological examination back in 1826, would no doubt have been equally pleased with this one.

But of all the obituaries that appeared in print on both sides of the Atlantic, probably none would have been more agreeable to Audubon than a long and laudatory essay written by Charles Wilkins Webber, celebrating Audubon as the "highest ideal of the Hunter-Naturalist." Webber, a Kentucky-born adventurer, Texas Ranger, and New York journalist, claimed he first encountered Audubon on a canal boat in Pittsburgh in the fall of 1843, when the aging naturalist was returning from his expedition to the Missouri River. As soon as Webber boarded the boat, he "heard above the buzz the name of Audubon spoken." Apparently already familiar with Audubon's reputation, Webber wrote that "there was one NAME that had so filled my life, that it alone would have been sufficient to inspire me."

> Audubon! Audubon! Delightful name! Ah, do I not remember well the hold it took upon my young imagination when I heard the fragmented rumor from afar, that there was a strange man aboard then, who lived in the wilderness with only his dog and gun, and did nothing by day, but follow up the birds; watching every thing they might do; keeping in sight of them all the time, wherever they went, while light lasted; then sleeping beneath the tree where they perched, to be up again to follow them again with the dawn, until he knew every habit and way that belonged to them.

And so Webber went on for a dozen exuberant pages, describing Audubon's "fine, classic head" and "patriarchal beard" and "hawk-like eyes," asserting that "the very hem of his garments—of that rusty and faded green blanket, ought to be sacred to all devotees of science."[6]

But Webber did not stop there, and his tribute to Audubon became a full-throated cry of scientific adulation. In Webber's eyes, Audubon represented much more than a master of ornithology or avian art: He embodied the "hero of the ideal," the supreme representation of the rugged American "pioneer hunters . . . who learned through starvation, and all the perils of savage warfare, and the inconsistent seasons, to know more accurately the habits, passions, transitions and localities of our animals . . . who have been able to unite scientific accuracy with the gleanings of their rude lore, who are to be depended upon as true delineators." Audubon might well have written essentially the same thing himself, of course, with his own self-described image of the American Woodsman merging easily with the magnified manliness of Webber's Hunter-Naturalist. Like Audubon, Webber used the supremely masculine image of the "pioneer" American scientist to create a dismissive, derisive contrast to the formulaic, effete, and implicitly feminine approach of European naturalists: Too many "scientific pedants in silk stockings" and "pur-blind Professors," Webber complained, had "technicalised" the study of nature "into what may almost be called a perfect whalebone state of sapless system . . . so heavily overlaid by the dry bones of Linnaean nomenclature as to become a veritable Golgotha of Science." As a result, he continued, ordinary people had become isolated from science, "repulsed, in dismay of its formidable hieroglyphics, from what is to them as a sealed book." By contrast, "Our glorious Audubon," the Hunter-Naturalist of the new nation, "lived and wrote like one of the people," and he thus represented a distinctly American approach to the art and science of natural history, a two-way relationship between the naturalist and the nation. Therefore, Webber declared, "We love and venerate him passed away."[7]

Audubon's Science, Audubon's Society

Audubon had indeed passed away, but some of the issues, even tensions, he embodied in the world of science would remain very much in focus over the years to come. Charles Wilkins Webber's brash portrayal of the "Hunter-

Naturalist" as a populist alternative to the "pur-blind Professors" might have been overblown, but it made an important point. As much as Audubon craved respectable standing in the scientific community—and he consistently sought and proudly accepted the accolades that came with membership in all the right scientific societies—he still had a self-conscious sense of reaching out to a broader range of people. His invitation to the "kind Reader" to assist him may have been a bit disingenuous at times, but it did take a significant rhetorical step toward acknowledging that science could be accessible to essentially everyone, that ordinary people had a place in ornithology. In the first half of the nineteenth century, nobody did more to promote that notion than Audubon, and it remained a central issue in the study of birds for decades after his death.

On one level, the professors seemed to be ascendant, and the move toward rigorous standards and disciplinary identity only increased in the years after Audubon's death.[8] The nation's first two organizations devoted expressly to ornithology—the Nuttall Ornithological Club, established in 1873, and the American Ornithologists' Union (commonly called the AOU), created a decade later—promoted a strong commitment to the exacting scientific study of birds, particularly with regard to systematic classification and nomenclature. The more Darwin began to replace the Deity as the primary source of scientific authority, the more taxonomy defined the focus of ornithological inquiry. The AOU also pushed for the professional certification of those who pursued such study, so that by the early decades of the twentieth century, people could actually get PhDs and paid positions, finding "reasonable employment prospects in colleges and universities, natural history museums, governmental wildlife agencies, and private conservation organizations."[9] Audubon may not have been the sort of ornithologist who could ever hold an indoor job, much less work for someone else, but he might have felt fortunate at least to ponder the professional opportunity.

At the same time, though, a passionate amateur could still have a respected, albeit sometimes contested, place in the field. While the AOU provided a reasonably exclusive enclave for scientists and other self-consciously "serious" students of ornithology, other bird-centered organizations offered a more open opportunity to a wide variety of enthusiasts, from casual watchers to committed, sometimes even obsessive, observers. The Audubon Society, founded in 1886, just three years after the AOU, is the most significant case in point. Audubon himself had nothing to do with starting the organization that bore his name—he was long dead, of course—but he did have at least an indirect connection to the founder, George Bird Grinnell, the young editor of *Forest and Stream*. Grinnell noted that his family lived near the Audubons in New York when he was a boy, and "I attended a school conducted by Madam Audubon in the Victor Audubon house, where she lived."[10] He had also read enough of *Ornithological Biography* to have been taught, so to speak, by Audubon himself.

In creating the Audubon Society, Grinnell intended it to be a complement, not a competitor, to the AOU, and he opened it to anyone who would adhere to its basic principles: protecting wild birds, their nests, and eggs; and preventing bird feathers from being collected for use in fashionable women's hats. (Like Audubon, Grinnell was a hunter, and he made an exception for killing birds for food, a position the modern-day manifestation of the organization still generally accepts.[11]) The Audubon Society provided a tent big enough for bird lovers of all sorts, offering them a sense of purpose and coherence they might not otherwise have discovered. One of the earliest and most active chapters, for instance, was organized by two undergraduates at Smith College, Florence Merriam and Fanny Hardy, who took several hundred female students not only into the field to see live birds, but also into the local milliners' shops to have dead birds removed from their hats. The rapid growth of the Smith College chapter provided a preview of the rapid growth of the Audubon Society: Within a year of its founding, the national organization had attracted upward of eighteen thousand members organized in three hundred local chapters.[12]

The Audubon Society, like the AOU, would have its ups and downs in membership and influence over the years, but the number of bird enthusiasts would continue to grow, seemingly exponentially, into the twentieth century. By the 1930s, some younger members of the AOU would occasionally express concern about the enduring influence of recreational bird-watchers on their organization, fearing that armchair amateurs could undermine the organization's professional respectability. The AOU's more astute leaders, however, recognized the potential symbiosis of the situation, accepting amateurs as occasional allies in field research, vocal advocates in policy issues, and, certainly not least, subscription-paying supporters of the AOU's major publication, *The Auk*. The relationship between professionals and amateurs would long remain a source of uneasiness, even occasional tension, but most people in the ornithological elite had to recognize that no one could claim true ownership of ornithology, and passionate amateurs had no intention of leaving the field. Beginning in the early years of the twentieth century, avid birders of all descriptions began to take part in such activities as the Christmas Bird Count, and the National Audubon Society now celebrates "Citizen Science" as a way of engaging active amateurs and recognizing their contributions to ornithological research.[13] Just as Audubon and other nineteenth-century naturalists had made good use of ordinary people in the past, so, too, have modern-day ornithologists benefited from all the additional eyes and ears so eagerly turned to the birds.

By the end of the twentieth century, the number of ornithologically inclined Americans had apparently grown well into the millions. In 2000, a government study—the National Survey on Recreation and the Environment (NSRE 2000), conducted by the U.S. Department of Agriculture and the National Oceanographic and Atmospheric Agency—asked a simple-seeming question: "During the past 12 months, did you view, identify, or photograph birds?" According to the study results, at least 67 million Americans answered "yes."[14] It would be wise, of course, to hedge one's bets on the accuracy, much less the real meaning of the number: that 67 million might include everyone from casual observers of birds in a backyard feeder, to more committed participants in one of the annual bird counts, to even more active travelers on eco-tourist trips to birding sites, to year-round birders who come to life during migration seasons, or even the hard-core competitors in the World Series of Birding or the Big Year. Still, whatever the degree of skill or commitment, the point is clear: Today, birders seem almost as ubiquitous as birds, and ordinary ornithologists of all stripes take to the field with considerable fervor, many of them compiling extensive life lists of the birds they eagerly (if often fleetingly) see.

And now, equipped with better binoculars around their necks and well-illustrated field guides in their hands, they can see birds in ways that Audubon himself would no doubt envy. Audubon and other ornithologists of his time relied on the gun as their main technology of taxonomy, and bird identification most often depended on bird destruction. In the post-Audubon era, improvements in optics and the rise of the field guide would change that dramatically. Beginning with basic and often bulky books in the late nineteenth century, most of them long on text and short on illustration, the field guide became both smaller and better in the twentieth century, and each new edition seemed to improve upon previous ones as a visual asset to bird-watching. From Roger Tory Peterson's pioneering work, which set the standards of the field guide from the 1930s onward, to David Allen Sibley's several books at the beginning of the current century (done, incidentally, under the auspices of the National Audubon Society), the best bird guides are works of art in their own right, with carefully detailed, full-color drawings of birds. One might consider them pocket-sized Audubons, in a sense, except for the most significant difference in composition: In the interest of accurate identification, modern field guides show each species frozen on the page in static profile, rather like the much earlier bird images of Mark Catesby, Alexander Wilson, and, in his adolescent years, even Audubon himself. To their credit, these guides include more birds than Audubon ever had in *The Birds of America*, but none of them would dare try to match the motion, much less emotion, that Audubon brought to avian art.[15]

Audubon's Birds Unbound

Audubon always knew that avian art would be his signature in life, and that his Great Work would truly be a lasting treasure. In the early days of its production, well before he had even completed the first volume, he confidently asserted that *The Birds of America* would be "a book that in fifty years will be sold at immense prices because of its rarity."[16] Later, after he had finished all four volumes, he wrote to Daniel Webster to remind him what a valuable asset he had purchased, hoping "with great sincerity, that you will never dispose of the 'Birds of America' until indeed you are reduced to the direst necessity."[17]

As it happened, Audubon's own family faced dire necessity first. After fire damaged some of Audubon's original copper plates in 1845, Audubon wrote his young friend Spencer Baird that he and his family were trying to restore some of the damaged plates, but he added a note of doubt about future prospects for *The Birds of America*, "if ever that work is republished in its original size at all."[18] In fact, it would not be, at least not by using the Havell plates. In 1858, seven years after Audubon died, his sons, Victor and John, attempted to generate additional business by selling subscriptions for a full-size but less expensive version of their father's work, this time produced with a new printer, Julius Bien, and a new technology, chromolithography. Bien, who had emigrated from Germany in 1848 and established a print shop in New York in 1850, had quickly become the country's leader in this complicated but cost-effective process. Bien and the workers in his shop would take wet-paper transfers of the images from Havell's copper plates onto lithographic stones, then print the images in several stages of coloring (with some hand-drawing added when necessary) and thus make full-size reproductions that preserved the detail of the original drawings. With the Bien edition of full-size plates printed on high-quality paper, the Audubon brothers hoped they had found a marketable mid-range option between the Double Elephant Folio and the Royal Octavo editions. Unfortunately, the project never worked out as well as the Audubon family had hoped. Victor died in August 1860, John less than two years later, and the Civil War brought the production of the plates to a halt. The Bien venture produced images of only 150 of Audubon's original 435 engravings and attracted probably no more than a hundred subscribers, and more likely fewer than that, thus failing to produce the income the Audubons needed.[19] To make up for the financial losses incurred, Lucy soon sold the copper plates to a New York firm, and some were eventually melted down into copper bars; she also sold 471 of Audubon's original watercolors to the New-York Historical Society, where they would occasionally be on display for viewers.[20]

When Lucy Audubon sold off the bulk of her husband's original watercolors and copper plates, she was only one participant in a process of breaking up *The Birds of America* that had begun much earlier in Audubon's era and endured well into the twentieth century. The greatest version of that Great Work, the Double Elephant Folio edition, has had a particularly perilous history of its own. It was from the beginning, by definition, a rare book, but it has since become, at least in the minds of some people, a work that stretches the very definition of what a book should be—if it should even be considered a book at all. Given the bulk and beauty of *The Birds*, most nineteenth-century owners of the complete work had custom-built stands and storage furniture made to house it.[21] Several decades before Thorstein Veblen coined the term "conspicuous consumption," they had a trophy book to be displayed—with their own status on display as well. As one mid-twentieth-century Audubon admirer put it, *The Birds of America* was "hardly a book that one can—or would want to—pick up and leaf through lightly as he lounges after luncheon in a favorite chair." But leafing through it may have not even been the point. "The proud possessor of a copy of the 1827–38 folio edition oft finds himself a bit embarrassed by the treasure he owns. . . . His nose tilts a little higher in the air as he walks the street, thinking of the men and institutions who would have an *elephanticus* if they could."[22]

Not many could, of course. Today, Audubon collectors and scholars offer varying estimates of the number of complete sets Audubon eventually delivered to his subscribers—perhaps as few as 175, maybe as many as 199—but now, at last count, apparently only 119 exist, all but a dozen or so of those in the possession of libraries and other institutions; many of the other original sets have been lost, destroyed, or, more commonly, broken up and sold off as individual plates.[23] That sort of destruction began with some of the very first subscribers, some of whom seemed to have been drawn largely by the decorative value of Audubon's work, caring less about scientific detail or collecting, but seeking some drama for their walls. In fact, one subscriber, the Marchioness of Hertford, the mistress of King George IV of England, even embedded them into her walls, cutting up the plates and making particular images part of her wallpaper. When Audubon discovered that another purchaser had allowed her plates to become "quite abused and tumbled," he at first became annoyed, but then he tried to tell himself, not very convincingly, that it was "not my concern."[24]

It did, however, become a considerable concern for later admirers of Audubon's art. As the naturalist Samuel N. Rhoads noted in 1916, "In the past twenty years it is probable that one New England print-dealer has broken up thirty or forty volumes of this magnificent work, selling the plates separately for framing and other illustrative purposes."[25] The breaking up of the full volumes continued well into the mid-twentieth century. Between 1931 and 1935, Macy's New York department store put four, perhaps five, full sets of the Double Elephant Folio on sale, plate by plate, and took in about $120,000.[26] Other book and print dealers followed suit, treating the images as some of Audubon's own nineteenth-century subscribers did, displaying them as individual works of art and not keeping them bound together in something that looked very much like a book—a very big book, at that. Breaking up complete sets of *The Birds of America* quite often proceeded without apology, sometimes even with a sense of moral superiority. What some Audubon aficionados might decry as an act of destruction, others celebrated as a form of artistic democratization. In 1947, for instance, the Old Print Shop in New York announced the sale of two hundred plates from the first two bound volumes of *The Birds of America*, and the prospectus for the sale made a preemptive strike against any criticism from book preservationists: "As is usual when we break up a bound volume of the Elephant Folio, the cry of 'Vandal, Vandal!' is heard, but we believe that *every* set should be broken up." The prospectus went on to argue that keeping such a rare book in "the reserve departments of public libraries" where it could be seen "only by a few so-called 'serious scholars'" would be almost the equivalent of subjecting it to a "public burning." Better to cut the volumes up and sell the individual prints, the Old Print Shop said, because "we feel sure that the actual presence of single impressions in individual homes is what has made Audubon the living force he is today." More to the point, the Old Print Shop continued, Audubon himself would want the volumes cut up: "It is not to be reconciled with the character of Audubon, his love of country, his love of birds, and his love of people, to think that he would have wanted to limit the circulation of the approximately 175 sets of his work to wealthy institutions or individuals." Some of those institutions could be research repositories, of course, "but it is our conviction that by far the greatest number of volumes should be broken up and the plates given the widest possible circulation." To facilitate this "widest possible circulation" to the "individual homes" of postwar America, the Old Print Shop offered individual plates of, say, the Ruffed Grouse at $375, or the Great Blue Heron for $325, or, for the more economy-minded, the Golden Plover for $15 and the Dusky Petrel for $10.[27]

This allegedly democratic destruction of *The Birds of America* did not stop with the Double Elephant Folio volumes. In 1955, Goodspeed's of Boston announced the sale of prints from the smaller Royal Octavo edition: "Since a complete set of the folio edition recently sold at auction for more than $25,000," Goodspeed's said, "and since the individual folio prints are becoming more expensive and less attainable, perhaps you will agree that a little bird in the hand is worth two big ones in the bush, and will settle for an original minia-ture."[28] That seemed to be the increasingly apparent postwar approach among other art dealers, who put a higher value (and thus a higher price) on the contents of the book, the individual plates, than on the book itself as a complete volume, or set of volumes. In a sense these modern dealers brought *The Birds of America* full circle, unbinding it—unbooking it, one might say—and taking it back to a collection of individual images to be exhibited on the walls of museums or galleries or even private homes.

The Birds Saved from Extinction

Still, what remains the fate of the original big book—or set of books? It may well be, as the Old Print Shop put it, that the "so-called 'serious scholar'" and "bibliographer" might be content with the complete copies that still exist in rare book collections of great libraries. But with the number of complete sets of the Double Elephant Folio edition now apparently down to 119 or so, the original manifestation of Audubon's Great Work is undeniably an endangered species. And as has often been the case with actual endangered species of birds, the closer they get to extinction, the more collectors scramble to get their hands on the last few specimens.[29] That has become markedly evident in the modern marketing of complete editions of *The Birds of America* in the past fifty years, as prices have soared higher than any bird ever dreamed of flying. Again, as Goodspeed's noted in 1955, a complete set of the large edition of *The Birds of America* had recently sold for $25,000. Then, less than fifteen years later, in 1969, another complete set sold at auction for $216,000, and in 1992, the auction price for another set jumped to $4.1 million, an increase of almost 2,000 percent in just over twenty years. Less than a decade later, in March 2000, even that big bid for the Double Elephant Folio edition was doubled, when a day of intense activity at the Christie's New York auction house ended with a purchase price of just over $8.8 million, setting a new world auction record for a printed book. (The previous recordholder had been a copy of Chaucer's *Canterbury Tales*, which fetched a little over $7.5 million at Christie's in London in 1998.[30]) The price of *The Birds of America* suddenly seemed so astronomical that it raised a serious question for those institutions that owned it: Would it make better sense to keep the book in the hand, or to put it on the auction block?

Two cases from the early twenty-first century offer contrasting approaches to the question. In early 2003, the directors of the Providence Athenaeum announced just before the organization's annual meeting that they had decided to sell their copy of *The Birds of America* to help shore up the institution's dwindling endowment—not to mention its sagging ceiling. If the auction at Christie's in 2000 had brought a price of $8.8 million, then a similar sale, they figured, could put the Athenaeum back into sound fiscal condition. The Athenaeum's leadership did not anticipate, however, the pro-Audubon anger of some of the institution's dues-paying members, who formed a protest organization called Save the Athenaeum Association (SAA) and filed suit to stop the sale. The dissident defenders of the Double Elephant Folio attacked the Athenaeum's leadership, charging them with financial incompetence and bibliophilic indifference toward Audubon's book—"They're treating this thing like a used car," said one outraged SAA member. Eventually, the executive director of the Athenaeum, fed up with the attacks and what he called the "lingering elitism among a small, but active, group of individuals," submitted his resignation and left in disgust.[31] The case became tied up in the courts, until a Rhode Island judge ruled that the Athenaeum's board had been within its rights, and the Audubon soon went on the block. In December 2005, it went for just over $5.6 million at Christie's—a good sum, but not the $7 million originally anticipated at the time, and not even close to the $11.5 million another complete set would fetch in 2010, just five years later. As the Christie's auctioneer explained after the Providence Athenaeum sale, "There's no way to know for sure, but it's possible that some people were put off by the way this particular copy of *Birds* came to the market."[32]

On the other hand, the recent case—and certainly a much less conflicted case—of the University of Pittsburgh provides an instructive lesson in preserving *The Birds of America*. The university has owned a Double Elephant Folio edition since 1918, when the family of William McCullough Darlington, a Pittsburgh lawyer and rare book collector, donated the late Mr. Darlington's collection, including the Audubon volumes, to the university library. (Darlington had apparently bought his Audubon at a bargain price of $400 in 1852, well below market even then and a good deal less than the $1,000 Audubon would have initially charged.) In 2000, the Hillman Library's Special Collections staff opened the four bound volumes for the first time in years, only to find a number of disturbing tears and stains, some of the latter apparently caused by the cigar ashes dropped by Mr. Darlington and his friends as they admired Audubon's birds. The university sent the four volumes away to be cleaned, which required undoing the original bindings so that each plate could be individually treated. When the process had been completed, though, the conservators decided not to rebind the plates into four huge volumes again, but to keep each plate in individual presentation folders with polyester cover sheets. As the university librarian explained, "This plan vastly improves access to the original plates for researchers, safeguards the plates for exhibition purposes, and eliminates nearly all physical stress to the individual plates when the plates are viewed for research or exhibition purposes."[33] In that regard, the University of Pittsburgh's decision reflects both an interest in public access and a concern for preservation. Unlike the postwar proprietors of the Old Print Shop and Goodspeed's, however, who had argued for breaking up the bound volumes and distributing them to the public by sale, the University of Pittsburgh has apparently made a commitment to keeping the Darlington copy of *The Birds of America* intact *as a collection*, but not as a book of four bound volumes. Given the ongoing tension between proper preservation and accessible use, the university may have hit upon a wise compromise that affords Audubon aficionados the opportunity to see *The Birds of America* and save it at the same time.

Finally, for those who still want to see the big edition of the Great Work in its fully bound, Double Elephant Folio form, several other institutions on both sides of the Atlantic have the ponderous volumes on display. Audubon's personal copy, which the family sold for $600 in 1861 to one of the founders of New York's Metropolitan Museum of Art, has now migrated southward to Orange, Texas, where it has come to rest in the H. J. Lutcher Stark Museum of Art. Audubon's own set is distinctive because he organized the images to be bound following the taxonomic system he defined in his *Synopsis of the Birds of North America* (1839), not in the order he originally established in 1827, and it also has thirteen additional images not found in almost all the other sets. A similarly organized set is the one that once belonged to Audubon's London physician and friend, Benjamin Phillips, which is now in Chicago's Field Museum. In Hartford, Connecticut, Trinity College has the volumes originally owned by Audubon's celebrated engraver, Robert Havell, who had the good fortune to be able to choose his own plates carefully, right from the shop, and thus possess "one of the finest sets in existence."[34]

Were Audubon truly able to come back to life and revisit *The Birds*, he might take special satisfaction in going to see Dartmouth College's set, which once belonged to Daniel Webster. Webster acquired only the first three volumes, and even though Audubon dogged him to make good on his subscription commitment, he may not have paid Audubon in full for them. Still, he followed Audubon's advice about their potential value and held onto his three volumes until he died. In later years they came into the possession of the college, where they remain in the Rauner Special Collections Library, accessible in a glass case for viewers. As the library puts it, with considerable pride, "we turn the page every week or two, so there is always something fresh to enjoy."[35] Given his long-term struggle in producing *The Birds of America*, not to mention tracking down as elusive a subscriber as Daniel Webster, Audubon would certainly find "something fresh to enjoy" in getting a view of his beloved *Birds* at Webster's beloved alma mater.

Audubon never doubted the significance of his Great Work, nor did he doubt his own greatness, even genius. Today, despite our doubts about Audubon as a flawed human being, we can hardly question the legacy of his work. Whatever the mistakes he made along the way, whatever the insults he suffered—and they sometimes seemed innumerable on both counts, destructive cuts to man and book alike—Audubon and *The Birds of America* have endured as much-celebrated American icons, each in its own way a remarkable work of art. In the end, both Audubon and *The Birds* stand essentially inseparable as products of the passion that drove one man to create an oversized book about every bird he could find in America and, in the process, to create an oversized identity of himself as the American Woodsman, the nation's first true human fusion of art and science.

Notes

～～～

Introduction

The epigraph to this chapter is from John James Audubon, *Ornithological Biography, Or, An Account of the Habits of the Birds of the United States of America: Accompanied by Descriptions of the Objects Represented in the Work Entitled "The Birds of America," and Interspersed with Delineations of American Scenery and Manners*, 5 vols. (Edinburgh, 1831–1839), I, v; hereafter cited as *OB*. In this and all subsequent quotations from Audubon, the punctuation, spelling, and emphasis reflect the original sources.

1. John James Audubon, *The Birds of America; From Original Drawings*, 4 vols. (London, 1827–1838).

2. Audubon to Lucy Bakewell Audubon, 1 October 1826, in *The 1826 Journal of John James Audubon*, edited by Alice Ford (New York, 1987), 238.

3. Three of Audubon's journals have been published in one work: Maria R. Audubon, *Audubon and His Journals; With Zoological and Other Notes by Elliott Coues*, 2 vols. (New York, 1897), which includes the European Journals (1826–1829), the Labrador Journal (1833), and the Missouri River Journals (1843). Unfortunately, Maria Audubon, his granddaughter, reworked all three journals, along with an extended autobiographical essay called "Myself," according to her late Victorian sensibilities and sense of her grandfather's place in history, which led her to bowdlerize them more than a bit. More recently, Audubon's Mississippi River Journal (1820–1821) and the Missouri River Journals (1843) have been published in *John James Audubon: Writings and Drawings*, edited by Christoph Irmscher (New York, 1999). The record of the first years of Audubon's time in Great Britain has been published in Ford, *1826 Journal*; and *John James Audubon's Journal of 1826: The Voyage to "The Birds of America,"* edited by Daniel Patterson and Patricio J. Serrano (Lincoln, NE, 2011). The journal for a later period in the United States is *Journal of John James Audubon Made While Obtaining Subscriptions to his "Birds of America" 1840–1843*, edited by Howard Corning (Boston, 1929). For a valuable overview of Audubon's fate at the hands of various editors, see Christoph Irmscher, "Audubon the Writer," *Eldridge-Audubon Royal Octavo Initiative: Preserving an Endangered Species*, http://www.audubonroyaloctavos.com/SITE/pages/Irmscher.html (accessed 26 February 2015).

4. I must here express my considerable gratitude to the Darlington Digital Library at the University of Pittsburgh, which has greatly facilitated the search for both words and birds in *Ornithological Biography* by making a digitized version of Audubon's work available online: http://digital.library.pitt.edu/d/darlington/ (accessed 26 February 2015).

5. Audubon, "The Ohio," in *OB*, I, 29.

6. Audubon, "The Pewee Flycatcher," in *OB*, II, 122.

7. For now, I will simply list the main Audubon biographies in chronological order, saving any specific comments for later, as needed: Mrs. Horace St. John, *Audubon, The Naturalist of the New World: His Adventures and Discoveries* (New York, 1856); Robert Buchanan, *The Life and Adventures of John James Audubon* (London, 1868); Francis Hobart Herrick, *Audubon the Naturalist: A History of His Life and Time*, 2 vols. (New York, 1917); Edward A. Muschamp, *Audacious Audubon: The Story of a Great Pioneer, Artist, Naturalist, and Man* (New York, 1929); Donald Culross Peattie, *Singing in the Wilderness: A Salute to John James Audubon* (New York, 1935); Constance Rourke, *Audubon* (New York, 1936); Stanley Clisby Arthur, *Audubon, an Intimate Life of the American Woodsman* (New Orleans, 1937); Alice Tyler, *I Who Should Command All* (New Haven, CT, 1937); Alice Ford, *John James Audubon: A Biography* (Norman, OK, 1964; rev. ed., New York, 1988); Alexander Adams, *John James Audubon: A Biography* (New York, 1966); John Chancellor, *Audubon: A Biography* (New York, 1978); Shirley Streshinsky, *Audubon: Life and Art in the American Wilderness* (New York, 1993); Ella M. Foshay, *John James Audubon* (New York, 1997); Duff Hart Davies, *Audubon's Elephant: America's Greatest Naturalist and the Making of "The Birds of America"* (New York, 2004); Richard Rhodes, *John James Audubon: The Making of an American* (New York, 2004); and William Souder, *Under a Wild Sky: John James Audubon and the Making of "The Birds of America"* (New York, 2004). This list does not include other works that focus on specific periods of his life, whether in Kentucky, Louisiana, Scotland, Florida, Labrador, or the American West, nor does it include biographical works intended for young readers or other works of fiction, some of which will be discussed in subsequent chapters.

8. Benjamin Smith Barton, *Fragments of the Natural History of Pennsylvania* (Philadelphia, 1799), viii.

9. With the expansion of newspapers and other forms of print media for a mass market, the notion of the "celebrity" was just beginning to take hold in the transatlantic cultural scene, and Audubon certainly gained a good measure of popular notoriety. By the 1830s, periodicals on both sides of the Atlantic not only reviewed Audubon's work ("The papers here have *blown me up* sky high," he crowed about the American press in 1831) but also took note of his movements from one place to the next ("This enthusiastic Naturalist is gone again to the woods," wrote a London newspaper in the same year), and he gained a following well beyond the scientific community; see Chapters 5 and 6. On the emergence of celebrity in the early decades of the nineteenth century, see Renee Sentilles, *Performing Menken: Adah Isaacs Menken and the Birth of American Celebrity* (New York, 2003); and *Constructing Charisma: Celebrity, Fame, and Power in Nineteenth-Century Europe*, edited by Edward Berenson and Eva Giloi (New York, 2010).

Celebrity goes only so far and lasts only so long in scientific circles. Audubon, like many of the other early nineteenth-century naturalists who studied and wrote about the flora and fauna of the new nation, fell far short of the professional scientific standards that would become commonplace later in the century. As the art historian Robert Hughes has observed in *American Visions: The Epic History of Art in America* (New York, 1999), 154, "Audubon had a vulnerable and sore point. He was not a scientist. He had no formal training, no degrees.... He could not compete on academic ground." Moreover, in George H. Daniels, *American Science in the Age of Jackson* (New York, 1968)—one of the standard works on American science in the first half of the nineteenth century, a book "intended as a study of key issues in the intellectual history of the American scientific community"—Audubon does not get a single mention.

10. Audubon, "Myself," in Irmscher, *Audubon: Writings and Drawings*," 765, 788; "Colonel Boon," in *OB*, I, 503; "Introduction," in *OB*, V, v.

Chapter 1

The epigraph for this chapter is from John James Audubon, "Myself," in *John James Audubon: Writings and Drawings*, edited by Christoph Irmscher (New York, 1999), 765.

1. John James Audubon to Lucy Bakewell Audubon, 12 March 1827, quoted in Francis Hobart Herrick, *Audubon the Naturalist: A History of His Life and Time*, 2 vols. (New York, 1917), I, 372.

2. On the self-invention prevalent in nineteenth-century autobiography, see Ann Fabian, *The Unvarnished Truth: Personal Narratives in Nineteenth-Century America* (Berkeley, 2000). Karen Haltunnen, in *Confidence Men and Painted Women: A Study of Middle-class Culture in America, 1830–1870* (New Haven, CT, 1986), argues that personal identity could often become wrapped in various layers of deceptive performance. In a study focused on the ambiguities of race and racial passing in American society, Allyson Hobbs points to a "much larger phenomenon that encompasses multiple disguises and forms of dissemblance" and notes that "particularly in societies with relatively open and fluid social orders, the permutations on passing were endless." See Hobbs, *A Chosen Exile: A History of Racial Passing in American Life* (Cambridge, MA, 2014), quotation on 19.

3. Audubon, *Ornithological Biography; or, An Account of the Habits of the Birds of the United States of America: Accompanied by Descriptions of Objects Represented in the Work Entitled "The Birds of America," and Interspersed with Delineations of American Scenery and Character*, 5 vols. (Edinburgh, 1831–1839), I, v–vi, viii; hereafter cited as *OB*, I, v–vi, viii.

4. Audubon, "Myself," 765; the autobiographical sketch was never published in Audubon's lifetime. Herrick, in *Audubon the Naturalist*, I, 16, explains that the manuscript was written in 1835, set aside, then lost, later found, and published by Audubon's granddaughter, Maria R. Audubon, keeper of his records and protector of his reputation, first in 1893 in *Scribner's* and subsequently in Maria R. Audubon, *Audubon and His Journals; With Zoological and Other Notes by Elliott Coues*, 2 vols. (New York, 1897). Among Audubon scholars, Maria Audubon has become infamous for freely and flagrantly changing the words and meanings of Audubon's writings, usually to make her grandfather look more dashing and distinctive and, on occasion, more discreet.

5. Audubon, "Myself," 765.

6. See, for instance, Wayne Hanley, *Natural History in America: From Mark Catesby to Rachel Carson* (New York, 1977), 66, who observes that "Audubon always gave his age as greater than it was, although the year of birth varied. Perhaps he had in mind misleading anyone who searched for the truth." Herrick notes that "the naturalist, in his letters and journals, made frequent allusions to his age, but, as his granddaughter remarked, with one exception, no two agree." The granddaughter, Maria Audubon, suggested that Audubon might have been born "anywhere from 1772 to 1783," eventually accepting the then-common consensus of May 5, 1780, only with some uncertainty. See Herrick, *Audubon the Naturalist*, I, 68.

7. On the inconsistencies and discrepancies in Audubon's official documents, see Francis James Dallett, "Citizen Audubon: A Documentary Discovery," *Princeton University Library Chronicle* 21, nos. 1–2 (Autumn 1959–Winter 1960), 89–93.

8. "Audubon's Ornithological Biography," *American Monthly Review* 8, no. 1 (June 1832), 349; "Audubon's Biography of Birds," *New American Review* 34 (April 1832), 387; "Audubon, the American Ornithologist," *Eastern Argus Tri-Weekly*, 30 October 1833, 2; "Audubon the Naturalist," *Burlington Free Press*, 8 September 1843, 2.

9. G. W. Curtis et al., *The Homes of American Authors: Comprising Anecdotical, Personal, and Descriptive Sketches* (New York, 1852), 7; "John James Audubon," *Spirit of the Times*, 15 March

1851, 45; "Audubon, the Naturalist," *Gleason's Pictorial Drawing Room Companion*, 25 September 1852, 196.

10. Herrick, *Audubon the Naturalist*, I, 52–53.

11. Herrick puts Anne Moynet's age at nine years older than Jean Audubon; see Herrick, *Audubon the Naturalist*, I, 32. The other indications of the age difference, twelve and fourteen years, come from, respectively, Alice Ford, *John James Audubon: A Biography* (New York, 1988), 15–16; and Shirley Streshinsky, *Audubon: Life and Art in the American Wilderness* (New York, 1993), 8.

12. For a brief overview of the sugar economy of eighteenth-century Saint-Domingue, see Doris L. Garraway, *Libertine Colony: Creolization in the Early French Caribbean* (Durham, NC, 2005), 8.

13. Herrick, *Audubon the Naturalist*, I, 39, 58.

14. The term comes from Garroway, *Libertine Colony*.

15. Herrick, *Audubon the Naturalist*, I, 30.

16. Ibid., 73–89. For a sustained study of the revolutionary-era conflict in the region, see Charles Tilly, *The Vendée* (Cambridge, MA, 1964), esp. chaps. 12 and 13. Audubon's comment on his father's naval service to the French republic is from "Myself," 767.

17. Margaret Kieran and John Kieran, *John James Audubon* (New York, 1954), 1–8.

18. Joan Howard, *The Story of John J. Audubon* (New York, 1954), 3–9.

19. Alexander B. Adams, *John James Audubon* (London, 1966), 19; Herrick, *Audubon the Naturalist*, I, 56.

20. Ira Berlin, *Many Thousands Gone: The First Two Centuries of Slavery in North America* (Cambridge, MA, 1998), 105. For a discussion of the uncertain implications of "creole" identity, see Christopher Iannini, *Fatal Revolutions: Natural History, West Indian Slavery, and the Routes of American Literature* (Chapel Hill, NC, 2012), 257–258; and Joseph G. Tregle Jr., "On That Word 'Creole' Again: A Note," *Louisiana History: The Journal of the Louisiana History Association* 23, no. 2 (Spring 1982), 193–198.

21. *The Negro Almanac: A Reference Work on the African American*, 5th ed., compiled and edited by Harry A. Ploski and James D. Williams (Detroit, 1989), 1035; for the Ebony Society of Philatelic Events and Reflections, see http://esperstamps.org/history1.htm (accessed 27 March 2012); on the same date, 27 March 2012, a telephone contact at the African American Cultural Center Library confirmed the Audubon listing in the Biography File.

22. Alice Ford, *John James Audubon: A Biography* (Norman, OK, 1964), 21.

23. Ford, *John James Audubon* (1988), 22.

24. Ibid., 29.

25. Hanley, *Natural History in America*, 67–68. The most recent biographies of Audubon follow Ford's lead on Rabin's identity, all describing her as a French chambermaid resident in Saint-Domingue; see Duff Hart Davies, *Audubon's Elephant: America's Greatest Naturalist and the Making of "The Birds of America"* (New York, 2004), 21; Richard Rhodes, *John James Audubon: The Making of an American* (New York, 2004), 4; and William Souder, *Under a Wild Sky: John James Audubon and the Making of "The Birds of America"* (New York, 2004), 18.

26. Audubon to Lucy Bakewell Audubon, 25 November 1827, in *Letters of John James Audubon, 1826–1840*, edited by Howard Corning, 2 vols. (Boston, 1930), I, 51. I address the masculine Americanness of Audubon's emerging identity in more detail in Chapter 6; see also Gregory

Nobles, "John James Audubon, the American 'Hunter-Naturalist,'" *Common-place* 12, no. 2 (January 2012).

27. Audubon, "Myself," 765.

28. Ibid.

29. Ibid., 768, 772.

30. Ibid., 769, 772.

31. Audubon, "My Style of Drawing Birds," in Irmscher, *Audubon: Writings and Drawings*, 760.

32. Audubon, "Myself," 766–767.

33. Ibid., 767. The more accurate account of Jean Audubon's experience during the American Revolution comes from Herrick, *Audubon the Naturalist*, I, 32–35.

34. Audubon, "Myself," 768.

35. Ibid., 768–769.

36. Ford, *John James Audubon* (1988), 34.

37. Audubon, "Myself," 769.

38. For a valuable account of the fate—and failure—of Napoleon's military venture in Saint-Domingue, see Laurent Dubois, *A Colony of Citizens: Revolution and Slave Emancipation in the French Caribbean, 1787–1804* (Chapel Hill, NC, 2004), 365–373, 402–404.

39. Robert Penn Warren, *Audubon: A Vision* (New York, 1969), 2.

40. Audubon to John Bachman, 22 March 1837, in Corning, *Letters*, II, 152.

41. Audubon, "Myself," 768, 772.

42. For a brief newspaper notice of Audubon's arrival, see *Poulson's Daily Advertiser*, 1 September 1803. Among the ships listed as "Arrived since our last" is "Brig *Hope,* Smith, from Nantz [*sic*]," with a passenger list of seven men, one of them "Audubon." I am grateful to Professor Billy G. Smith of Montana State University for this reference. It is worth noting that when Audubon wrote about his later return to France, in 1805, he indicated, "My passage was taken on board the brig 'Hope,' of New Bedford"—perhaps the same ship, or perhaps a conflation of vessels in Audubon's memory. See Audubon, "Myself," 779.

43. For information about the probable Saint-Domingue origins of New York's 1803 yellow fever outbreak, see Billy G. Smith, *Ship of Death: A Voyage That Changed the Atlantic World* (New Haven, CT, 2013), which provides an outstanding account of the arrival of yellow fever in Philadelphia a decade earlier but also notes the subsequent spread of the disease to other cities on the East Coast of the United States. See also Ronald Angelo Johnson, *Diplomacy in Black and White: John Adams, Toussaint Louverture, and Their Atlantic World Alliance* (Athens, GA, 2014), 72–77. For contemporary accounts of the epidemic and debates about its origins, see James R. Manley, *An Inaugural Dissertation on the Yellow Fever* (New York, 1803); and James Hardie, *An Account of the Yellow Fever, Which Occurred in the City of New York, in the Year 1822 ... [and] In the Years 1798, 1799, 1803, & 1805* (New York, 1822).

Chapter 2

The epigraph for this chapter is from Audubon, "Account of the Method of Drawing Birds Employed by J. J. Audubon, Esq. F. R. S. E.," in *John James Audubon: Writings and Drawings*, edited by Christoph Irmscher (New York, 1999), 753–758, quotation on 753.

1. Audubon, "Myself," in Irmscher, *Audubon: Writings and Drawings*, 774.

2. David Jaffee, "'A Correct Likeness': Culture and Commerce in Nineteenth-Century Rural America," in *Reading American Art*, edited by Marianne Doezema and Elizabeth Milroy (New Haven, CT, 1998), 109–127; see also Jaffee, "One of the Primitive Sort: Portrait Makers of the Rural North, 1760–1860," in *The Countryside in the Age of Capitalist Transformation: Essays in the Social History of Rural America*, edited by Steven Hahn and Jonathan Prude (Chapel Hill, NC, 1985), 103–138.

3. On the relationship between painter and patron, see Paul Straiti, "Character and Class: The Portraits of John Singleton Copley," and Alan Wallach, "Thomas Cole and the Aristocracy," in Doezema and Milroy, *Reading American Art*, 31–37, 79–108.

4. Audubon, "Account of the Method of Drawing Birds," 754.

5. Audubon, "Small-Headed Flycatcher," in *Ornithological Biography; or, An Account of the Habits of the Birds of the United States of America: Accompanied by Descriptions of Objects Represented in the Work Entitled "The Birds of America," and Interspersed with Delineations of American Scenery and Character*, 5 vols. (Edinburgh, 1831–1839), V, 291; hereafter cited as *OB*. This bird, incidentally, is one that Audubon claimed to have first encountered in Kentucky in 1808, and one he later alleged Alexander Wilson had copied from Audubon's drawing; there is also some suspicion that Audubon copied Wilson's. In any event, this bird seems to be a now-vanished—or perhaps spurious—species: as one of today's leading ornithologists explains, "Nothing matching the drawings and descriptions has ever been seen again." See Scott Weidensaul, *Of a Feather: A Brief History of Birding* (Orlando, FL, 2007), 73.

6. Audubon, "Account of the Method of Drawing Birds," 754.

7. Audubon, "Myself," 786.

8. For a valuable perspective on Audubon's involvement in, even embrace of, the world of commerce, see Jennifer L. Roberts, *Transporting Visions: The Movement of Images in Early America* (Berkeley, CA, 2014), 73, 110–112.

9. Scott Weidensaul, "Introduction: The Birthplace of American Birding," in F. Brock, S. Fordyce, D. Kunkle, and T. Fenchel, *Eastern Pennsylvania Birding and Wildlife Guide* (Harrisburg, PA, 2009), 8; Weidensaul, *Of a Feather*, 54–56; Scott Weidensaul, *Living on the Wind: Across the Hemisphere with Migratory Birds* (New York, 1999), 34. I am also grateful to Weidensaul for personal correspondence, 28 June 2012, regarding the greater abundance of birds in Audubon's era.

10. *Travels of William Bartram*, edited by Mark van Doren (New York, 1928), 235.

11. Benjamin Smith Barton, *Fragments of the Natural History of Pennsylvania* (Philadelphia, 1799), x–xi.

12. Audubon, "Myself," 774.

13. Ibid., 775.

14. Quoted in Francis Hobart Herrick, *Audubon the Naturalist: A History of His Life and Time*, 2 vols. (New York, 1917), I, 112.

15. Audubon, "Myself," 783.

16. Ibid.

17. W. J. Rohrbach estimates that between 1800 and 1830 per capita consumption of alcoholic spirits reached five gallons, "a rate nearly triple that of today's consumption." See *The Alcoholic Republic: An American Tradition* (New York, 1979), 8.

18. Audubon, "Myself," 783.

19. Audubon, "The Pewee Flycatcher," in *OB*, II, 123–126.

20. One additional step he might well have taken, however, was back toward Philadelphia, where the world's then most notable naturalist, Alexander von Humboldt, made two short visits in May and June 1804. In his defense, one must suspect that Audubon apparently never knew of von Humboldt's arrival in the region, and it could have been presumptuous for a teenager—even a brash and self-promoting young man like Audubon—to have imposed himself on the eminent visitor without an invitation. (By comparison, fifteen-year-old John Bachman, who would later become one of Audubon's closest friends in the field, did receive an invitation to attend a dinner with von Humboldt in Philadelphia in 1804, and he remembered the experience for the rest of his life.) Over two decades later, in 1826, Audubon did get a letter of introduction to von Humboldt, but there is no indication that he ever used it to arrange a personal meeting. In fact, Audubon never mentioned von Humboldt at all in the five volumes of *Ornithological Biography*, and when he did, it was only fleetingly in his correspondence.

On von Humboldt's visit to Philadelphia, see Gerhard Casper, "A Young Man from 'Ultima Thule' Visits Jefferson: Alexander von Humboldt in Philadelphia and Washington," *Proceedings of the American Philosophical Society* 155, no. 3 (September 2011), 247–262; and Andrea Wulf, *The Invention of Nature: Alexander von Humboldt's New World* (New York, 2015), 96–98. Bachman's boyhood meeting with von Humboldt is noted in Herrick, *Audubon the Naturalist*, II, 284. On Audubon's letter of introduction to von Humboldt, see Audubon to Victor Gifford Audubon, 1 September 1826, in *The Letters of John James Audubon, 1826–1840*, 2 vols., edited by Howard Corning (Boston, 1930), 4; and Alice Ford, *John James Audubon: A Biography* (New York, 1988), 183.

21. Audubon, "My Style of Drawing Birds," in Irmscher, *Audubon: Writings and Drawings*, 760; Audubon, "Myself," 784.

22. Audubon, "My Style of Drawing Birds," 760–761.

23. Audubon, "Account of the Method of Drawing Birds," 756.

24. Audubon, "Myself," 769.

25. Ibid., 776, 784.

26. Ibid., 778.

27. Jean Audubon to Francis Dacosta, n.d., 1804–1805, quoted in Herrick, *Audubon the Naturalist*, I, 118.

28. Audubon, "Myself," 779–780.

29. Ibid., 780.

30. Ibid., 781.

31. Ibid., 781–782.

32. Herrick, *Audubon the Naturalist*, I, 170–172.

33. Lucy Bakewell Audubon to Euphemia Gifford, 27 May 1808, in *The Audubon Reader*, edited by Richard Rhodes (New York, 2006), 15.

34. Alexander Wilson, *American Ornithology, or The Natural History of the Birds of the United States*, 9 vols. (Philadelphia, 1808–1814); hereafter cited as *AO*. By the time Wilson arrived in Louisville, he had completed and published volumes I (1808) and II (1810).

35. For brief descriptions of Wilson's early background, see Herrick, *Audubon the Naturalist*, I, 202–220, quotation on 211; Weidensaul, *Of a Feather*, 41–63; and William Souder, *Under a Wild Sky: John James Audubon and the Making of "The Birds of America"* (New York, 2004), 21–28, 45–64. The standard modern biography of Wilson is Robert Cantwell, *Alexander Wilson, Naturalist and*

Pioneer (Philadelphia, 1961), and the most recent book-length study of Wilson is Edward H. Burtt Jr. and William E. Davis Jr., *Alexander Wilson: The Scot Who Founded American Ornithology* (Cambridge, MA, 2013).

36. Herrick, *Audubon the Naturalist*, I, 212–213. For an insightful analysis of Wilson and his work, see Laura Rigal, "Empire of Birds: Alexander Wilson's *American Ornithology*," *Huntington Library Quarterly* 59, nos. 2 and 3 (1996), 232–268; and "Feathered Federalism: Alexander Wilson's *American Ornithology, 1807–1814,*" in *The American Manufactory: Art, Labor, and the World of Things in the Early Republic* (Princeton, NJ, 1998), 145–178.

37. Wilson, *AO*, I, 1.

38. Wilson, *AO*, IX, xxxv–xxxvi. Wilson himself became extinct, so to speak, before seeing *American Ornithology* through to completion, and the frustrating search for sympathetic subscribers dogged him to the end of his days.

39. Herrick, *Audubon the Naturalist*, I, 204–205, 223. Wilson did not stumble upon Audubon completely by accident. On his way westward from Philadelphia, he stopped in Pittsburgh, where he met Audubon's kinsman and former employer, Benjamin Bakewell, now the proprietor of the region's leading glass factory. Bakewell apparently didn't purchase Wilson's work—his name does not appear on the list of subscribers—but he did offer Wilson a letter of introduction and, equally important, the useful suggestion that he look up this young man who dabbled in drawing birds.

40. Audubon, "Louisville in Kentucky," in *OB*, I, 438–439.

41. Quoted in George Ord, *Sketch of the Life of Alexander Wilson, Author of the American Ornithology* (Philadelphia, 1828), cxlvi–cxlvii. Ord's always-partisan approach to the Audubon-Wilson relationship renders questionable, if not suspect, his quotations from Wilson's diary.

42. Audubon, "Louisville in Kentucky," in *OB*, I, 440. Burtt and Davis Jr., in *Alexander Wilson*, 333–337, offer the most recent account of the Audubon-Wilson meeting in Louisville, stating that "the two men appear to have parted on good terms and never met again."

43. Wilson's kingfisher would appear in print in less than a year, as plate 23 of the third volume of *American Ornithology*, which was published in February 1811.

44. See, for instance, Sacheverell Sitwell, Handasyde Buchanan, and James Fisher, *Fine Bird Books, 1700–1900* (New York, 1990), for the opinion that the standard profile presentation "becomes monotonous," 8. Christoph Irmscher notes, a bit more charitably, that while Audubon, like Wilson, "adhered to the conventions of natural history illustrations" by showing his birds "motionless and in profile" in his early work, he later began to "introduce drama and movement into his compositions" in *The Birds of America*; see Irmscher, *The Poetics of Natural History: From John Bartram to William James* (New Brunswick, NJ, 1999), 195. See also Linda Dugan Partridge, "By the Book: Audubon and the Tradition of Ornithological Illustration," *Huntington Library Quarterly* 59, nos. 2 and 3 (1996), 269–301.

45. As one modern-day observer has put it, "Audubon . . . had by this time reached Wilson's level. If he wasn't yet the elegant pencil draftsman that Wilson was, he was superior in terms of composition and color, in drawing birds to scale, and in depicting their natural habitats." See Theodore E. Stebbins Jr., "Audubon's Drawings of American Birds, 1805–38," in *John James Audubon: The Watercolors for "The Birds of America,"* edited by Annette Blaugrund and Theodore E. Stebbins Jr. (New York, 1993), 3–26, quotation on 9.

46. Burtt and Davis, *Alexander Wilson*, 339–340.

47. A more detailed discussion of the battles between Audubon's and Wilson's allies, along with their implications in the larger transatlantic scientific community, follows in Chapter 5.

48. John Wilson, *Critical and Miscellaneous Essays of Christopher North (Professor Wilson)*, 3 vols. (Philadelphia, 1842), II, 148.

49. Audubon, "Myself," 785. Herrick describes Audubon at the time as "still more of a sportsman than a naturalist, and when not occupied with drawing, he spent most of his time in the forest, to the neglect of his trade." Herrick, *Audubon the Naturalist*, I, 232.

50. Audubon, "Myself," 786.

51. Ibid., 785.

52. Thomas Ashe, *Travels in America, Performed in 1806 . . .* (London, 1808), 248, quoted in Alexander B. Adams, *John James Audubon: A Biography* (London, 1967), 112. On the early history of Henderson, see Edmund Lyne Starling, *History of Henderson County, Kentucky* (Henderson, KY, 1887), 35, 157. Starling estimated that the population would rise to just over a thousand by 1820, but Federal Census figures put the 1820 population considerably lower, at 532, but still a decent increase. See also Carolyn E. DeLatte, *Lucy Audubon: A Biography* (Baton Rouge, FL, 1982; updated ed., 2008), 73–74.

53. Audubon, "Fishing in the Ohio," in *OB*, III, 122–123.

54. Ibid.

55. Audubon, "Journey up the Mississippi," in *The Winter's Wreath: A Collection of Original Contributions in Prose and Verse* (Liverpool, UK, 1829), 104–127, quotation on 118. A more readily accessible version is in Rhodes, *Audubon Reader*, 21–36, quotation on 23.

56. Audubon, "Journey up the Mississippi," in Rhodes, *Audubon Reader*, 29.

57. Audubon's characterizations of Ste. Genevieve and its inhabitants come from "Journey up the Mississippi," in Rhodes, *Audubon Reader*, 33; and Audubon, "Myself," 787.

58. For a brief sketch of Rozier's subsequent success in Ste. Genevieve, see Herrick, *Audubon the Naturalist*, I, 245–246.

59. For Audubon's efforts at collecting from Rozier, see Herrick, *Audubon the Naturalist*, I, 242–243, 249–250; for the description of Audubon's long walk in a short time, see Audubon, "Myself," 787–788.

60. Audubon, "Myself," 788; DeLatte, *Lucy Audubon*, 67–71.

61. Starling, *Henderson County*, 150.

62. Audubon, "Myself," 789. For a survey of Audubon's growing prosperity in Henderson, see Starling, *Henderson County*, 124; and DeLatte, *Lucy Audubon*, 74–78. Audubon's position on slavery and abolition will be discussed in more detail in Chapter 7.

63. Audubon, "Myself," 789. The description of the steam mill is in Starling, *Henderson County*, 794.

64. On the place of Henderson within the wider economy of the Green River region of Kentucky, see Stephen Aron, *How the West Was Lost: The Transformation of Kentucky from Daniel Boone to Henry Clay* (Baltimore, MD, 1996), 166–167.

65. *Henderson, A Guide to Audubon's Home Town in Kentucky, Compiled by Workers of the Writers' Program, Works Project Administration in the State of Kentucky* (Northport, NY, 1941), 55.

66. Starling, *Henderson County*, 148.

67. Audubon, "Myself," 790.

68. DeLatte, *Lucy Audubon*, 90–91. Audubon later wrote, rather charitably, that Bakewell eventually wound up in Cincinnati, "where he has made a large fortune, and glad I am of it." Audubon, "Myself," 791.

69. Audubon, "Myself," 790–791; Herrick, *Audubon the Naturalist*, I, 258–259.

70. For the Panic of 1819, see Murray N. Rothbard, *The Panic of 1819: Reactions and Policies* (New York, 1962), and, more recently, Daniel Dupre, "The Panic of 1819 and the Political Economy of Sectionalism," in *The Economy of Early America: Historical Perspectives and New Directions*, edited by Cathy Matson (University Park, PA, 2006), 263–297, esp. 271–272. On the rise and fall of the Bank of Henderson, when "paper money of all kinds and denominations began to flood the country, worthless bank-notes, private bills, and other shin-plasters, seemed determined to crowd out the specie currency," see Starling, *Henderson County*, 148, 150–152; and also *Henderson, A Guide to Audubon's Home Town in Kentucky*, 34.

71. Starling, *Henderson County*, 152.

72. Audubon, "Myself," 791.

73. William Charvat, *The Profession of Authorship in America, 1800–1870* (Columbus, OH, 1968), 48.

74. DeLatte, *Lucy Audubon*, 98. The inventory of the family's goods, compiled by Audubon himself, is in Rhodes, *Audubon Reader*, 93–94. The seven slaves are not listed here, but they were conveyed to Berthoud in a separate transaction.

75. Audubon, "Myself," 791.

76. Audubon, "Little Sandpiper," in *OB*, IV, 184. Speculating about the avian population in Audubon's era, the nineteenth-century historian of the Henderson region observed that "birds were far more plentiful and of a greater variety in those days than they have ever been since to woodsman commenced clearing the country." See Starling, *Henderson County*, 794–795. The current-day count, according to the pamphlet "Birds of Audubon State Park and Wolf Hills," compiled by Nancy Richardson and Don Dodson, lists 169 species but notes that "whereas Audubon saw such birds as whooping cranes and ivory-billed woodpeckers, we no longer can see these birds here." Audubon also wrote about other birds he saw in Henderson that are not on the current list, including the Wild Turkey and the now-extinct Passenger Pigeon.

77. Audubon, "Introductory Address," in *OB*, I, xiii–xiv. As one modern student of Audubon's early art has noted, "We can surmise that Audubon was much less active as an artist during these years, due to the press of business and personal affairs," and, given the demands of both, "the quality of his work varied as a result." Stebbins, "Audubon's Drawings," 8–10.

78. Audubon, "Myself," 792.

79. Ibid., 792–793.

80. On Daniel Drake's life and career, see Emanuel D. Rudolph, "Daniel Drake as a Nineteenth Century Educational Reformer," *Ohio Journal of Science* 85, no. 4 (September 1985), 146–151.

81. For a very useful exploration of Peale's Philadelphia "Repository," see David R. Brigham, "'Ask the Beasts, and They Shall Teach Thee': The Human Lessons of Charles Willson Peale's Natural History Displays," *Huntington Library Quarterly* 59, nos. 2 and 3 (1996), 183–206.

82. On the founding and early purpose of the Western Museum, see M. H. Dunlop, "Curiosities Too Numerous to Mention: Early Regionalism and Cincinnati's Western Museum," *American Quarterly* 36, no. 4 (Autumn 1984), 524–548.

83. Daniel Drake, *An Anniversary Discourse, on the State and Prospects of the Western Museum Society* (Cincinnati, 1820), 5–6.

84. Audubon, "Myself," 793.

85. Ibid.

86. Elijah Slack to Reverends Veil and Chapman, 10 October 1820, quoted in Audubon, "Mississippi River Journal," in Irmscher, *Audubon: Writings and Drawings*, 43.

87. Drake, *Anniversary Discourse*, 9–10.

88. Audubon to Henry Clay, 12 August 1820, quoted in Audubon, "Mississippi River Journal," 154–155.

89. Henry Clay to Audubon, 23 August 1820, quoted in ibid., 44.

Chapter 3

The epigraph for this chapter is from Audubon, "Mississippi River Journal," in *John James Audubon: Writings and Drawings*, edited by Christoph Irmscher (New York, 1999), 3. The Irmscher transcription is one of two published versions of Audubon's journal, the other being *Journal of John James Audubon Made During His Trip to New Orleans in 1820–1821*, edited by Howard Corning (Boston, 1929). I chose to use the Irmscher version here for the sake of accessibility and fidelity to Audubon's original (and often idiosyncratic) spelling.

1. Audubon, "Mississippi River Journal," 32.

2. Ibid., 3.

3. Ibid., 32. At the close of his unpublished memoir, "Myself," which was also intended for his sons, Audubon notes that after leaving Cincinnati in October 1820, "my journals are kept with fair regularity, and if you read them you will easily find all that followed afterward." See Audubon, "Myself," in Irmscher, *Audubon: Writings and Drawings*, 793.

4. Henry Clay to Audubon, 23 August 1820, in Audubon, "Mississippi River Journal," 43.

5. J. H. B. Latrobe, *The First Steamboat Voyage on the Western Waters* (Baltimore, MD, 1871), 3. For the connection between Audubon and Nathaniel Roosevelt, see Mary Helen Dohan, *Mr. Roosevelt's Steamboat* (New York, 1981), 70–71. For the more general history of Mississippi River boats at the time, see Erik F. Haites, James Mak, and Gary M. Walton, *Western River Transportation: The Era of Early Internal Development, 1810–1860* (Baltimore, MD, 1975), 19–33.

6. In his Mississippi River Journal, Audubon described young Mason as being "about eighteen," but Mason was born in 1808 and not quite thirteen when he set off downriver with Audubon. See Alice Ford, *John James Audubon: A Biography* (New York, 1988), 112; and Irving T. Richards, "Audubon, Joseph R. Mason, and John Neal," *American Literature* 6, no. 2 (May 1934), 122–140, in which Mason is described as Audubon's "inseparable and loved companion, prized as much for his skill as an assistant as for his good character and good-fellowship." Mason would remain with Audubon until 1822, but he would soon become resentful about not getting enough credit for his contribution to Audubon's work.

7. Haites, Mak, and Walton, *Western River Transportation*, 14, 21.

8. Harry N. Scheiber, *Ohio Canal Era: A Case Study of Government and the Economy, 1820–1861* (Athens, OH, 1969), 9.

9. Michael Allen, *Western Rivermen, 1763–1861: Ohio and Mississippi Boatmen and the Myth of the Alligator Horse* (Baton Rouge, LA, 1990), 105.

10. Ibid., 4, 7, 8, 10.

11. Ibid., 7.

12. Carl von Linne (Linnaeus), *A General System of Nature, Through the Three Grand Kingdoms of Animals, Vegetables, and Minerals...*, edited by William Turton (London, 1806). On Audubon's copying the bird image from Turton, see Linda Dugan Partridge, "By the Book: Audubon and the Tradition of Ornithological Illustration," *Huntington Library Quarterly* 59, nos. 2 and 3 (1998), 276–278.

13. Audubon, "Mississippi River Journal," 16–19.

14. Ibid., 17, 25, 26, 34–36.

15. Ibid., 41. Audubon's later encounter with the Golden Eagle is in Chapter 6.

16. Audubon, "Mississippi River Journal," 21.

17. Ibid., 22–23; see also his journal entry for Sunday, 3 December 1820, 37–38.

18. Ibid., 28.

19. Ibid., 47, 58.

20. Ibid., 58, 59, 65.

21. Ibid., 23–24.

22. Audubon, "The Earthquake," in *Ornithological Biography, Or, An Account of the Habits of the Birds of the United States of America: Accompanied by Descriptions of the Objects Represented in the Work Entitled "The Birds of America," and Interspersed with Delineations of American Scenery and Manners*, 5 vols. (Edinburgh, 1831–1839), I, 241; hereafter cited as *OB*. For a recent study of the economic and cultural impact of the New Madrid earthquakes, see Conevery Bolton Valencius, *The Lost History of the New Madrid Earthquakes* (Chicago, 2013).

23. Audubon, "Mississippi River Journal," 26, 27, 51.

24. Ibid., 24, 48, 49.

25. Ibid., 72, 73, 74.

26. On the early development and diversity of New Orleans, see Richard Campanella, *Geographies of New Orleans: Urban Fabrics Before the Storm* (Lafayette, LA, 2006); Richard Campanella, "An Ethnic Geography of New Orleans," *Journal of American History* 94 (December 2007), 704–715; Paul F. LaChance, "The 1809 Immigration of Saint-Domingue Refugees to New Orleans: Reception, Integration, and Impact," *Louisiana History* 29, no. 2 (Spring 1988), 109–141; and Peter J. Kastor, *The Nation's Crucible: The Louisiana Purchase and the Creation of America* (New Haven, CT, 2004), esp. 19–34.

27. Benjamin Henry Boneval Latrobe, *Impressions Respecting New Orleans: Diary and Sketches, 1818–1820*, edited by Samuel Wilson Jr. (New York, 1951), 21–22.

28. Ibid., 33–34.

29. Gary A. Donaldson, "A Window on Slave Culture: Dances at Congo Square in New Orleans, 1800–1862," *Journal of Negro History* 69, no. 2 (Spring 1984), 63–72, quotation on 65. (The site is now named Armstrong Square, for Louis Armstrong, still in keeping with the musical roots deep in the city's culture.) See also Mary Gehman, *The Free People of Color of New Orleans: An Introduction* (New Orleans, 1994), 25.

30. Latrobe, *Impressions Respecting New Orleans*, 49–51.

31. Ibid., 76–77.

32. Audubon, "Mississippi River Journal," 77–78.

33. Again, see Audubon, "Myself," 765.

34. Audubon, "Mississippi River Journal," 75–76, 93.

35. Ibid., 76, 82, 86, 100.

36. Ibid., 103, 124.

37. Ibid., 104.

38. Ibid., 125–127. For a fuller description of Audubon's time at Oakley, see Danny Heitman, *A Summer of Birds: John James Audubon at Oakley House* (Baton Rouge, LA, 2008).

39. Audubon, "Mississippi River Journal," 131, 143–144, 145–148.

40. One of Audubon's earlier biographers, Stanley Clisby Arthur, conducted a close examination of Audubon's original watercolor drawings and concluded that "at least, 167 of Audubon's original bird drawings were drawn in Louisiana: that 56 were drawn in New Orleans; 40 were made at Percy's Beech Woods plantation; and 26 while Audubon was with the Pirrie family at Oakley plantation in West Feliciana parish." Suspecting that these numbers were probably a bit low, Arthur later increased his Louisiana count to 175. See Arthur, *Audubon: An Intimate Life of the American Woodsman* (New Orleans, 1937), 500–506, quotation on 506; and Richards, "Audubon, Joseph R. Mason, and John Neal," 129, fn. 28. For the Louisiana-based stages of Audubon's drawing of the Carolina Parakeet, see *John James Audubon: The Watercolors for "The Birds of America,"* edited by Annette Blaugrund and Theodore E. Stebbins Jr. (New York, 1993), 154. At the end of his written description in *Ornithological Biography,* Audubon notes that "the specimen from which this drawing was taken was shot at Bayou Sara, in Louisiana." Audubon, "Carolina Parrot," in *OB,* I, 140.

41. Audubon, "Purple Gallinule," in *OB,* IV, 38.

42. Audubon, "Mississippi River Journal," 77, 81–83.

43. Ibid., 91–92.

44. Ibid., 131, 134–136, 149–154.

45. Alexander Wilson, *American Ornithology, or The Natural History of the Birds of the United States,* 9 vols. (Philadelphia, 1808–1814), I, 9; hereafter cited as *AO.*

46. On Wilson's discoveries, see George Ord, *Sketch of the Life of Alexander Wilson, Author of the American Ornithology* (Philadelphia, 1828), clxiii–clxiv. Scott Weidensaul notes somewhat lower numbers, 268 species in all, of which 26 were new, the result of misidentifications of birds and prior discoveries by other naturalists; see Scott Weidensaul, *Of a Feather: A Brief History of Birding* (Orlando, FL, 2007), 50–51.

47. Audubon to John Bachman, 22 January 1836, in Irmscher, *Audubon: Writings and Drawings,* 838.

48. Ibid., 3, 49.

49. Ibid., 61–63.

50. Ibid., 63, 71.

51. Ibid., 79–80, 83.

52. Ibid., 89.

53. Ibid., 81.

54. Ibid., 98, 105.

55. Ibid., 114.

56. Ibid., 146–148.

57. For Audubon's short-lived partnership with Steen, see Ford, *John James Audubon* (1988), 139–140.

58. For a brief account of Audubon's several dead ends before going to Philadelphia, see Ford, *John James Audubon* (1988), 140–144.

59. Audubon, "Introductory Address," in *OB,* I, x.

60. Gary Nash, *First City: Philadelphia and the Forging of Historical Memory* (Philadelphia, 2006), 108; Nash attributes the "Athens" reference to Benjamin Henry Latrobe. The "Mecca"

reference comes from Francis Hobart Herrick, *Audubon the Naturalist*, 2 vols. (New York, 1917), I, 327.

61. This discussion of Peale's "Repository of Natural Curiosities" relies on David R. Brigham, "'Ask the Beasts, and They Shall Teach Thee': The Human Lessons of Charles Willson Peale's Natural History Displays," *Huntington Library Quarterly* 59, nos. 2 and 3 (1998), 183–206. For suggestions that Audubon may have drawn from specimens in the museum, see Partridge, "By the Book," 298–301; Richard Rhodes, *John James Audubon: The Making of an American* (New York, 2004), 93; and Roberta J. M. Olson, "The 'Early Birds' of John James Audubon," *Master Drawings* 50, no. 4 (Winter 2012), 454–456.

62. Brigham, "'Ask the Beasts, and They Shall Teach Thee,'" 197–203. There is, however, precious little evidence to indicate that Peale's museum—or Cincinnati's Western Museum or any other for that matter—ever truly succeeded in taming American society and imposing a sense of hierarchical order over an often raucous populace. In this as in so many other hopeful plans in the early republic, expectation much exceeded actual experience.

63. This brief survey of the scientific variety available to early nineteenth-century Philadelphians stems from unpublished papers and personal correspondence from Susan Branson, to whom I am grateful.

In the case of phrenology, even a self-conscious man of science like Audubon could find such study fascinating and, in his own experience, certainly satisfying. In 1826, while living in Edinburgh, he became friends with the well-known phrenologist George Combe, and he submitted to several examinations of the bumps on his head, all of which indicated that he was, indeed, a man of artistic talent and even much more: "I was astounded when they ... said I must be a strong and constant lover and affectionate father, that I had great veneration for high, talented men, that I would have made a great general, that music was not to be compared with painting in me, that I was extraordinarily generous, &c." Audubon had to agree that if phrenology could get it so right in his case, it must be good, and he "came off full of wonder at the singularity of this science." For Audubon's encounters with Combe and his friends in phrenology, see *The 1826 Journal of John James Audubon*, edited by Alice Ford (New York, 1987), 354–356, 402–403.

64. On Audubon's connections in Philadelphia, see Audubon, "Introductory Address," in *OB*, I, x–xi; and Ford, *John James Audubon* (1988), 145–147.

65. Ford, *John James Audubon* (1988), 147–148.

66. Nash, *First City*, 142. On the relationship between Lawson and Wilson, see Bayard H. Christy, "Alexander Lawson's Bird Engravings," *The Auk* 43, no. 1 (January–March 1926), 47–61.

67. For a sketch of the Lawson sisters, see Jordan D. Marche II and Theresa A. Marche, "A 'Distinct Contribution': Gender, Art, and Scientific Illustration in Antebellum America," *Knowledge and Society* 12 (2000), 77–106, an essay that focuses primarily on Helen Elizabeth Lawson, Orra White Hitchcock, and Lucy Sistare Say. Although they had received little formal training in either science or drawing—much less engraving, a craft their father thought suitable only for his son—the Lawson sisters learned their craft in Peale's museum, which they and other artists, male and female, used as an informal studio. Still, Philadelphia's world of print remained a largely gendered craft, and, even with Ord's support, the Lawson sisters never experienced the sort of success that men gained in the print profession—nor did they, as one study of female illustrators has pointedly put it, "undertake self-promotion in the manner of ornithological artist Audubon." See also Sally Gregory Kohlstedt, "Nature by Design: Masculinity and Animal Display in

Nineteenth-Century America," in *Figuring It Out: Science, Gender, and Visual Culture*, edited by Ann B. Shteir and Bernard V. Lightman (Hanover, NH, 2006), 114.

68. Audubon, "Introductory Address," in *OB*, I, xiv.

69. Audubon, "Louisiana Hawk," in *OB*, V, 34. Audubon also named a species of woodpecker "Harris's Woodpecker," but that was an ornithological error, the result of a spurious identification of a bird that turned out to be a western version of the Hairy Woodpecker. I have addressed in more detail the misidentification of this and other birds, which Audubon knew only from skins sent back from the West by Thomas Nuttall and John K. Townsend, in Gregory Nobles, "Audubon's Western Woodpeckers: Specious Species and Expansionist Science in the Pacific Northwest," *Columbia: The Magazine of Northwest History* (October 2008), 18–23. Harris did have another bird named for him, Harris's Sparrow, but that species was not identified and named until 1931, almost a century too late for Audubon; see http://www.audubon.org/field-guide/bird/harriss-sparrow (accessed 20 April 2015).

70. Quoted in Edward A. Muschamp, *Audacious Audubon: The Story of a Great Pioneer, Artist, Naturalist, and Man* (New York, 1929), 241.

71. For a fuller discussion of Audubon's commission for the New Jersey banknote, see Robert M. Peck and Eric P. Newman, "Discovered! The First Engraving of an Audubon Bird," *Journal of the Early Republic* 30, no. 3 (Fall 2010), 443–461.

72. Audubon, "Introductory Address," in *OB*, I, xi.

73. Audubon to Reuben Haines III, 25 December 1825, in Matthew R. Halley, "The Heart of Audubon," *Common-place* 16, no. 1 (Fall 2015), http://common-place.org/issue/vol-16-no-1/ (accessed 21 March 2016).

74. Audubon, "Introductory Address," in *OB*, I, xi–xii.

75. Audubon to Thomas Sully, 14 August 1824, quoted in Herrick, *Audubon the Naturalist*, I, 339.

76. Ford, *1826 Journal*, 71.

77. On Audubon's shipboard inebriation, see Ford's editorial notes in *1826 Journal*, 61, 71, 77.

Chapter 4

The epigraph for this chapter is from Audubon, 23 June 1826, in *The 1826 Journal of John James Audubon*, edited by Alice Ford (New York, 1987), 40–41. The date of the journal entry is 22 June, but within the text, Audubon adds a section for 23 June. For the sake of accessibility and accuracy, I have relied for the most part on Ford's version of the journal rather than the edition published almost a century earlier by Audubon's granddaughter, Maria R. Audubon, *Audubon and His Journals; With Zoological and Other Notes by Elliott Coues*, 2 vols. (New York, 1897), in which, according to Ford, Maria Audubon "cut, censored, paraphrased, bowdlerized, and even rewrote at will." See Ford, *1826 Journal*, 10. Ford's *1826 Journal* also includes letters addressed to specific individuals, which will be cited as such below. For Audubon's general journal entries, such as the one here, the citation will include only the date.

1. The best basic works about *The Birds of America* as a book are Waldemar H. Fries, *The Double Elephant Folio: The Story of Audubon's "Birds of America"* (Chicago, 1973); and Susanne M. Low, *An Index and Guide to Audubon's "Birds of America"* (New York, 1988). The quotation from Audubon about the weight of his work is in Audubon, "Small-Headed Flycatcher," in *Ornithological Biography*,

Or, An Account of the Habits of the Birds of the United States of America: Accompanied by Descriptions of the Objects Represented in the Work Entitled "The Birds of America," and Interspersed with Delineations of American Scenery and Manners, 5 vols. (Edinburgh, 1831–1839), V, 291; hereafter cited as *OB*.

2. Henry Clay to Rufus King, 24 February 1826, in Ford, *1826 Journal*, 17.

3. Vincent Nolte to Richard Rathbone, 13 May 1826, in ibid., 18.

4. Audubon to Lucy Bakewell Audubon, 7 August 1826, in ibid., 130.

5. 27 May 1826, in ibid., 30.

6. 27 May 1826, in ibid. The reference to "finny tribes" is on 23, and the story of the dying bunting on 28.

7. 26 June 1826, in ibid., 41.

8. 18 July 1826 and 24 July 1826, in ibid., 75 and 80, respectively.

9. 21–27 July 1826, in ibid., 81–99, quotation on 99.

10. 31 July 1826, in ibid., 111.

11. Audubon to Nicholas Berthoud, 7 August 1826, in ibid., 135. On the same day, Audubon wrote essentially the same thing to Lucy Audubon, in ibid., 130–134.

12. 1 August 1826, in ibid., 114.

13. 16 August 1826, in ibid., 156.

14. Audubon to Victor G. Audubon, 1 September 1826, in *Letters of John James Audubon, 1826–1840*, 2 vols., edited by Howard Corning (Boston, 1930), I, 3.

15. Liverpool *Mercury*, 11 August 1826.

16. Ibid., 1 September 1826.

17. Audubon to Lucy Bakewell Audubon, 1 September 1826, in Corning, *Letters*, I, 5–6.

18. Audubon to Lucy Bakewell Audubon, 10 December 1826, in Ford, *1826 Journal*, 381–382. Howard Corning dates the letter a little later, on 21 December; see Corning, *Letters*, I, 7–13.

19. 12 September 1826, in Ford, *1826 Journal*, 190.

20. 14 September 1826, in ibid., 195.

21. 9 August 1826, in ibid., 139.

22. 25 September 1826, in ibid., 222.

23. Audubon to Lucy Bakewell Audubon, 22 October 1826, in ibid., 293; Audubon to Lucy, 21 December 1826, in *The Audubon Reader*, edited by Richard Rhodes (New York, 2006), 175.

24. Audubon, "Introduction," in *OB*, II, vii.

25. The portrait by John Syme (1795–1861) was commissioned by William Home Lizars, Audubon's Edinburgh engraver.

26. Audubon to Victor G. Audubon, 1 September 1826, in Corning, *Letters*, I, 4; Audubon to Mrs. Richard Rathbone, 29 November 1826, in Ford, *1826 Journal*, 361. In a later letter to Lucy, Audubon noted that he did make a change to his hairstyle in early 1827: "I cut it off and I am now *rather more* of an Englishman but in good Circles and all societies Appearance has very little to do." Audubon to Lucy Bakewell Audubon, 16 May 1827, in Corning, *Letters*, I, 27. He would eventually grow his hair back to its former frontier appearance.

27. 1 December 1826, in Ford, *1826 Journal*, 365.

28. 22 September 1826, in ibid., 215.

29. 23 September 1826, in ibid., 218.

30. Audubon to Lucy Bakewell Audubon, 1 October 1826, in ibid., 238.

31. 15 December 1826, in ibid., 394. The exhibit ran from 14 November to 23 December, and it garnered £152.18.10 in admissions and £20.12.6 in catalog sales; see John Chalmers, *Audubon in Edinburgh and His Scottish Associates* (Edinburgh, 2003), 45, 59.

32. Edinburgh *Caledonian Mercury*, 16 November 1826, 27 November 1826.

33. Mark Catesby, *The Natural History of Carolina, Florida and the Bahama Islands*, 2 vols. (London, 1731, 1743); John Abbot and James Edward Smith, *The Natural History of the Rarer Lepidopterous Insects of Georgia*, 2 vols. (London, 1797); Alexander Wilson, *American Ornithology, or The Natural History of the Birds of the United States*, 9 vols. (Philadelphia, 1808–1814).

34. Until very recently, in fact, it remained so, but the title now goes to a big, five-by-seven-foot, 133-pound volume, Michael Hawley, *Bhutan: A Visual Odyssey Across the Last Himalayan Kingdom* (Charlestown, MA, 2003).

35. Audubon, "Introductory Address," in *OB*, I, xvi–xvii.

36. Jennifer L. Roberts, *Transporting Visions: The Movement of Images in Early America* (Berkeley, CA, 2014), 90.

37. Richard Harlan, one of Audubon's Philadelphia allies, warned him that "I know of no society here likely to subscribe to so costly a work, [because] they have been too accustomed to enlarge their libraries by presents, and begging.—besides Men of Nat. Science are scarce here, and those not able to purchase many expensive books on their own account." Harlan to Audubon, 19 November 1828, in Audubon Box, New-York Historical Society.

38. Audubon to William Rathbone, 24 November 1826, in Ford, *1826 Journal*, 347.

39. Audubon to Nicholas Berthoud, 7 August 1826, in ibid., 136.

40. 24 November 1826, in ibid., 347.

41. 16 December 1826, in ibid., 395.

42. Edinburgh *Journal*, 15 February 1827, quoted in Roberts, *Transporting Visions*, 71; 181, fn. 5.

43. On the larger context for private collections, see *Collecting Across Cultures: Material Exchanges in the Early Modern Atlantic World*, edited by Daniela Bleichmar and Peter C. Mancall (Philadelphia, 2011), 1–4.

44. Audubon to William Rathbone, 24 November 1826, in Ford, *1826 Journal*, 346.

45. Audubon to Lucy Bakewell Audubon, 15 May 1827, John James Audubon Collection, Princeton University Library Department of Rare Books and Special Collections, Box 2: Letters, Folder 6. He wrote almost the same thing a little over a month later, assuring Lucy that "if I can procure as many as 500 subscribers it will be an immense revenue, quite sufficient to make us all comfortable for the rest of our days." Audubon to Lucy Bakewell Audubon, 20 June 1827, in Rhodes, *Audubon Reader*, 207.

46. Prideaux John Selby, *Illustrations of British Ornithology*, 2 vols. (Edinburgh, 1819–1834); Sir William Jardine and P. J. Selby, *Illustrations of Ornithology*, 4 vols. (Edinburgh, 1826).

47. For Audubon's early meeting with Lizars (1788–1859), see Francis Hobart Herrick, *Audubon the Naturalist: A History of His Life and Time*, 2 vols. (New York, 1917), I, 358–359.

48. Audubon to William Rathbone, 24 November 1826, in Ford, *1826 Journal*, 346.

49. 27 November 1826, in ibid., 356.

50. Unfortunately, most of the original copper plates were destroyed after being sold for scrap, in 1870, when Lucy Audubon needed money two decades after her husband's death: see Chapter 10. Today, only eighty plates remain, and one, of the American Bittern, is now on display at the Audubon Museum and Nature Center in Henderson, Kentucky.

51. Audubon, "The Prospectus of *The Birds of America*" (Edinburgh, 1827). At the rate of two guineas, or one pound one shilling, per number, the total cost for the original complete set of 300 images, as originally promised, would have been 120 guineas, or about £132, for a British buyer; in the end, the total cost of the final four-volume, 435-plate edition would have been 170 guineas, or £187. See Chalmers, *Audubon in Edinburgh*, 55.

52. Quoted in Herrick, *Audubon the Naturalist*, II, 387.

53. Audubon, "Introductory Address," in *OB*, I, xi.

54. Audubon to Lucy Bakewell Audubon, 16 May 1827, in Corning, *Letters*, I, 23–24.

55. Ibid.

56. Quoted in Chalmers, *Audubon in Edinburgh*, 123. For a description of Wilson's difficulties in selling subscriptions to *American Ornithology*, see William M. Souder, *Under a Wild Sky: John James Audubon and the Making of "The Birds of America"* (New York, 2004), 104–117. For a longer view of the importance of subscriptions to support works of natural history, see David R. Brigham, "Mark Catesby and the Patronage of Natural History in the First Half of the Nineteenth Century," in *Empire's Nature: Mark Catesby's New World Vision,* edited by Amy R. W. Meyers and Margaret Beck Pritchard (Chapel Hill, NC, 1998), 91–146.

57. Audubon to Victor G. Audubon, 22 December 1828, in Corning, *Letters*, I, 76.

58. For estimates of middle-class and working-class annual incomes in early nineteenth-century America, see Bruce Laurie, *Artisans into Workers: Labor in Nineteenth-Century America* (New York, 1989), 56–61. Today, of course, the full, four-volume Double Elephant Folio set of *The Birds of America* goes for much, much more, now fetching well into the millions. For a discussion of the twentieth-century sales of *The Birds of America*, see Chapter 10.

59. Audubon to Lucy Bakewell Audubon, 21 December 1827, Audubon Collection, Princeton University Library Department of Rare Books and Special Collections, Box 2: Letters, Folder 9.

60. Audubon to Lucy Bakewell Audubon, 25 November 1827, in Corning, *Letters*, I, 47. Unfortunately for Audubon, not to mention the monarch himself, George IV died in 1830, and his subscription then lapsed; see Chalmers, *Audubon in Edinburgh*, 123.

61. 8 November 1826, in Audubon, *Audubon and His Journals*, I, 267.

62. 25 March, 1827, in ibid., I, 292.

63. Quoted in Duff Hart-Davies, *Audubon's Elephant: America's Greatest Naturalist and the Making of "The Birds of America"* (New York, 2004), 123.

64. Low, *Index and Guide*, 31. See also David F. Bottjer, "Robert Havell and Creating 'The Birds of America,'" *Journal of the American Historical Print Collectors Society* 37, no. 1 (Spring 2012), 25–39.

65. Audubon, "Havell's Tern," in *OB*, V, 122.

66. Audubon to Victor Woodhouse Audubon, 10 November 1828, in Corning, *Letters*, I, 72.

67. Roberts, *Transporting Visions*, 80.

68. Audubon to Robert Havell Jr., 31 December 1827, New-York Historical Society, Audubon Box.

69. For examples of Audubon's complaints to Havell about the quality of workmanship, see Audubon to Robert Havell Jr., 29 June 1830 and 31 October 1830, both in Corning, *Letters*, I, 112, 120.

70. 21 January 1827, in Audubon, *Audubon and His Journals*, 278.

71. Audubon to Robert Havell Jr., 14 March 1831, New-York Historical Society, Audubon Box.

72. Audubon to Victor G. Audubon, 15 September 1833, in Corning, *Letters*, I, 249.

73. Audubon to Robert Havell Jr., 20 April 1833, in ibid., 212–213.

74. Lucy Bakewell Audubon to Euphemia Gifford, 10 June 1831, in Audubon-Bakewell and Gifford Correspondence, Stark Museum of Art, SMA ABGC 0011.103.011.

75. Audubon to Lucy Bakewell Audubon, 10 December 1826, in Ford, *1826 Journal*, 383.

76. Lucy Bakewell Audubon to Victor Gifford Audubon, 15 June 1828, in Rhodes, *Audubon Reader*, 217.

77. Audubon to Lucy Bakewell Audubon, 24 March 1827, in Corning, *Letters*, I, 18.

78. Audubon to Lucy Bakewell Audubon, 6 August 1827 and 6 February 1828, in ibid., 31, 58. For other letters to Lucy expressing a similar hope for support from "some great Nabob" or prominent institution, see ibid., 18, 55.

79. For a sympathetic treatment of Lucy Audubon's side of the marriage, see Carolyn DeLatte, *Lucy Audubon: A Biography* (Baton Rouge, LA, 1982; updated ed., 2008). DeLatte's narrative goes only to 1830, with the last twenty years of the Audubons' married life covered in only a fifteen-page epilogue; still, her extensive coverage of the period of Audubon's separation from Lucy while he was in Great Britain recounts the many difficulties of an estranged and increasingly strained relationship.

80. Audubon to Lucy Bakewell Audubon, 21 December 1826, in Rhodes, *Audubon Reader*, 176–177.

81. Audubon to Lucy Bakewell Audubon, 15 May 1827, Audubon Collection, Princeton University Library Rare Books and Special Collections, Box 2: Letters, Folder 6.

82. Audubon to Lucy Bakewell Audubon, 16 May 1827, in Corning, *Letters*, I, 25.

83. Audubon to Lucy Bakewell Audubon, 5 December 1827, in Rhodes, *Audubon Reader*, 213; on Audubon's being "fatigued of our separation" and generally unhappy in London, see also 23 December 1828, in ibid., 225.

84. Audubon to Lucy Bakewell Audubon, 8 August 1828, in Corning, *Letters*, I, 69.

85. Audubon to Lucy Bakewell Audubon, 2 November 1828, John James Audubon Collection, Princeton University Library Rare Books and Special Collections, Box 2: Letters, Folder 10.

86. DeLatte, *Lucy Audubon*, 191.

87. Lucy Bakewell Audubon to Victor Gifford Audubon, 15 June 1828; 19 January 1829; 30 January 1829, in Rhodes, *Audubon Reader*, 217, 227, 232.

88. Audubon to Lucy Bakewell Audubon, 20 January 1829, in ibid., 229–230. Given the vagaries of transatlantic mail traffic, Lucy also wrote to Audubon just a few weeks later, before she received his letter, saying that she had "come to the resolution of giving up my occupation and joining you." Before she would leave her teaching job in Louisiana for England, though, she still needed some assurance of his intentions toward her: "I beg of you, my husband, to be plain and clear in your reply; if any circumstance has occurred to change your affection for me . . . be explicit and let me remain where I am." Lucy Bakewell Audubon to Audubon, 8 February 1829, in ibid., 234–235.

89. Undated journal entry, in ibid., 282.

90. Audubon to Charles Lucien Bonaparte, 5 May 1830, in ibid., 288. According to Dr. Daniel Feller, editor of the Papers of Andrew Jackson, there is, unfortunately, no record of Jackson's account of the meeting or his observations about Audubon (email correspondence from Feller, 22 October 2013).

91. Lucy Bakewell Audubon to Euphemia Gifford, 19 July 1831, in Audubon-Bakewell and Gifford Correspondence, Stark Museum of Art, SMA ABGC 0011.103.003.

92. Audubon to John Woodhouse Audubon, 28 October 1826, in Ford, *1826 Journal*, 313.

93. Audubon to Lucy Bakewell Audubon, 8 August 1828, in Corning, *Letters*, I, 67. Two days later, he wrote directly to John, telling the teenager that "I would give you 500 dollars per annum were you able to make for me such drawings as I will want." See Audubon to John Woodhouse Audubon, 10 August 1828, in Rhodes, *Audubon Reader*, 219.

94. John Woodhouse Audubon to Victor Gifford Audubon, 5 November 1833, in Audubon Papers, American Philosophical Society.

95. Audubon to Victor Gifford Audubon, 21 December 1833, in Corning, *Letters*, I, 274.

96. Audubon to Victor Gifford Audubon, 22 December 1828, in ibid., I, 75.

97. Audubon to Victor Gifford Audubon, 18 October 1834, Audubon Collection, Princeton University Library Rare Books and Special Collections, Box 2: Letters, Folder 29.

98. Victor Gifford Audubon to Robert Havell Jr., 29 April 1832; see also subsequent letters, 21 July 1832 and 18 February 1833, all in John James Audubon Papers (MS Am 1842), Houghton Library, Harvard University.

99. Audubon to Victor Gifford Audubon, quoted in Stanley Clisby Arthur, *Audubon, an Intimate Life of the American Woodsman* (New Orleans, 1937), 421.

100. For the background of the British book law, see Fries, *Double Elephant Folio*, 47.

101. Audubon to Robert Havell Jr., 23 March 1831, in Corning, *Letters*, I, 132.

102. Audubon to Lucy Bakewell Audubon, 1 October 1826, in Ford, *1826 Journal*, 238.

103. Audubon to Lucy Bakewell Audubon, 16 October 1830, quoted in Arthur, *Audubon, An Intimate Life*; and Herrick, *Audubon the Naturalist*, I, 438.

104. Audubon, "Introductory Address," in *OB*, I, xix. On the early career and scientific adventures of MacGillivray, see David Elliston Allen, *The Naturalist in Britain* (Princeton, NJ, 1976), 67–68.

105. Herrick, *Audubon the Naturalist*, I, 438; Audubon, "Introductory Address," in *OB*, I, xvii.

106. Edward Dwight notes that Audubon completed forty-two paintings during his trip to New Jersey and Pennsylvania in 1829 and "fully half of all the paintings for *The Birds of America*" between 1831 and 1838; see Dwight, *The Original Watercolor Paintings for "The Birds of America*,*"* 2 vols. (New York, 1966), II, xxxv.

Chapter 5

The epigraph for this chapter is from John James Audubon, "Anhinga or Snake-Bird," in *Ornithological Biography, Or, An Account of the Habits of the Birds of the United States of America: Accompanied by Descriptions of the Objects Represented in the Work Entitled "The Birds of America,"* *and Interspersed with Delineations of American Scenery and Manners*, 5 vols. (Edinburgh, 1831–1839), IV, 136–169, quotations on 136, 154; hereafter cited as *OB*.

1. John Wilson, *Critical and Miscellaneous Essays. By Christopher North, (Professor Wilson) In Three Volumes* (Philadelphia, 1842), II, 122.

2. John T. Battalio, *The Rhetoric of Science in the Evolution of American Ornithological Discourse* (Stamford, CT, 1998), 17.

3. Audubon, "Introductory Address," in *OB*, I, xvii–xviii.

4. Audubon, "The Eccentric Naturalist," in *OB*, I, 455–460.

5. Ibid., 456.

6. Ibid., 457, 460.

7. James Fenimore Cooper, *The Prairie: A Tale* (New York, 1827).

8. On Rafinesque's background, see George H. Daniels, *American Science in the Age of Jackson* (New York, 1968), 220–221; see also Joseph Kastner, *A Species of Eternity* (New York, 1977), 240–253.

9. On the full extent of Audubon's deceptions and Rafinesque's gullibility, see Jason Daley, "Audubon Pranked Fellow Naturalist by Making Up False Rodents," Smithsonian.com, 27 April 2016, http://www.smithsonianmag.com/smart-news/audubon-pranked-fellow-naturalist-making-fake-rodents-180958907/?no-ist (accessed 15 June 2016).

10. Alexis de Tocqueville, *Democracy in America*, 2 vols. (London, 1835, 1840), I, xxxvi.

11. Tocqueville's observations on American science appeared in his second volume of *Democracy in America*, 35–48, published a year after the final volume of Audubon's *Ornithological Biography* appeared in print, and therefore too late for Audubon to have referenced Tocqueville there. By the 1840s, Audubon was in declining health and had essentially finished his best scientific work, and I have found no subsequent reference to Tocqueville in Audubon's later writings.

12. The phrase quoted comes from the title of Tocqueville's chapter X.

13. Tocqueville, *Democracy in America*, II, 41.

14. The discussion of the early years of the American Philosophical Society in this and the following paragraph stems largely from Tom Shachtman, *Gentleman Scientists and Revolutionaries: The Founding Fathers in the Age of Enlightenment* (New York, 2014), 38–44. On the subject of transatlantic networks of correspondence and exchange, I have also benefited from the work of Susan Scott Parrish, *American Curiosity: Cultures of Natural History in the Colonial British Atlantic World* (Chapel Hill, NC, 2006), 103–173; and Robyn Davis McMillin, "Science in the American Style" (PhD dissertation, University of Oklahoma, 2009), 189–194.

15. Benjamin Franklin, "A Proposal for Promoting Useful Knowledge among the British Plantations in America" (Philadelphia 1743), 1–2. On Franklin's criticism of the "idle Gentlemen" in the APS, see Gordon S. Wood, *The Americanization of Benjamin Franklin* (New York, 2005), 49.

16. Chandos Michael Brown, "A Natural History of the Gloucester Sea Serpent: Knowledge, Power, and the Culture of Science in Antebellum America," *American Quarterly* 42, no. 3 (September 1990), 406–407.

17. Georges-Louis LeClerc, Comte de Buffon, *Histoire naturelle, générale et particulière*, 36 vols. (Paris, 1749–1788). Buffon has been widely discussed in works on early American science and natural history, but my summary of his work here relies largely on two works by Paul Lawrence Farber: *The Emergence of Ornithology as a Scientific Discipline, 1760–1850* (Dordrecht, The Netherlands, 1982); and *Finding Order in Nature: The Naturalist Tradition from Linnaeus to E. O. Wilson* (Baltimore, MD, 1994), 13–21.

18. Buffon completed thirty-six volumes while he lived, and *Histoire naturelle* eventually ran to forty-four volumes in all, the last eight produced posthumously by assistants. Farber, in *Emergence of Ornithology*, 25, notes that some people have considered Buffon's prose style to be a bit too literary for scientific writing, but he gives Buffon credit for his "penchant for *la grande vue* and his aesthetic sensitivity that prompted him to ask fundamental questions about his data."

19. Farber, *Finding Order in Nature*, 20.

20. Buffon, *Histoire naturelle*, V, 237.

21. Jefferson's rebuttal of Buffon has been widely described, and the brief account here relies largely on I. Bernard Cohen, *Science and the Founding Fathers: Science in the Political Thought of*

Jefferson, Franklin, Adams, and Madison (New York, 1995), 72–88. See also Pamela Regis, *Describing Early America* (De Kalb, IL, 1992), 91–93. For a discussion of Jefferson and the role of the mastodon and moose skeletons, see Lee Alan Dugatkin, *Mr. Jefferson and the Giant Moose: Natural History in Early America* (Chicago, 2009); Paul Semonin, *American Monster: How the Nation's First Prehistoric Creature Became a Symbol of National Identity* (New York, 2000); and an abbreviated essay drawn from that book, "Peale's Mastodon: The Skeleton in Our Closet," *Common-place* 4, no. 2 (2004). For the lingering uneasiness about Buffon, see Andrew J. Lewis, *A Democracy of Facts: Natural History in the Early Republic* (Philadelphia, 2011), 19–20.

22. Elliott Coues, *Birds of the Colorado Valley: A Repository of Scientific and Popular Information Concerning North American Ornithology* (Washington, DC, 1878), 601. Coues's estimation of Wilson came in an appendix in which he compiled a very extensive "List of Faunal Publications Relating to North American Ornithology," beginning with John Smith, *A Map of Virginia* (Oxford, 1612), and running up to L. P. Vieillot, *Histoire naturelle des oiseaux de L'Amerique Septentrionale* (Paris, 1807), just preceding the citation of Wilson's *American Ornithology*. Coues's comment on Audubon's *Birds of America* is on 612.

23. Henry Adams, *History of the United States During the First Administration of Thomas Jefferson* (New York, 1889), I, 124.

24. Alexander Wilson, "Wood Thrush," in *American Ornithology, or The Natural History of the Birds of the United States,* 9 vols. (Philadelphia, 1808–1814), I, 34 (hereafter cited as *AO*); "Canada Jay," in *AO*, II, 33; "Downy Woodpecker," in *AO*, I, 155; "Sea Eagle," in *AO*, VII, 17–18.

25. George Ord, "Life of Wilson," in *AO*, IX, xliii, xliv.

26. Audubon, "Swainson's Warbler," in *OB*, II, 563; "The Hemlock Warbler," in *OB*, II, 205; "The Red-Eyed Vireo," in *OB*, II, 289; "Harris's Woodpecker," in *OB*, V, 191; "Introduction," in *OB*, III, 9; "The Sora Rail," in *OB*, III, 251.

27. Audubon, "Wilson's Plover," in *OB*, III, 73–74.

28. Audubon, "The Rough-Legged Falcon," in *OB*, II, 378; "The Great Red-Breasted Rail," in *OB*, III, 27.

29. See, for instance, Audubon, "American Widgeon," in *OB*, IV, 338; "Sharp-Shinned or Slate-Coloured Hawk," in *OB*, IV, 526; "Little Screech Owl," in *OB*, I, 486; "Knot or Ash-Coloured Sandpiper," in *OB*, IV, 132; "Connecticut Warbler," in *OB*, V, 82.

30. Audubon, "Smew or White Nun," in *OB*, IV, 350; "Appendix," in *OB*, V, 458.

31. Audubon, "Anhinga or Snake-Bird," in *OB*, IV, 154. It's been suggested that Peale's collection also had an influence on Audubon and that he may also have worked from specimens there. See Roberta J. M. Olson, "The 'Early Birds' of John James Audubon," *Master Drawings* 50, no. 4 (Winter 2012), 454–456. If so, Audubon would never have admitted that.

32. Audubon, "Introduction," in *OB*, V, vii–viii.

33. Audubon, "Swainson's Warbler," in *OB*, II, 563. With perhaps false modesty, Audubon excluded himself in his reference to those extending the list of American birds, but listed "our BONAPARTES, our NUTTALLS, our BACHMANS, our COOPERS, PICKERINGS, TOWNSENDS, PEALES, and other zealous naturalists."

34. Francis Hobart Herrick, *Audubon the Naturalist: A History of His Life and Time,* 2 vols. (New York, 1917), I, 328–329; Alice Ford, *John James Audubon: A Biography* (New York, 1988), 147; William Souder, *Under a Wild Sky: John James Audubon and the Making of "The Birds of America"* (New York, 2004), 13; Richard Rhodes, *John James Audubon: The Making of an American* (New York, 2004), 221.

35. This biographical sketch of Ord stems primarily from a more positive profile by Al Dorof, "George Ord (1781–1866)," Southwark Historical Society, January 2012, http://www.qvna.org/qvna/georgeord/ (accessed 16 June 2016).

36. Robert Henry Welker, *Birds and Men: American Birds in Science, Art, Literature, and Conservation, 1800–1900* (Cambridge, MA, 1955), 62.

37. Ibid., 399.

38. Herrick, *Audubon the Naturalist*, I, 399–400. When the publisher of the *Magazine of Natural History*, John C. Loudon, kept pressing for an article, Audubon finally agreed to do one on the Bird of Washington, which appeared in the July 1828 issue.

39. Audubon to Richard Harlan, 29 November 1830, in Arthur E. Lownes, "Ten Audubon Letters," *The Auk* 52, no. 2 (April 1935), 156.

40. Audubon to Robert Havell Jr., 20 September 1831, in *Letters of John James Audubon, 1826–1840*, edited by Howard Corning, 2 vols. (Boston, 1930), I, 136; Audubon to Lucy Bakewell Audubon, 7 November 1831, in ibid., 148; Audubon should have said "Academy of Natural Sciences," but it was a minor mistake. Audubon's election to the American Philosophical Society, 15 July 1831, is recorded in the *Early Proceedings of the American Philosophical Society . . . from 1744 to 1838* (Philadelphia, 1884), 618.

41. *Vermont Gazette*, 2 June 1829, 1; Boston *Daily Advertiser and Patriot*, reprinted in New York *American*, 1 March 1833, 2.

42. Charleston (SC) *Courier*, 20 July 1831, 2; Bucks County (PA) *Intelligencer*, reprinted in *Eastern Argus Tri Weekly*, 30 October 1833, 2.

43. New York *Mercury Advertiser*, reprinted in Savannah (GA) *Daily Georgian*, 20 August 1831, 3; Philadelphia *National Gazette*, 4 May 1833, 1.

44. Charleston (SC) *City Gazette and Commercial Daily Advertiser*, 21 October 1831, 2; New York *American*, 3 April 1833, 2.

45. Alexandria (VA) *Gazette*, 8 December 1831, 2.

46. Boston *Columbian Centinel*, 23 March 1833, 2.

47. Hampshire *Gazette*, 27 March 1833, 3.

48. Charles Waterton, *Essays on Natural History: Chiefly Ornithology* (London, 1845), 496.

49. Brian W. Edginton, *Charles Waterton: A Biography* (Cambridge, UK, 1996), 131.

50. Norman Moore, "Life of the Author," in Waterton, *Essays on Natural History* (1845), 127–128.

51. Charles Waterton to George Ord, 16 February 1832, quoted in Ford, *John James Audubon* (1988), 300; Ord to Waterton, 23 April 1832, George Ord Papers, American Philosophical Society.

52. George Ord to Charles Waterton, 23 April 1832; 28 April 1833; 15 April 1835; 16 April 1835; 29 September 1835, George Ord Papers, American Philosophical Society.

53. George Ord to Charles Waterton, 17 April 1835, in ibid.

54. George Ord to Charles Waterton, 29 September 1835, in ibid.; Charles Waterton to George Ord, 7 April 1836, quoted in Ford, *John James Audubon* (1988), 349.

55. William Swainson, "Some Account of the Work Now Publishing by Mr. Audubon," *Loudon's Magazine of Natural History* 1 (May 1828), 48–52, quoted in Herrick, *Audubon the Naturalist*, I, 404.

56. Audubon to William Swainson, 13 August 1828, quoted in Herrick, *Audubon the Naturalist*, I, 404.

57. William Swainson to Audubon, 18 January 1829, in Ruthven Deane, "William Swainson to John James Audubon," *The Auk* 22 (July 1905), 250–251.

58. William Swainson to Audubon, 10(?) May 1830, quoted in Herrick, *Audubon the Naturalist*, II, 97.

59. Audubon to William Swainson, 22 August 1830, in ibid., II, 101–102.

60. William Swainson to Audubon, 24–28 August 1830, 2 October 1830, in ibid., II, 104–108.

61. Audubon to Charles Lucien Bonaparte, 2 January 1831, in *The Audubon Reader*, edited by Richard Rhodes (New York, 2006), 313–314.

62. William Swainson, *Natural History of Birds* (London, 1836–1837), 210, 217–218.

63. Audubon to John Bachman, 1 December 1835, in Corning, *Letters*, II, 106.

64. Charles Waterton, "An Ornithological Letter to William Swainson, Esq FRS" (Wakefield, UK, 1837); quoted in Charles Waterton, *Essays on Natural History*, edited by Norman Moore (New York, 1871), 511–523.

65. Audubon, "Wilson's Petrel," in *OB*, III, 486.

66. For a very useful overview of the history of classification in the eighteenth and early nineteenth centuries, see Farber, *Finding Order in Nature*, 6–42. Natural history, Michel Foucault has observed, "is nothing more than the nomination of the visible." But as he has also noted, observing means being "content with seeing a few things systematically." See Foucault, *The Order of Things: An Archaeology of the Human Sciences* (New York, 2002), 144, 146. Beginning in the eighteenth century, seeing things systematically became the scientific standard among European naturalists and, later, their counterparts in the United States: They shared a passion for classification, for both revealing the underlying order in nature and then imposing an overarching order on their revelations. They did not yet have complete order in that effort itself, because they had several competing classification systems to choose from. Peter Kalm's mentor, Carolus Linnaeus, had gained the greatest prominence in the field with the publication of his *Systema Naturae* in 1735, but later in the century, Georges-Louis Leclerc, Comte de Buffon, published another influential work, *Histoire naturelle de oiseaux* (1781). Two of Buffon's fellow Frenchmen, Georges Cuvier and Etienne Geoffroy Saint-Hilaire, soon followed with books of their own, and by the early years of the nineteenth century, the various approaches to classification seemed to be multiplying with every succeeding decade.

67. Audubon to [no name or address], 4 October 1837, in Corning, *Letters*, II, 183.

68. John James Audubon, *A Synopsis of the Birds of North America* (Edinburgh, 1839). When Audubon later bound his own copies of the plates into four volumes, he rearranged the order of the plates in accordance with the ornithological classification scheme that he had developed in his *Synopsis*. The plates in Audubon's personal copy (which is now in the H. J. Lutcher Stark Museum in Orange, Texas) do not follow the standard order in which Audubon produced them, beginning with the famous image of the Great American Cock, or Wild Turkey, which was the original plate 1; rather, Audubon starts with the Turkey Vulture, plate 426, which came toward the very end of the standard order in *The Birds of America*. (The same is true for those copies owned by two of his close friends, Edward Harris and Dr. Benjamin Phillips.) In a sense, Audubon ultimately chose to impose his authority as scientist over his own authority as artist.

69. Audubon to Dr. Richard Harlan, 16 November 1834, in Rhodes, *Audubon Reader*, 485; Audubon to Victor G. Audubon, 21 December 1833, in Corning, *Letters*, I, 272; Audubon to John

Bachman, 3 December 1834, in Corning, *Letters*, II, 53–54; Audubon to John Bachman, 10 December 1834, in Rhodes, *Audubon Reader*, 486.

70. Audubon to Richard Harlan, 20 January 1834, in Audubon Collection, Princeton University Library Department of Rare Books and Special Collections, Box 2, Folder 90 (transcript of original in Gilbert Collection, College of Physicians, Philadelphia).

71. Audubon to John Bachman, 13 March 1834, in Audubon Collection, Princeton University Library Department of Rare Books and Special Collections, Box 2, Folder 27; Audubon to Bachman, 27 October 1838, in Corning, *Letters*, II, 207.

72. George Ord to Charles Waterton, 7 October 1852, in George Ord Papers, American Philosophical Society.

Chapter 6

The epigraph for this chapter is from Audubon, "Great Horned Owl," in *Ornithological Biography, Or, An Account of the Habits of the Birds of the United States of America: Accompanied by Descriptions of the Objects Represented in the Work Entitled "The Birds of America," and Interspersed with Delineations of American Scenery and Manners*, 5 vols. (Edinburgh, 1831–1839), I, 315; hereafter cited as *OB*.

1. London *Morning Chronicle*, 16 June 1831.

2. The passage from the Philadelphia *Gazette* was taken from a subsequent issue of the New York *Commercial Advertiser*, 15 September 1831, 2. See also the Washington, DC, *Daily National Journal*, 10 October 1831, 3; and the Charleston (SC) *City Gazette*, 18 October 1831, 2, both of which used identical language about the "tedious, difficult, and perilous journey" Audubon was about to take.

3. Audubon to Lucy Bakewell Audubon, 25 November 1827, in *Letters of John James Audubon, 1826–1840*, edited by Howard Corning, 2 vols. (Boston, 1930), I, 51.

4. Audubon, "Introduction," in *OB*, V, v.

5. Audubon, "The Wood Pewee," in *OB*, II, 93.

6. Audubon, "Great Pine Swamp," in *OB*, I, 57. Audubon, as we saw in Chapter 2, had been no stranger to fine fashion, even a bit of foppery, when he was a young man himself, sometimes going out hunting wearing silk and satin and ruffles; by the 1830s, though, he had perfected quite another sartorial style.

7. The best biographical source for Nuttall is Jeannette E. Graustein, *Thomas Nuttall, Naturalist: Explorations in America, 1808–1841* (Cambridge, MA, 1967).

8. Ibid., 32–41.

9. Benjamin Smith Barton to Thomas Nuttall, 7 April 1810, in Benjamin Smith Barton Papers, American Philosophical Society.

10. Ibid.

11. Quoted in *Nuttall's Travels into the Old Northwest: An Unpublished 1810 Diary*, edited by Jeannette E. Graustein (Waltham, MA, 1951), 21, 25.

12. Ibid., 33.

13. Ibid., 37.

14. Graustein, *Thomas Nuttall, Naturalist*, 42–77.

15. Ibid., 146.

16. Ibid., 151.

17. Alexander Wilson to David Wilson, 6 June 1811, quoted in *The Poems and Literary Prose of Alexander Wilson: The American Ornithologist*, edited by Alexander B. Groshart, 2 vols. (Paisley, Scotland, 1876), I, 227.

18. Alexander Wilson, "Pied Oyster-catcher," in *American Ornithology, or The Natural History of the Birds of the United States*, 9 vols. (Philadelphia, 1808–1814), VIII, 18; hereafter cited as *AO*.

19. Wilson, "Introduction," in ibid., I, 4.

20. George Ord, *Sketch of the Life of Alexander Wilson, Author of the American Ornithology* (Philadelphia 1828), clxii.

21. Audubon, "Great Horned Owl, in *OB*, I, 315.

22. Audubon, "Kentucky Sports," in *OB*, I, 285.

23. Ibid., 394.

24. Audubon, "Colonel Boon," in *OB*, I, 503.

25. John Mack Faragher, *Daniel Boone: The Life and Legend of an American Pioneer* (New York, 1993), 308. Faragher adds that when Audubon later painted a portrait of Boone from memory, the result "was as inaccurate as his pen picture" (309).

26. "American Ornithology," *Western Monthly Magazine* 3, no. 19 (July 1834), 349–350.

27. Audubon to G. W. Featherstonehaugh, 7 December 1831, quoted in Francis Hobart Herrick, *Audubon the Naturalist: A History of His Life and Time*, 2 vols. (New York, 1917), II, 9. Born in upstate New York, Bachman received his adolescent education in Philadelphia between 1804 and 1815, where he came to know William Bartram, George Ord, and Alexander Wilson. He moved to the warmer climes of Charleston, South Carolina, for his health in 1815 and took up a pastorate that he would hold for fifty-six years.

28. Audubon to Lucy Bakewell Audubon, 23 October 1831, in Corning, *Letters*, I, 142.

29. On the enduring Audubon-Bachman relationship, see Jay Shuler, *Had I the Wings: The Friendship of Bachman and Audubon* (Athens, GA, 1995).

30. Audubon, "The Long-Billed Curlew," in *OB*, III, 241–243.

31. John Bachman to Audubon, 22 January 1836, in *The Audubon Reader*, edited by Richard Rhodes (New York, 2006), 498.

32. John Bachman to Audubon, 1 April 1833, quoted in Alice Ford, *John James Audubon: A Biography* (New York, 1988), 306.

33. Audubon to John Bachman, 27 December 1837, quoted in ibid., 337; see also Corning, *Letters*, II, 108.

34. Audubon to John Bachman, 22 January 1836, in *John James Audubon: Writings and Drawings*, edited by Christoph Irmscher (New York, 1999), 838–839. Whether Audubon paid her anything in monetary terms, in fact, remains unclear. He did give her thanks a dozen times in *Ornithological Biography*, almost always identified as the sister or sister-in-law of Bachman. His most appreciative gesture came when he named a bird for her, Maria's Woodpecker, writing, "I feel bound to make some ornithological acknowledgement for the aid she has on several occasions afforded me in embellishing my drawings of birds, by adding to them beautiful and correct representations of plants and flowers." Audubon, "Maria's Woodpecker," in *OB*, V, 181. Unfortunately, the bird turned out to be a form of the Hairy Woodpecker, not a distinct species, so the name has not stuck; see Gregory Nobles, "Audubon's Western Woodpeckers: Specious Species and Expansionist Science in the Pacific Northwest," *Columbia: The Magazine of Northwest History* (October 2008), 18–23.

35. Katherine Govier, *Creation: A Novel* (Woodstock, NY, 2002), 199.

36. Audubon, "Bewick's Wren," in *OB*, I, 96.

37. Ibid.

38. Audubon, "Introductory Address," in *OB*, I, xii.

39. Audubon, "The White-Headed Eagle," in *OB*, II, 160–161.

40. Ibid.

41. Audubon to G. W. Featherstonhaugh, 31 December 1831, quoted in Herrick, *Audubon the Naturalist*, II, 16–17.

42. Audubon, "The Night Heron," in *OB*, III, 276.

43. Audubon, "The Sandwich Tern," in *OB*, III, 531.

44. Audubon, "The Golden Eagle," in *OB*, II, 464–470.

45. Audubon to Victor Gifford Audubon, 24 February 1833, in Corning, *Letters*, I, 200.

46. Wilson, "Ivory-Billed Woodpecker," in *AO*, IV, 22.

47. Audubon, "The Trumpeter Swan," in *OB*, IV, 541; "The Swallow-Tailed Hawk," in *OB*, I, 369; "Bonaparte's Fly-Catcher," in *OB*, I, 27.

48. Audubon, "The Golden Eagle," in *OB*, II, 464.

49. Audubon to Victor Gifford Audubon, 24 February 1833, in Corning, *Letters*, I, 201.

50. Audubon, "The Golden Eagle," in *OB*, II, 465.

51. Ibid.

52. Ibid., 465–466.

53. Christoph Irmscher, *The Poetics of Natural History: From John Bartram to William James* (New Brunswick, NJ, 1999), 227.

54. Audubon to Victor Gifford Audubon, 5 February 1833, in Herrick, *Audubon the Naturalist*, II, 35.

55. Audubon to Dr. Richard Harlan, 20 March 1833, in Rhodes, *Audubon Reader*, 372–373.

56. Audubon to Victor Gifford Audubon, 19 March 1833, in Audubon Papers, American Philosophical Society. The following day, he also wrote his friend Richard Harlan to say that "the attack seized on my mouth & particularly my lips, so much so that I neither could articulate or hold anything. My good dear wife was terribly frightened, and yet acted with prudence & knowledge that I was relieved, as I already said, in about one hour." Audubon to Richard Harlan, MD, 20 March 1833, in Rhodes, *Audubon Reader*, 372.

57. Philadelphia *Gazette*, quoted in the Newark *Daily Advertiser*, 30 April 1833, 2.

58. Audubon, "The Golden Eagle," in *OB*, II, 464, 465.

59. Audubon even wrote his own wife out of the story. Although Lucy was with Audubon in Boston, she does not appear in Audubon's account; the only other family member mentioned is his son John Woodhouse. When Katherine Govier recounts the tale of the Golden Eagle in her novel *Creation*, she does put Lucy in the scene, engaging in a quarrelsome exchange about the bird's tenacity and her husband's vulnerability: "You've met your match. You'll do yourself in, Fougère, if you're not careful." See Govier, *Creation*, 67–69.

60. Audubon, "The Golden Eagle," in *OB*, II, 467–468.

61. Ella M. Foshay, for instance, has emphasized Audubon's use of the imagery of "heroic pursuit," in which Audubon "identifies himself as an artist embarked on a dangerous journey that requires courage, skill, and ingenuity to survive." She points out, to be sure, that the heroism depicted in the painting was "entirely false," given Audubon's actual indoor encounter with the eagle: The image only "typifies the tenuous balance between fact and fiction that runs through

John James Audubon's life and work." Foshay, *John James Audubon* (New York, 1997), 9. Amy R. W. Meyers notes that Audubon equates himself with the eagle and "portrays himself as being equally tenacious and daring, braving all dangers to possess his quarry. His presence in the piece also suggests that he is equally ruthless." Meyers, "Observations of an American Woodsman: John James Audubon as Field Naturalist," in *John James Audubon: The Watercolors for "The Birds of America,"* edited by Annette Blaugrund and Theodore E. Stebbins Jr. (New York, 1993), 48. Christoph Irmscher likewise stresses the connection between the artist and his prey and suggests that "we might even read the majestic bird in the foreground as an elaborate metaphor for the artist's struggle with this composition." Irmscher, "Violence and Artistic Representation in John James Audubon," *Raritan: A Quarterly Review* 15, no. 2 (Fall 1995), 31.

62. Irmscher, *Poetics of Natural History*, 204, 225. See, for instance, Audubon's image of the Great White Heron, with Key West, Florida, in the background; or of the Golden Plover, with a view of Charleston, South Carolina.

63. Audubon, "The Golden Eagle," in *The Birds of America*, 7 vols. (New York, 1840), I, 51.

64. The extensive journal that Audubon kept on the voyage was edited by Audubon's granddaughter, Maria R. Audubon, who, as Richard Rhodes notes in *Audubon Reader,* "refined, formalized and expurgated her grandfather's more colorful prose." That is the only extant version of the journal, however, and Rhodes has reproduced it in *Audubon Reader,* 374–450; it will be cited here as "Labrador Journal."

65. Audubon, "Labrador Journal," 383–384.

66. Audubon, "The Arctic Tern," in *OB,* III, 367.

67. Audubon, "The Black Guillemot," in *OB,* III, 151.

68. Audubon, "Labrador Journal," 411.

69. Audubon, "The Great Black-Backed Gull," in *OB,* III, 305.

70. Audubon, "Labrador Journal," 394.

71. Audubon, "The Eggers of Labrador," in *OB,* III, 83–84, 86.

72. Audubon, "Labrador Journal," 401.

73. Audubon, "The Eggers of Labrador," in *OB,* III, 85–86.

74. Audubon, "Labrador Journal," 420.

75. Govier, *Creation,* 25.

76. Audubon, "Labrador Journal," 437.

77. Audubon to Richard Harlan, 26 May 1833, in Audubon Collection, Princeton University Library Department of Rare Books and Special Collections, Box 2, Folder 20.

Chapter 7

The epigraph for this chapter is from Audubon, "The Ohio," in *Ornithological Biography, Or, An Account of the Habits of the Birds of the United States of America: Accompanied by Descriptions of the Objects Represented in the Work Entitled "The Birds of America," and Interspersed with Delineations of American Scenery and Manners,* 5 vols. (Edinburgh, 1831–1839), I, 29; hereafter cited as *OB.*

1. The only image in *The Birds of America* to include a human figure is the Snowy Egret (plate 44), which shows a small image of a man in the background on the right. Another picture of a human being, the struggling woodsman originally painted into the watercolor version of the Golden Eagle, was, as we saw in Chapter 6, removed before final engraving.

2. Audubon to Richard Harlan, 29 November 1830, in Albert E. Lownes, "Ten Audubon Letters," *The Auk* 52, no. 2 (April 1935), 154.

3. Audubon to William Swainson, 22 August 1830, in Francis Hobart Herrick, *Audubon the Naturalist: A History of His Life and Time*, 2 vols. (New York, 1917), II, 102.

4. He wrote in the introduction to the fourth volume of *Ornithological Biography* that he had been "obliged to exclude" the Episodes "in order to make room for anatomical notices, of more interest to the scientific reader"—for the most part, drawings of digestive systems—and even then, he could not get in everything he wanted; *OB*, IV, xxiv.

5. Herrick offered this observation in the introduction to a single-volume collection of Audubon's episodes, *Delineations of American Scenery and Character* (New York, 1926), x–xi, and he might be forgiven for trying to make a case for the importance of the book.

6. Audubon, "Ruddy Duck," in *OB*, IV, 326.

7. Audubon, "The Carolina Parrot," in *OB*, I, 135; "The Wood Wren," in *OB*, I, 428.

8. Audubon, "The House Wren," in *OB*, I, 428.

9. Audubon, "The Great-Footed Hawk," in *OB*, I, 85.

10. Audubon, "The White-Headed Eagle," in *OB*, I, 160.

11. *Cookery as It Should Be: A New Manual of the Dining Room and Kitchen, for Persons in Moderate Circumstances* (Philadelphia, 1856), 55. See also Caroline F. Sloat, "Pigeons and Their Cuisine," *Common-place* 11, no. 3 (April 2011).

12. Maria Eliza Ketelby Rundell, *A New System of Domestic Cookery* (London, 1819), 112–114.

13. Audubon, "The Purple Finch," and "The Tyrant Fly-Catcher," in *OB*, I, 25, 406. On the "delicate flavor" of the Purple Finch, see also P. J. Giraud, *The Birds of Long Island* (New York, 1844), 127.

14. Audubon, "The Golden-Winged Woodpecker," in *OB*, I, 194.

15. Audubon, "The American Snipe," in *OB*, III, 325.

16. Audubon, "The Marsh Blackbird," in *OB*, I, 251.

17. Audubon, "The Florida Cormorant," in *OB*, III, 393.

18. Audubon, "The Great Northern Diver, or Loon," in *OB*, IV, 52–53.

19. He also thought the Ruffed Grouse a much better bargain, going for about seventy-five cents in urban markets in the winter (although he once bought a pair at twelve and a half cents in Pittsburgh), as opposed to the Pinnated Grouse: "I have no reason to regret my inability to purchase Prairie Hens for eating at five dollars the pair." Audubon, "The Ruffed Grouse," in *OB*, I, 211, 218; "The Pinnated Grouse," in *OB*, II, 501–502.

20. Audubon, "The American Woodcock," in *OB*, III, 480.

21. Ibid., 475.

22. Audubon, "The White-Throated Sparrow," in OB, I, 43.

23. Audubon, "The Golden Plover," in *OB*, III, 624–625.

24. Audubon, "The Pigeon Hawk," in *OB*, I, 466.

25. Alexander Wilson, "Passenger Pigeon," in in *American Ornithology, or The Natural History of the Birds of the United States*, 9 vols. (Philadelphia, 1808–1814), III, 7; hereafter cited as *AO*.

26. Audubon, "The Passenger Pigeon," in *OB*, I, 322.

27. James Fenimore Cooper, *The Pioneers, or, The Sources of the Susquehanna* (New York, 1823 [1988]), 242–250, quotations on 247, 250; Wilson, "Passenger Pigeon," in *AO*, III, 9; Audubon, "The Passenger Pigeon," in *OB*, I, 321–324.

28. Audubon, "The Passenger Pigeon," in *OB*, I, 325. For a long-term exploration of the eventual fate of the Passenger Pigeon, see Christopher Cokinos, *Hope Is the Thing with Feathers* (New York, 2000); and Joel Greenberg, *A Feathered River Across the Sky: The Passenger Pigeon's Flight to Extinction* (New York, 2014).

29. *New York Sporting Magazine* 1, no. 4 (June 1833), 181–182; the attendance figure of two thousand comes from the subsequent issue, no. 5 (222).

30. "Pigeon Matches Near Baltimore," *American Turf Register and Sporting Magazine* 1, no. 7 (September 1829), 359.

31. "Great Pigeon Shooting—Maryland Beaten by New York," *American Turf Register and Sporting Magazine* 1, no. 10 (June 1830), 500.

32. "Shooting Extraordinary," *American Turf Register and Sporting Magazine* 1, no. 10 (June 1830), 495.

33. "A Gentleman of Philadelphia County" [Jesse Y. Kester], *The American Shooter's Manual* (Philadelphia, 1827), ix, 14, 23, 35–36.

34. J. Cypress Jr., *Sporting Scenes and Sundry Sketches*, 2 vols. (New York, 1842), I, 22; John Krider, *Krider's Sporting Anecdotes, Illustrative of the Habits of Certain Varieties of American Game*, edited by H. Milnor Klapp (Philadelphia, 1853), 174–175, 263–264.

35. "Gentleman of Philadelphia County," *American Shooter's Manual*, ix–x, 144–145.

36. Giraud, *Birds of Long Island*, 87; "Frank Forester" [Henry William Herbert], "The Game of North America," *United States Magazine and Democratic Review* 16 (December 1845), 461; "Nature and Habits of the Woodcock," *New York Sporting Magazine* II, no. 2 (August 1834), 30.

37. E. J. Lewis, *Hints to Sportsmen* (Philadelphia, 1851), inscription on title page, viii, 53.

38. Herbert, "Game of North America," 461.

39. See, for instance, Audubon's reliance on David Eckleiy, Esq., of Boston, in his chapter on "The Pinnated Grouse," in *OB*, II, 498–500, and a seven-plus-page published article by J. J. Sharpless, of Philadelphia, in the chapter on "The Canvas-back Duck," in *OB*, IV, 3–10.

40. Audubon, "The Virginian Partridge," in *OB*, I, 390,391; "The Passenger Pigeon," in *OB*, I, 325; "The American Snipe," in *OB*, III, 326; "The Painted Finch," in *OB*, I, 279.

41. Audubon, "The Canada Jay," in *OB*, II, 53.

42. Audubon, "The Ohio," in *OB*, I, 29–32.

43. Audubon, "A Maple-Sugar Camp," in *OB*, III, 438–444; "The Florida Keys," in *OB*, II, 312–316; "A Long Calm at Sea," in *OB*, III, 491–494; "The Niagara," in *OB*, I, 362–363.

44. Audubon, "A Flood," in *OB*, I, 155–159; "The Earthquake," in *OB*, I, 239–241; "The Hurricane," in *OB*, I, 262–264; "The Force of the Waters," in *OB*, II, 97–101; and "Breaking Up of the Ice," in *OB*, III, 408–410.

45. Audubon, "Fishing in the Ohio," in *OB*, III, 122–127; on the portrayal of avian violence in Audubon's images, see Christoph Irmscher, "Violence and Artistic Representation in John James Audubon," *Raritan* 15, no. 2 (September 1995), 1–34.

46. Audubon, "The Lost One," in *OB*, II, 69–73.

47. Audubon, "The Regulators," in *OB*, I, 105.

48. Audubon, "The Prairie," in *OB*, I, 82, 84.

49. Ibid., 84.

50. Audubon, "The Ohio," in *OB*, I, 31–32.

51. Audubon, "Kentucky Sports," in *OB*, I, 290, 292.

52. Audubon, "The Pinnated Grouse," in *OB*, II, 491.

53. Audubon, "St. John's River in Florida," in *OB*, II, 293. Audubon does, however, revisit and revise his ideas about Indians when he goes to the West in 1843, a topic that will be discussed in Chapter 9.

54. Audubon, "The Runaway," in *OB*, II, 27–32.

55. Lucy Bakewell Audubon to Victor Gifford Audubon, 18 October 1834, in Audubon Collection, Princeton University Library Rare Books and Special Collections, Box 2: Letters, Folder 29.

56. See Chapter 3; see also Audubon, "Mississippi River Journal," in *John James Audubon: Writings and Drawings*, edited by Christoph Irmscher (New York, 1999), 77–78.

57. Audubon, "Mississippi River Journal," in Irmscher, *Audubon: Writings and Drawings*, 71–72.

58. Audubon to G. W. Featherstonhaugh, 31 December 1831, quoted in Herrick, *Audubon the Naturalist*, II, 15, 18.

59. Audubon, "Myself," in Irmscher, *Audubon: Writings and Drawings*, 789.

60. Audubon's biographers tend to be as circumspect on the issue of slavery as he was himself. The few references to his buying and selling slaves can be found in Shirley Streshinsky, *Audubon: Life and Art in the American Wilderness* (New York, 1993), 82, 253; Richard Rhodes, *John James Audubon: The Making of an American* (New York, 2004), 115, 141; and William Souder, *Under a Wild Sky* (New York, 2004), 135. See also Carolyn DeLatte, *Lucy Audubon: A Biography* (Baton Rouge, LA, 1982; updated ed., 2008), 81, 98.

61. Audubon to Victor G. Audubon, 23 September 1833; Audubon to Lucy Bakewell Audubon, 5 September 1834, in *Letters of John James Audubon 1826–1840*, edited by Howard Corning, 2 vols. (Boston, 1930), I, 255; II, 34.

62. Streshinsky, *Audubon*, 82.

63. Audubon, "The Runaway," in *OB*, II, 27–28.

64. Ibid., 28–29.

65. Ibid., 29–30.

66. Ibid., 30–31.

67. Ibid., 31.

68. Ibid., 32.

69. The Louisiana Slave Code of 1824, enacted in the year that Audubon left Louisiana for Philadelphia, said nothing about such restrictions on breaking up families and, indeed, made clear that slave marriages and the ownership of children remained under the control of the masters: Article 182 declared that "slaves cannot marry without the consent of their masters, and their marriages do not produce any of the civil effects which result from such contract," and Article 183 followed by saying that "children born of a mother then in a state of slavery, whether married or not, follow the condition of their mother; they are consequently slaves and belong to the master of their mother." See *A Documentary History of Slavery in North America*, edited by Willie Lee Nichols Rose (Oxford, UK, 1976; Athens, GA, 1999), 175–178.

70. "The Runaway Slave," *The Liberator*, 4 April 1835.

71. Newark (NJ) *Daily Advertiser*, 24 March 1835. In Ireland, the Belfast *News-Letter*, 23 January 1835, also ran the story in its entirety.

72. In more recent times, the tale of "The Runaway" has been largely ignored in most biographical accounts of Audubon's life, apparently buried for being too implausible, too embarrassing, or both. Even those works that do address this episode fail to grasp its biographical implications. Duff Hart-Davies, in *Audubon's Elephant: America's Greatest Naturalist and the*

Making of "The Birds of America" (New York, 2004), makes only a brief reference to the story on page 220 and then comes to the naïve-seeming conclusion that "even if there is a saccharine taste in Audubon's telling of the saga, it is easy enough to believe that he was deeply moved by it." William Souder, in *Under a Wild Sky*, 262, surmises that the story "could have been true" but adds that the happy ending "seems dubious." His reason for making that judgment, however, stems from his notion of Audubon's social status as a mere painter confronting a wealthier planter, not from any consideration of Audubon's attitudes toward slavery and abolition.

Chapter 8

The epigraph for this chapter is from Audubon, "Introductory Address," in *Ornithological Biography, Or, An Account of the Habits of the Birds of the United States of America: Accompanied by Descriptions of the Objects Represented in the Work Entitled "The Birds of America," and Interspersed with Delineations of American Scenery and Manners*, 5 vols. (Edinburgh, 1831–1839), I, ix; hereafter cited as *OB*.

1. Peter Kalm, *Travels into North America*, 3 vols. (Warrington, UK, 1770), II, 144–146, 194.

2. Alexander Wilson, "Preface," in *American Ornithology, or The Natural History of the Birds of the United States*, 9 vols. (Philadelphia, 1808–1814), II, vii; hereafter cited as *AO*.

3. William Swainson, *A Preliminary Discourse on the Study of Natural History* (London, 1834), 130.

4. For the origins of the term "scientist," see the first edition of the *Oxford English Dictionary* (Oxford, 1884–1928), which notes that the term first came into use in 1840, apparently coined by William Whewell. The second edition of the *OED* (1989) puts Whewell's use of the word in 1833.

5. Thomas Dunlap, in *Nature and the English Diaspora: Environment and History in the United States, Canada, Australia, and New Zealand* (Cambridge, 1999), offers a good illustration of the relationship between folkbiology and natural history in the early part of the nineteenth century; 21–45, quotation on 23. For very valuable analyses of folkbiology in the modern world, see Scott Atran, *Cognitive Foundations of Natural History: Towards an Anthropology of Science* (Cambridge, 1990); and Douglas L. Medin and Scott Atran, eds., *Folkbiology* (Cambridge, 1999).

6. Audubon, "The Pipiry Flycatcher," in *OB*, II, 394. For a similar story about Audubon's getting assistance from ordinary people—in this case, soldiers at the Key West garrison—see "The Mangrove Cuckoo," in *OB*, II, 390.

7. Audubon, "Chuck-Will's-Widow," in *OB*, I, 275–276.

8. See, for instance, Audubon, "The Mocking Bird," in *OB*, 1, 113; "The Spotted or Canada Grous," in *OB*, II, 440; "Great Northern Diver or Loon," in *OB*, IV, 48.

9. Audubon, "The Purple Martin," in *OB*, I, 118–119. The image of the Purple Martin in *The Birds of America* (plate 22) shows four of the birds surrounding a calabash house hanging on a bare branch, thus drawing attention to the custom he describes for Native Americans.

10. Ibid., 115–116.

11. James Madison Meteorological Journals, 1784–1793, 1798–1802, American Philosophical Society, vol. II, n.p.

12. Jeremy Belknap, *The History of New Hampshire*, 3 vols. (Boston, 1792; Dover, NH, 1812), III, 56.

13. Samuel Williams, *Natural and Civil History of Vermont* (Walpole, NH, 1794), I, 144.

14. Gilbert Imlay, *A Topographical Description of the Western Territory of North America*, 2nd ed. (London, 1793), 96.

15. Kalm, *Travels*, I, 291–292; II, 65, 76, 79. For an overview of Kalm's references to birds in his *Travels*, see Spencer Trotter, "Notes on the Ornithological Observations of Peter Kalm," *The Auk* 20, no. 3 (July 1903), 249–262.

16. Wilson, "Purple Grakle," in *AO*, III, 47.

17. Wilson, "Turkey Vulture," in *AO*, IX, 97. The words here may be those of George Ord, who edited and completed the final volumes of Wilson's work after the ornithologist's death in 1813.

18. Audubon, "The Red-Winged Starling, or Marsh Blackbird," in *OB*, I, 349.

19. Audubon, "The Tyrant Fly-Catcher," in *OB*, I, 403.

20. Audubon, "The American Crow," in *OB*, II, 318; "The Purple Grakle or Common Crow-Blackbird," in *OB*, I, 35.

21. Audubon, "The American Crow," in *OB*, II, 317–318.

22. Ibid., 322.

23. Audubon, "The Wild Turkey," in *OB*, I, 11–12.

24. Audubon, "The American Robin or Migratory Thrush," in *OB*, II, 190.

25. Thomas Morton, *New English Canaan* (Amsterdam, 1637; Boston, 1883 [reprint]), 193.

26. Jean Bossu, *Travels Through That Part of North America Formerly Called Louisiana*, 2 vols. (London, 1771), I, 288–290. On the role of Native Americans as valuable informants in eighteenth-century American natural history, see Susan Scott Parrish, *American Curiosity: Cultures of Natural History in the Colonial British Atlantic World* (Chapel Hill, NC, 2006), 247–256; and Kathleen S. Murphy, "Portals of Nature: Networks of Natural History in Eighteenth-century British Plantation Societies" (PhD dissertation, Johns Hopkins University, MD, 2008), chapter 4. For a larger exploration of Native American knowledge of the avian world, see Shepard Krech III, *Spirits of the Air: Birds and American Indians in the South* (Athens, GA, 2009).

27. John Kirk Townsend, *Sporting Excursions in the Rocky Mountains, Including a Journey to the Columbia River, and a Visit to the Sandwich Islands, Chili, &c.*, 2 vols. (London, 1840), II, 106. Townsend's observations about the Arkansas Flycatcher and the Canada Jay are quoted in Audubon's passages on those birds in *OB*, IV, 423, 124, respectively.

28. Wilson, "Carolina Parrot," in *AO*, III, 96.

29. Wilson, "Chuck-Will's-Widow," in *AO*, VI, 96.

30. Jared Diamond and K. David Bishop, "Ethno-ornithology of the Ketengban People, Indonesian New Guinea," in Medin and Atran, *Folkbiology*, 18.

31. Bossu, *Travels*, I, 187–188.

32. Jonathan Carver, *Travels Through the Interior Parts of North America, in the Years 1766, 1767, and 1768* (London, 1778), 468.

33. Wilson, "Tell-Tale Godwit, or Snipe," in *AO*, VII, 63; the comment is Ord's addition to Wilson's description.

34. Wilson, "Night Heron, or Qua-Bird," in *AO*, VII, 106; "White-eyed Flycatcher," in *AO*, II, 166; "Night Hawk," in *AO*, V, 59.

35. Kalm, *Travels*, II, 152.

36. Audubon, "Audubon's Warbler," in *OB*, V, 52. *Sylvia Audubonii* and *Sylvia coronata* are the western and eastern versions, respectively, of the same species, the Yellow-rumped Warbler. For the current classification of Audubon's Warbler with the Yellow-rumped Warbler (*Dendroica*

coronata), see David Allen Sibley, *National Audubon Society: The Sibley Guide to Birds* (New York, 2000), 436.

37. Andrew Lewis, *A Democracy of Facts: Natural History in the Early Republic* (Philadelphia, 2011), 5.

38. This story is quoted in "Wilson, the Ornithologist," *American Quarterly Review* VIII, no. 16 (September–December 1830), 377–378.

39. Benjamin Smith Barton, *Fragments of the Natural History of Pennsylvania* (Philadelphia, 1799). On the ornithological significance of Barton's pamphlet, see Robert Henry Welker, *Birds and Men: American Birds in Science, Art, Literature, and Conservation, 1800–1900* (Cambridge, MA, 1955), 17.

40. Barton, *Fragments*, vi.

41. Ibid., xii. Indeed, the swallow question had become accepted as something of a scientific given among many students of nature in Barton's time. Respected authors such as Jeremy Belknap, in *History of New Hampshire*, and Samuel Williams, in *Natural History of Vermont*, added their authority to the swallow submersion tales. For a full discussion of the history of the swallow immersion theory, see Lewis, *Democracy of Facts*, 13–45.

42. Barton, *Fragments*, xvii, 24.

43. Wilson, "Barn Swallow," in *AO*, V, 35–36.

44. Audubon, "Mississippi River Journal," in *John James Audubon: Writings and Drawings*, edited by Christoph Irmscher (New York, 1999), 132.

45. Wilson, "Carolina Parrot," in *AO*, III, 93, 99.

46. Audubon, "Mississippi River Journal," 54.

47. *American Turf Register and Sporting Magazine* 5, no. 1 (September 1833), 42–44.

48. Audubon, "The Black Vulture, or Carrion Crow," in *OB*, II, 46–47.

49. Audubon, "The Sora Rail," in *OB*, III, 256.

50. Audubon, "The Great Crested Flycatcher," in *OB*, II, 176.

51. On the correspondence between William H. Edwards and Audubon, see Margaret Curzon Welch, "John James Audubon and His American Audience: Art, Science, and Nature, 1830–1860" (PhD dissertation, University of Pennsylvania, 1988), 84. On the friendship between Audubon and Baird during the last years of Audubon's life, see Chapter 9. Unlike Baird, Edwards did not develop a sustained relationship with Audubon, but he later went on to become one of the leading American entomologists of the nineteenth century.

52. For a discussion of Audubon's network of informants, see Welch, "Audubon and His Audience," 78. The letter from Audubon to R. O. Anderson, 16 May 1836, is in the Audubon Collection, Princeton University Library Department of Rare Books and Special Collections, Box 2, Folder 32.

Chapter 9

The epigraph for this chapter is from Audubon, "Missouri River Journals," in *Audubon: Writings and Drawings*, edited by Christoph Irmscher (New York, 1999), 726.

1. "Art. IV—1. *History of British Birds*, by William MacGillivray . . . 2. *Ornithological Biography*, by John James Audubon," *North American Review* 50, no. 107 (April 1850), 381–404, quotations on 388–389, 403.

2. Audubon to John Bachman, 16 July 1837, in *Letters of John James Audubon, 1826–1840*, edited by Howard Corning, 2 vols. (Boston, 1830), II, 172.

3. Audubon, "Surf Duck," in *Ornithological Biography, Or, An Account of the Habits of the Birds of the United States of America: Accompanied by Descriptions of the Objects Represented in the Work Entitled "The Birds of America," and Interspersed with Delineations of American Scenery and Manners*, 5 vols. (Edinburgh, 1831–1839), IV, 161; hereafter cited as *OB*.

4. Audubon, "Introduction," in *OB*, V, viii.

5. London *Morning Chronicle*, 15 November 1837, 3. A later report, written in the aftermath of Audubon's death, in 1851, looked back at "the convulsions of 1837" and noted that "nearly one half of the subscribers upon whom he had depended to enable him to sustain the great expenses of his work, availed themselves of the crisis to withdraw their subscriptions." See Boston *Daily Atlas*, 2 March 1851, 2.

6. Audubon to Robert Havell Jr., 8 July 1837, in Corning, *Letters*, II, 167–168.

7. *Recollections of Samuel Breck; with Passages from His Note-Books (1771–1862)*, edited by H. E. Scudder (Philadelphia, 1877), 260–261.

8. Audubon to "My Dearest Friends," 1 July 1838, in R. W. Shufeldt and M. R. Audubon, "The Last Portrait of Audubon, Together with a Letter to His Son," *The Auk* 11, no. 4 (October 1894), 311. For Audubon's list of subscribers, see *OB*, V, 647–651.

9. Newburyport (MA) *Herald*, 15 March 1839, 1.

10. Audubon to Victor G. Audubon, 15 September 1833, in Corning, *Letters*, I, 247.

11. *The Birds of America, from Drawings made in the United States and its Territories. With 500 plates*, Royal Octavo ed., 7 vols. (New York and Philadelphia, 1840–1844).

12. For a detailed discussion of the production of this smaller version of *The Birds of America*, see Ron Tyler, *Audubon's Great National Work: The Royal Octavo Edition of "The Birds of America"* (Austin, TX, 1993), esp. 47–72. For a broader overview of the role of lithography in ornithological illustration in antebellum Philadelphia, see also Jonathan David Grunert, "Aesthetics for Birds: Institutions, Artist-Naturalists, and Printmakers in American Ornithologies, from Alexander Wilson to John Cassin" (MS thesis, Virginia Polytechnic Institute and State University, 2015), esp. chap. 3.

13. Robert Buchanan, *Life and Adventures of Audubon the Naturalist* (London, 1869; reprint, New York, 2005), 392. See also Francis Hobart Herrick, *Audubon the Naturalist: A History of His Life and Time*, 2 vols. (New York, 1917), II, 151–152.

14. *Journal of John James Audubon: Made While Obtaining Subscriptions to his "Birds of America" 1840–1843*, edited by Howard Corning (Boston, 1929), 8, 35–36, 40–41.

15. Corning, *Journal . . . 1840–1843*, 55–57. A more readily accessible source for Audubon's Worcester visit is *Audubon's America: The Narratives and Experiences of John James Audubon*, edited by Donald Culross Peattie (Boston, 1940), 267–268. Audubon's hopes about getting Burritt and Davis to subscribe to the larger version of *The Birds of America* were not to be fulfilled: Both men appear on a list of nineteen Worcester residents who subscribed to the smaller Royal Octavo edition. See Audubon, "List of Subscribers Since the Publication of the First Volume," in *Birds of America* (Royal Octavo), II, 205.

16. *Under Its Generous Dome: The Collections and Programs of the American Antiquarian Society*, edited by Marcus McCorison, 2nd ed. (Worcester, MA, 1992), 23. On a happier note, the society does have a copy of Audubon's *Ornithological Biography*, the Royal Octavo version of *The Birds of America*, and 105 plates from the later Bien edition.

17. Corning, *Journal . . . 1840–1843*, 62–63.

18. Ibid., 82.

19. G. W. Curtis et al., *Homes of American Authors: Comprising Anecdotal, Personal, and Descriptive Sketches, by Various Writers* (New York, 1852), 4–5.

20. Ibid., 6.

21. Audubon to John Bachman, 2 January 1840, in Irmscher, *Audubon: Writings and Drawings*, 855. The work that would eventually result would come into print as *The Viviparous Quadrupeds of North America (with John Bachman). With 150 plates. Issued in 30 parts*, Imperial folio (New York, 1845–1846); as *The Viviparous Quadrupeds of North America Text to the plates noted above*, Royal Octavo ed., 3 vols. (New York, 1846–1854); *The Quadrupeds of North America, With 155 Plates*, Royal Octavo ed., 3 vols. (New York, 1849–1854).

22. Ruthven Deane, "Unpublished Letters of Introduction Carried by John James Audubon on His Missouri River Expedition," *The Auk* 25, no. 2 (April 1908), 170–173.

23. Audubon, "Missouri River Journals," in Irmscher, *Audubon: Writings and Drawings*, 572, 684.

24. Ibid., 606.

25. Ibid., 657. Other journal entries supplied, if not written, by men whom Audubon met out West appear on 640–641, 714–717.

26. John Kirk Townsend, *The Narrative of a Journey Across the Rocky Mountains to the Columbia River and a Visit to the Sandwich Islands* (Philadelphia, 1839). In the same year, Townsend also published another book that Audubon might well have looked upon with some measure of competitive concern, *Ornithology of the United States of North America, or, Descriptions of the Birds Inhabiting the States and Territories of the Union with an Accurate Figure of Each, Drawn and Coloured from Nature* (Philadelphia, 1839). See Tyler, *Audubon's Great National Work*, 48–49.

27. Audubon, "Missouri River Journals," 566, 572–573.

28. See, for instance, the Indian references in "The Purple Martin," in *OB*, I, 119, and "The Raven," in *OB*, II, 1.

29. Audubon, "Missouri River Journals," 564, 577, 588.

30. Ann Fabian, *The Skull Collectors: Race, Science, and America's Unburied Dead* (Chicago, 2010), offers the most extensive study of craniology, both its scientific intentions and racial implications. Audubon himself had some earlier acquaintance with skull collecting. In late April 1837, when he arrived in Galveston, Texas, on a bird-hunting trip along the Gulf Coast, he met a beef contractor for Sam Houston's army who "promised me some skulls of Mexicans." The nearby battlefield at San Jacinto, the scene of Houston's decisive victory over Santa Anna's troops a year earlier, still remained littered with the unburied remains of Mexican soldiers, whose skulls were readily available for the taking. Audubon took some. See Buchanan, *Life and Adventures of Audubon*, 307.

31. Audubon, "Missouri River Journals," 664.

32. For a description of Catlin's work in the Upper Missouri region, see Kathryn S. Hight, "'Doomed to Perish': George Catlin's Depiction of the Mandan," in *Reading American Art*, edited by Marianne Doezema and Elizabeth Milroy (New Haven, CT, 1998), 150–163.

33. Audubon, "Missouri River Journals," 584, 615, 628.

34. Ibid., 618.

35. Ibid., 604, 690, 696.

36. Ibid., 712.

37. Ibid.

38. Ibid., 633, 580, 582, 591, 660, 708, 726.

39. Ibid., 731–749.

40. Spencer Fullerton Baird to Audubon, 24 November 1843, quoted in Deane, "Unpublished Letters of John James Audubon and Spencer F. Baird," *The Auk* 24, no. 1 (January 1907), 55.

41. Audubon to Spencer Fullerton Baird, 3(?) November 1843, quoted in William Healey Dall, *Spencer Fullerton Baird: A Biography* (Philadelphia, 1915), 93. The 3 November date Dall gives must be an error, because Audubon refers to Baird's letter of 24 November; the more likely date is 3 December 1843.

42. Spencer Fullerton Baird to Audubon, 4 June 1840, quoted in Ruthven Deane, "Unpublished Letters," *The Auk* 23, no. 2 (April 1906), 199.

43. Audubon to Spencer Fullerton Baird, 13 June 1840, quoted in ibid., 200.

44. Baird to Audubon, 20 June 1840, quoted in Deane, "Unpublished Letters," *The Auk* 23, no. 3 (July 1906), 318–320.

45. On the friendship between Audubon and Baird during the last years of Audubon's life, see Herrick, *Audubon the Naturalist*, II, 218–252, 272–280; and Dall, *Baird: A Biography*, chaps. 2–4. A more general work on early American naturalists, Joseph Kastner, *A Species of Eternity* (New York, 1977), gives some attention to the Audubon-Baird relationship, but only briefly on pages 306 and 316–317. Similarly—and surprisingly—most of the modern Audubon biographies pay scant, if any, attention to the Audubon-Baird connection: see, for instance, Alice Ford, *John James Audubon: A Biography* (New York, 1988), 376, 385, 388–389, 415, 421–422; Shirley Streshinsky, *Audubon: Life and Art in the American Wilderness* (New York, 1993), 336, 344–345, 356; Duff Hart-Davies, *Audubon's Elephant: America's Greatest Naturalist and the Making of "The Birds of America"* (New York, 2004), 260–261. There are no Audubon-Baird references at all in Alexander B. Adams, *John James Audubon: A Biography* (London, 1967); William Souder, *Under a Wild Sky: John James Audubon and the Making of "The Birds of America"* (New York, 2004); or Richard Rhodes, *John James Audubon: The Making of an American* (New York, 2004).

46. See, for instance, Baird's first letter to Audubon, 4 June 1840, quoted in Deane, "Unpublished Letters," *The Auk* 23, no. 2 (April 1906), 199, and the shift to "Your affectionate Pupil," in a letter dated 8 February 1842, quoted in Deane, "Unpublished Letters," *The Auk* 23, no. 3 (July 1906), 322–324.

47. Dall, *Baird: A Biography*, 46–47.

48. Baird to Audubon, 4 March 1842, quoted in Deane, "Unpublished Letters," *The Auk* 23, no. 3 (July 1906), 328.

49. Spencer Fullerton Baird to Samuel Baird, 8 July 1831, quoted in Dall, *Baird: A Biography*, 28.

50. The basic background on Baird's early years comes from Dall, *Baird: A Biography*, chaps. 1 and 2.

51. William M. Baird to Spencer F. Baird, 23 November 1842, quoted in Dall, *Baird: A Biography*, 82.

52. Spencer F. Baird to William M. Baird, 20 December 1841, quoted in ibid., 58.

53. Baird to Audubon, 8 February 1842, quoted in Deane, "Unpublished Letters," *The Auk* 23, no. 2 (July 1906), 322–324.

54. Audubon to Spencer Fullerton Baird, 10 February 1842, quoted in Deane, "Unpublished Letters," *The Auk* 23, no. 3 (July 1906), 325.

55. Baird to Audubon, 27 July 1842, quoted in Deane, "Unpublished Letters," *The Auk* 21, no. 2 (April 1904), 256.

56. Audubon to Spencer Fullerton Baird, 30 July 1842, quoted in ibid., 258.

57. Audubon to Spencer Fullerton Baird, 31 January 1843, quoted in Dall, *Baird: A Biography*, 89.

58. Audubon to Spencer Fullerton Baird, 23 February 1843, quoted in ibid., 92.

59. Ibid., 100.

60. Ibid., 99.

61. Baird to Audubon, 4 November 1846, quoted in Deane, "Unpublished Letters," *The Auk* 24, no. 1 (January 1907), 65–66.

62. Ibid., 66.

63. Report to Dickinson College as Curator of Museum, 8 July 1846, Dickinson College Archives.

64. Baird to Audubon, 8 February 1847, quoted in Deane, "Unpublished Letters," *The Auk* 24, no. 1 (January 1907), 69.

65. Dall, *Baird: A Biography*, 159.

66. C. C. Jewett to Spencer F. Baird, 5 July 1850; Joseph Henry to Spencer F. Baird, 8 July 1850, quoted in ibid., 211–212.

67. For Baird's role as a mentor to younger men of science, see Daniel Lewis, *The Feathery Tribe: Robert Ridgway and the Modern Study of Birds* (New Haven, CT, 2012), 35–38.

68. "Audubon and Washington Irving—The Plague of Railroads," *International Monthly Magazine of Literature, Science & Art* 1, no. 8 (August 1850), 232.

69. Park Benjamin, "Audubon's Blindness," *Graham's Magazine* (September 1850), 169. On Audubon's failing eyesight and his trouble with wearing glasses, see Herrick, *Audubon the Naturalist*, II, 288.

70. Herrick, *Audubon the Naturalist*, II, 288–290.

Chapter 10

The epigraph for this chapter is from Audubon, "Introduction," in *Ornithological Biography, Or, An Account of the Habits of the Birds of the United States of America: Accompanied by Descriptions of the Objects Represented in the Work Entitled "The Birds of America," and Interspersed with Delineations of American Scenery and Manners*, 5 vols. (Edinburgh, 1831–1839),V, xxv; hereafter cited as *OB*.

1. New York *Evening Post*, 28 January 1851, 2; see also reprinted versions in Albany (NY) *Evening Journal*, 29 January 1851, 2; Philadelphia *Public Ledger*, 29 January 1851, 3; and Savannah (GA) *Republican*, 3 February 1851, 2.

2. Derby (England) *Mercury*, 9 June 1852, 3; "New Books, Literature, and the Fine Arts in New York," Washington, DC *Daily National Intelligencer*, 31 January 1851, 3.

3. "Death of Audubon, the Ornithologist," Boston *Evening Transcript*, 28 January 1851, 2; "J. J. Audubon," Litchfield (CT) *Republican*, 30 January 1851, 3.

4. "The Death of Audubon," New York *Evening Post*, 28 January 1851, 2.

5. "John James Audubon," *American Phrenological Journal & Repository of Science, Literature & General Intelligence* 13, no. 4 (April 1851), 77–80, quotation on 77.

6. Charles Wilkins Webber, *Romance of Sporting; Or, Wild Scenes and Wild Hunters* (Philadelphia, 1859), 88, 92–94. Webber's reference to the "rusty and faded green blanket" no doubt refers to the "green blanket coat with fur collar and cuffs" Audubon was wearing on the day he came home to Minnie's Land, and in which his son John later painted his portrait; see Francis Hobart Herrick, *Audubon the Naturalist: A History of His Life and Time*, 2 vols. (New York, 1917), II, 258.

7. Webber, *Romance of Sporting*, 31–33.

8. The discussion of the post-Audubon path of ornithology owes much to Mark V. Barrow Jr., *A Passion for Birds: American Ornithology After Audubon* (Princeton, NJ, 1998); and Daniel Lewis, *The Feathery Tribe: Robert Ridgway and the Modern Study of Birds* (New Haven, CT, 2012).

9. Barrow, *Passion for Birds*, 207.

10. George Bird Grinnell, "Some Audubon Letters," *The Auk* 33 (April 1916), 119.

11. Barrow, *Passion for Birds*, 118–119. Today, the Audubon Society states that it "has never been opposed to the hunting of game species if that hunting is done ethically and in accordance with laws and regulations designed to prevent depletion of the wildlife resource. We have made this clear repeatedly in official statements of policy and it remains Audubon policy." The society's statement goes on to say, though, that the organization does advocate restrictions on hunting "whenever we are convinced that the welfare of the species requires it." See, for example, Washingtonians for Wildlife Conservation, "What They Say About Hunting," www.w4wc.org/what -they-say-about-hunting.html (accessed 21 September 2015).

12. On the participation of women in the early Audubon Society, see Carolyn Merchant, "George Bird Grinnell's Audubon Society: Bridging the Gender Divide in Conservation," *Environmental History* 15, no. 1 (January 2010), 3–30. On Florence Merriam and the origins of the Audubon Society chapter at Smith College, see Madelyn Holmes, *American Women Conservationists: Twelve Profiles* (Jefferson, NC, 2004), 38–41. For a lively discussion of the opposition to birds in women's hats, see Jennifer Price, *Flight Maps: Adventures with Nature in Modern America* (New York, 1999), 57–109.

13. Barrow, *Passion for Birds*, 209. The National Audubon Society now takes pride in having "pioneered the idea of citizen science with the first Christmas Bird Count" over a century ago, and it now "has helped transform citizen science into an everyday activity for tens of thousands of birders . . . [who] provide an ongoing assessment of bird populations that is fast becoming an invaluable resource for conservation." See https://www.audubon.org/conservation/science (accessed 21 September 2015).

14. See http://www.srs.fs.usda.gov/trends/Nsre/nsre2.html (accessed 21 September 2015).

15. On the development of American bird guides, see Thomas R. Dunlap, *In the Field Among the Feathered: A History of Birders and Their Guides* (New York, 2011).

16. Quoted in Waldemar Fries, *The Double Elephant Folio: The Story of Audubon's "Birds of America"* (Chicago, 1973), 13.

17. Arthur E. Lownes, "Ten Audubon Letters," *The Auk* 52, no. 2 (April 1935), 166.

18. Audubon to Spencer Fullerton Baird, 7 August 1845, quoted in William Healey Dall, *Spencer Fullerton Baird: A Biography* (Philadelphia, 1915), 124.

19. On the production of the Bien edition, see Daniel Lewis, "Night and Day: Revisiting Audubon's Birds of America," *Huntington Frontiers* (Fall/Winter 2006), 21–23. Also useful is the information provided by the American Antiquarian Society, the leading scholarly repository of American print culture, where I had an opportunity to see several plates of the society's Bien

holdings. See http://www.americanantiquarian.org/annualreports/2012.pdf (accessed 6 October 2014). I am also grateful to Nan Wolverton of AAS for personal email correspondence, 31 March 2015.

20. Herrick, *Audubon the Naturalist*, II, 304–308. One of those viewers, a "Long Island Farmer" named Bloodgood H. Cutter, was moved to verse when he beheld the Audubon paintings one night in November 1879. He wrote an eight-stanza poem that began with great appreciation of Audubon's skill and closed with even greater appreciation for the rigors of his research; an excerpt is included here:

> Audubon's birds that hang 'round here
> So very life like do appear
> And when I did them closely view
> Was astonished at what Man can do. . . .
> How many years of toil and pain
> He camped in Woods, in cold and rain,
> To achieve all this did persevere,
> And for th' same, all should prize him dear.

Artistically, Cutter's poetry seemed no match for Audubon's painting, but had Audubon been alive to read this celebration of the "toil and pain" he had been through as the American Woodsman, he would have been pleased to see that someone understood his suffering. Bloodgood H. Cutter, "Long Island Farmer on Seeing the Birds in the *New York Historical Society*," New-York Historical Society, Audubon Box.

21. For an example of the special furniture built to house and display the Double Elephant Folio edition of *The Birds of America*, see Benjamin W. Williams, "Audubon's 'The Birds of America' and the Remarkable History of Field Museum's Copy," *Field Museum of Natural History Bulletin* (June 1986), 10–11.

22. Henry Lyttleton Savage, "John James Audubon: A Backwoodsman in the Salon," *Princeton University Library Chronicle* 5, no. 4 (June 1944), 129–130.

23. Suzanne M. Low, compiler, *Catalogue of the New Birds of America Section of the Audubon Archives* (New York: Department of Ornithology, American Museum of Natural History, 1991), 3. Low counted 107 copies in the possession of institutions and 12 in the hands of individuals. When the Providence Athenaeum sold its copy at auction to an anonymous buyer in 2005, the sale may have raised the number of individual owners to 13. Writing almost two decades earlier, in 1973, Waldemar Fries counted 134 complete copies of the Double Elephant Folio edition, all but 6 of which he said he had examined personally; see Fries, *Double Elephant Folio*, 196. The difference may be simply a discrepancy of counting, but it more likely stems from the continuing destruction of folio volumes for the sake of selling individual plates.

24. Robert M. Peck, "Cutting Up Audubon for Science and Art." *Antiques* (October 2003), 104–113, quotation on 112–113.

25. Samuel N. Rhoads, "More Light on Audubon's Folio 'Birds of America," *The Auk* 33 (April 1916), 130.

26. Fries, *Double Elephant Folio*, 203–204.

27. *The Old Print Shop Portfolio* 6, no. 5 (January 1947), 100–101, in John James Audubon Collection, Princeton University Library Department of Rare Books and Special Collections, Box 3, Folder 6.

28. *The Month at Goodspeed's* 26, no. 9 (June 1955), 211, in John James Audubon Collection, Princeton University Library Department of Rare Books and Special Collections, Box 3, Folder 6.

29. On the race to collect specimens of near-extinct birds, see Scott Weidensaul, *The Ghost with Trembling Wings: Science, Wishful Thinking, and the Search for Lost Species* (New York, 2002), 8–9; and Christopher Cokinos, *Hope Is the Thing with Feathers: A Personal Chronicle of Vanished Birds* (New York, 2000), 35.

30. Fries, *Double Elephant Folio*, xxii. The 1992 price resulted from the auction of the University of Edinburgh's copy of *The Birds of America*, the sole complete set in that city where Audubon first began production of his Great Work with the engraver William Home Lizars. For a report on the 2000 auction that fetched twice that amount, see *"Birds of America* Takes Off," *Maine Antiques Digest* (May 2000). The winning offer came in over the phone from Saud bin Mohammed bin Ali Al-Thani, the culture minister of the Persian Gulf nation of Qatar, who at the time had already spent over a quarter of a billion dollars acquiring collections for a natural history museum he was building in his homeland.

31. "Bookish Contretemps in Providence," *New York Times*, 15 December 2003.

32. "Providence Athenaeum Audubon Sells for $5 Million," *American Libraries*, http://www .americanlibrariesmagazine.org/archive/2005abc/december2005ab/audubon (accessed 6 October 2014).

33. "Fabulous Audubon Illustrations Take Flight," PittsburghLIVE.com, 9 December 2003; Dr. Rush Miller, "Foreword," and Michael Lee, "The Conservation of the Darlington Library's Double Elephant Folio *Birds of America*," in *Taking Flight: Selected Prints from John James Audubon's "Birds of America,"* edited by Josienne N. Piller (Pittsburgh: University Art Gallery, University of Pittsburgh, 2003), 3, 19. The Darlington set of *The Birds of America* has also been digitized and made accessible and searchable online, a great boon to Audubon scholarship.

34. On the different organizational approach of the copies of *The Birds of America* owned by Audubon and Phillips, see Williams, "Audubon's 'The Birds of America,'" 7–8. A third copy organized in the same fashion was initially owned by Audubon's good friend Edward Harris, but the location of that set is now unknown. The evaluation of Havell's set is in Herrick, *Audubon the Naturalist*, II, 203.

35. "Webster's Birds of America," http://raunerlibrary.blogspot.com/2010/07/websters-birds-of-america.html (accessed 3 October 2014).

Index

~~~

Bird names in parentheses indicate Audubon's nomenclature. Page numbers in italics indicate figures; *pl.* indicates color plates.

Acknowledgments

In the course of researching and writing this book, I developed quite a life list of friends and colleagues who gave me steady support and showed generous indulgence. One friend, in fact, played a very important role long before I even thought about the book. Back in the 1980s, Tom Strikwerda first introduced me to the joys (and occasional frustrations) of birding, and I realize I might not have developed an interest in Audubon had it not been for his excellent example. The first thanks go to him.

Some years later, when I actually started work on Audubon, the circle of gratitude began to expand. I went first to a favorite refuge for research, the American Antiquarian Society (AAS), the historian's haven to which I have returned time and time again, for visits both short and long. I want to thank several early allies at AAS, especially Gigi Barnhill, Joanne Chaison, Ellen Dunlap, John Hench, Thomas Knoles, Marie Lamoureux, Jim Moran, and Caroline Sloat, for years of friendship and assistance. I was honored to be invited to give my first public talk on Audubon at AAS, the J. Russell Wiggins Lecture on the History of the Book in American Culture, which was subsequently published in the AAS *Proceedings*. More recently, I am grateful to Paul Erickson, Lauren Hewes, and especially Nan Wolverton at AAS for their wonderful hospitality and help when I directed the Center for the History of American Visual Culture (CHAViC) Summer Seminar, "The Art of Science and Technology."

I have also had the good fortune to spend time at several other research institutions where the staff members and other research fellows have proven to be invaluable in their support. At the Huntington Library, Alan Jutze, Dan Lewis, and Roy Ritchie were gracious and engaging colleagues during my fellowship period there, and I particularly enjoyed the Wednesday-morning bird walks Romaine Ahlstrom led on the Huntington grounds for staff and readers. Jim Grossman and Sara Austin helped me have an excellent six-month stay at the Newberry Library, and I much appreciate the strenuous workout my fellow Newberry fellows gave me in my seminar presentation, which helped me sharpen my thinking. I felt fortunate that my longtime friend and mentor, the late Alfred Young, was still the resident eminence at the Newberry at the time. Even though Al sometimes seemed puzzled that I was working on American ornithology rather than the American Revolution, he read my work carefully and astutely, and as always, I took his comments very much to heart. While I was at the Newberry, I also took the opportunity to venture southward several times to the Field Museum, where Ben Williams gave me free rein in the library and even let me hold some bird skins from the Field's collection (which I confess I did a bit tentatively). Gretchen Oberfranc made me feel intellectually welcome at the Princeton University Library Department of Rare Books and Special Collections, and I can now say that I've finally done serious scholarly work at my undergraduate alma mater. George Dillman was also very kind during my too-brief visit to the Adelson Library at the Cornell Lab of Ornithology, a place that offers a wonderfully distracting overlook of the pond at Sapsucker Woods; it is the only research facility I've visited that makes binoculars available to help patrons take in the view.

The McNeil Center for Early American Studies (MCEAS) at the University of Pennsylvania has been especially generous. Most important, MCEAS committed financial support for the publication of this book at the outset, making more feasible the inclusion of color images. The genial leader of MCEAS, Daniel Richter, not only offered me work space on occasional research visits to Philadelphia but also asked me to give a talk in the center's summer seminar series. In each instance, Dan's administrative colleagues Amy Baxter-Bellamy and Barbara Natello cheerfully and capably helped with logistical details. Simon Middleton and Simon Newman likewise invited me to present a paper to their MCEAS-sponsored conference, "On the Anvil of Labor History in the Revolutionary Era: Billy G. Smith and Fellow Artisans," where I was happy both to honor Billy and to try to make a case for labor relations in early American science.

At Georgia Tech, nine years of administrative duties as the director of the Georgia Tech Honors Program certainly slowed the writing process, but I came to count on my beloved colleagues, Monica Halka and Nicole Leonard, to take care of business when I tried to find the occasional morning to hide out and write. The Honors Program also gave me the opportunity to be the teaching wingman for my professorial pal Hugh Crawford in a first-year writing class on "Birds and Trees." Back in my academic department, the School of History and Sociology, my faculty colleagues gave me very smart questions and suggestions in a departmental seminar and, more important, helped keep our corner of the campus sane and humane.

I have had numerous other opportunities to talk about Audubon in scholarly conferences and public lectures—too numerous, in fact, to list here. I will note that of all the scholarly comment I've received, none has been more sustained than the feedback from my fellow participants in the Workshop on Early American Biographies sponsored by the Omohundro Institute for Early American History and Culture, the *William and Mary Quarterly*, and the University of Southern California-Huntington Early Modern Studies Institute. Held in the comfortable confines of the Huntington Library and ably led by Annette Gordon-Reed, Christopher Grasso, and Peter Mancall, the workshop brought together ten authors of biographical works in progress for two days of perceptive but supportive analysis. I am grateful to my fellow workshop participants for their insights into Audubon, particularly to Martha Jones for her subsequent follow-up with very helpful sources on eighteenth-century Saint-Domingue.

Several other friends have been generous in reading and commenting on individual chapters or other parts of the manuscript, and I particularly want to acknowledge Susan Branson, Robyn Lily Davis, Ronald Angelo Johnson, Catherine Kelly, Kenneth Lockridge, Jonathan Schneer, and Paul Sivitz. Ann Fabian and Dan Lewis read the entire manuscript for the University of Pennsylvania Press, both of them offering very smart feedback. Anne Harper also read various parts of the book and took part in any number of dinner-table discussions about its direction, but even more, she has remained a reliable source of energy and encouragement, especially as this book came closer to completion and she could see a new one in the future.

This book is certainly better (and shorter) because of the good work and good sense of Robert Lockhart, my editor. I got to know Bob several years before I actually began to work with him, and I quickly realized I liked his style—good-humored, well read, and, above all, smart. I knew he was the editor for me. Once he had done what he could do on the book, I was put into the hands of a careful copy editor, Christine Dahlin, and an excellent managing editor at the press, Erica Ginsburg, who guided the book through production.

I save the final word of thanks for my former college roommate, fellow college professor, and loyal and long-enduring friend, Philip Terrie. On one of our annual low-budget birding trips, we visited Audubon's haunts in Louisiana, where Phil claimed to have seen the Painted Bunting fly by before I did. He was (and doggedly remains) very much wrong about that, but I can now overlook one minor birding error to offer him this book's dedication.